# Applied Chemical Process Design

# Applied Chemical Process Design

**FRANK AERSTIN AND GARY STREET**

*Dow Chemical*
*Midland, Michigan*

**With a Foreword by**
**K. D. Timmerhaus**

**PLENUM PRESS · NEW YORK AND LONDON**

Library of Congress Cataloging in Publication Data

Aerstin, Frank.
  Applied chemical process design.

  Includes index.
  1. Chemical processes. 2. Chemical engineering. I. Street, Gary, joint author. II. Title.
TP155.7.A35                    660.2'81                    78-9104
ISBN 0-306-31088-0

First Printing — November 1978
Second Printing — May 1980

© 1978 Plenum Press, New York
A Division of Plenum Publishing Corporation
227 West 17th Street, New York, N.Y. 10011

Printed in the United States of America

# Contents

# Foreword

Development of a new chemical plant or process from concept evaluation to profitable reality is often an enormously complex problem. Generally, a plant-design project moves to completion through a series of stages which may include inception, preliminary evaluation of economics and market, data development for a final design, final economic evaluation, detailed engineering design, procurement, erection, startup, and production.

The general term *plant design* includes all of the engineering aspects involved in the development of either a new, modified, or expanded industrial plant. In this context, individuals involved in such work will be making economic evaluations of new processes, designing individual pieces of equipment for the proposed new ventures, or developing a plant layout for coordination of the overall operation. Because of the many design duties encountered, the engineer involved is many times referred to as a design engineer. If the latter specializes in the economic aspects of the design, the individual may be referred to as a cost engineer. On the other hand, if he or she emphasizes the actual design of the equipment and facilities necessary for carrying out the process, the individual may be referred to as a process design engineer. The material presented in this book is intended to aid the latter in developing rapid chemical designs without becoming unduly involved in the often complicated theoretical underpinnings of these useful notes, charts, tables, and equations.

The authors have attempted to emphasize those areas most often encountered in chemical process design, namely heat transfer, mass transfer, fluid flow, and mixing. Other design areas considered, but to a lesser extent, include cooling towers, liquid–liquid separations, gas–solid separations, vapor–liquid separations, pumps, safety valves and rupture disks, steam ejectors, and vessel design. These design procedures are supplemented with information on the thermal and transport properties of many materials and chemicals needed in the design of such process equipment, the mechanical properties of a host of metals commonly used in their construction, and the dimensions and properties of steel piping and tubing. In addition, two measures of economic profitability have been included to assist the process design engineer in justifying a specific design or process to management.

In heat transfer, the authors have considered heat exchangers and condensers normally used in the chemical process industry. This has involved presenting the basic concept of heat transfer coefficients with a tabulation of coefficients for a variety of applications and a simple adjustment for the effect of velocity on the heat transfer rates. The user is then provided with design procedures to estimate heat losses from tanks and insulated pipelines, calculate heat requirements of process piping and vessels, and specify double-pipe heat exchangers, shell and tube heat exchangers, reboilers, vaporizers, water-cooled condensers, and air-cooled heat exchangers. (The extensive material for air-cooled heat exchangers is particularly welcome in those areas where cooling water is at a premium.) Much of the material is presented in easy-to-use tabular and graphical form. Additionally, the authors have provided numerous example problems outlining understandable step-by-step design procedures for much of this heat transfer equipment.

The authors have approached the subject of distillation in a similar manner. The most widely used simple shortcut methods for determining the optimum number of trays and the optimum reflux ratio are outlined in easy-to-use graphs. A step-by-step procedure is outlined to obtain these two key items in a simple distillation process. These procedures are supplemented with information on flash vaporization and how to determine column diameter and tray efficiency. For smaller volume separations, design procedures are also outlined for packed columns.

The section on fluid flow provides useful tabular data and graphs to obtain pressure drops through pipes, elbows, tees, bends, reducers, and valves for water, steam, and air. Simple corrections for the pressure drop due to viscosity and density changes increase the usable range of the data and correlations. Design flow limits for weirs, flumes, orifices, and vertical pipes are also included in this survey. The section concludes with a brief calculation procedure for two-phase flow using the Lockhart–Martinelli approach.

All in all, the material presented in *Applied Chemical Process Design* should prove to be a handy reference for the process design engineer. The emphasis throughout has been to provide concise design procedures which can be used in conjunction with a small hand calculator and which will be able to handle about 80 to 90% of the engineering design problems that are encountered in the field. The methods summarized are useful for quick checks of prior designs and in some cases are sufficiently accurate for a final design.

K. D. Timmerhaus

# Preface

*Applied Chemical Process Design* was prepared to give the chemical process engineer a ready reference that can be used at the office, in the field, or while on business travel.

After spending several years in the chemical industry, we had found that each of us had a rather scattered collection of useful notes, charts, tables, articles, etc. The need to organize and consolidate these references was obvious.

This book has been intentionally kept concise, to maintain its usefulness while in the field. Theory has been virtually eliminated. However, the material presented is adequate to solve many design and/or plant problems. Those wishing to learn more of the background or theory behind the methods presented should consult the references and selected readings given at the end of each chapter.

The areas given the highest priority are those encountered most often: agitation, distillation, heat transfer, and fluid flow.

The book is intended to help students, process design engineers, pilot plant engineers, and production engineers. It is hoped that it will be of particular value to younger engineers in bridging the gap between theory and application.

## Acknowledgments

We would like to express our thanks to our colleagues at Dow Chemical, USA, whose constructive comments have been very helpful. In particular, the help of Lanny Robbins, Bruce Lovelace, Clarence Voelker, James Huff, Gerald Geyer, Douglas Leng, Thomas Tefft, Leo Schick, Jay Bleiweiss, James May, Paul Handt, and Kenneth Coulter has been appreciated. We would also like to thank Dr. James Pfafflin (Stevens Institute of Technology) and Dr. Harold Donnelly (Wayne State University) for their comments and help. Finally, we would like to thank the department secretaries (Susan Krantz, Erna Nash, Barbara Talicska, Anne Marie Duranczyk, and Nancy Roop), whose patience and perseverance have been greatly appreciated.

Frank Aerstin
Gary Street

# Figures

# Tables

**XV**

# Conversion Tables

| To convert from | To | Multiply by |
|---|---|---|
| Å | in. | $3.937 \times 10^{-9}$ |
| Å | m | $1 \times 10^{-10}$ |
| Å | microns | $1 \times 10^{-4}$ |
| ampere-hour (absolute) | coulombs (absolute) | 3,600 |
| atm | mm Hg (32°F) | 760 |
| atm | dyn/cm² | $1.0133 \times 10^6$ |
| atm | N/m² | 101,325 |
| atm | feet of water (39.1°F) | 33.90 |
| atm | g/cm² | 1,033.3 |
| atm | in. Hg (32°F) | 29.921 |
| atm | lb/ft² | 2,116.3 |
| atm | lb/in.² | 14.696 |
| barrels (oil) | m³ | 0.15899 |
| barrels (oil) | gal | 42 |
| barrels (US liquid) | m³ | 0.11924 |
| barrels (US liquid) | gal | 31.5 |
| bar | atm | 0.9869 |
| bar | N/m² | $1 \times 10^5$ |
| bar | lb/in.² | 14.504 |
| boiler horsepower | Btu/hr | 33,480 |
| boiler horsepower | kW | 9.803 |
| Btu | cal (g) | 252 |
| Btu | centigrade heat units (c.h.u. or p.c.u.) | 0.55556 |
| Btu | ft-lb | 777.9 |
| Btu | horsepower-hr | $3.929 \times 10^{-4}$ |
| Btu | J | 1,055.1 |
| Btu | kW-hr | $2.930 \times 10^{-4}$ |
| Btu/hr | W | 0.29307 |
| Btu/min | horsepower | 0.02357 |
| Btu/lb | cal/g | 0.5556 |
| Btu/lb | J/kg | 2,326 |
| Btu/lb-°F | cal/g-°C | 1 |
| Btu/lb-°F | J/kg-°K | 4,186.8 |
| Btu/sec | W | 1,054.4 |
| Btu/ft²-hr | J/m²-sec | 3.1546 |
| Btu/ft²-min | kW/ft² | 0.1758 |
| cal (g) | Btu | $3.968 \times 10^{-3}$ |
| cal (g) | ft-lb | 3.087 |

1

## Conversion Tables—(continued)

| To convert from | To | Multiply by |
|---|---|---|
| cal (g) | J | 4.1868 |
| cal (g) | horsepower-hr | $1.5591 \times 10^{-6}$ |
| cal (g)/g-$^\circ$C | J/kg-$^\circ$K | 4,186.8 |
| cal (kg) | kW-hr | 0.0011626 |
| cal (kg)/sec | kW | 4.185 |
| cal/g | Btu/lb | 1.8 |
| centigrade heat units | Btu | 1.8 |
| curies | disintegrations/min (dpm) | $2.2 \times 10^{12}$ |
| curies | coulombs/min | $1.1 \times 10^{12}$ |
| cm | Å | $1 \times 10^{8}$ |
| cm | ft | 0.03281 |
| cm | in. | 0.3937 |
| cm | m | 0.01 |
| cm | $\mu$ | 10,000 |
| $cm^2$ | $ft^2$ | 0.0010764 |
| $cm^3$ | $ft^3$ | $3.532 \times 10^{-5}$ |
| $cm^3$ | gal | $2.6417 \times 10^{-4}$ |
| $cm^3$ | oz (US fluid) | 0.03381 |
| $cm^3$ | quarts (US fluid) | 0.0010567 |
| cm Hg ($0^\circ$C) | atm | 0.013158 |
| cm Hg ($0^\circ$C) | ft $H_2O$ ($39.1^\circ$F) | 0.4460 |
| cm Hg ($0^\circ$C) | $N/m^2$ | 1,333.2 |
| cm Hg ($0^\circ$C) | $lb/ft^2$ | 27.845 |
| cm Hg ($0^\circ$C) | $lb/in.^2$ | 0.19337 |
| cm $H_2O$ ($4^\circ$C) | $N/m^2$ | 98.064 |
| cP | lb/ft-sec | $6.72 \times 10^{-4}$ |
| cP | lb/ft-hr | 2.42 |
| cP | P | 0.01 |
| cP | $\mu$P | $1 \times 10^{4}$ |
| cP | kg/hr-m | 3.60 |
| cSt | $m^2$/sec | $1 \times 10^{-6}$ |
| cSt | cP | fluid density, $g/cm^3$ |
| degrees | radians | 0.017453 |
| dyn | N | $1 \times 10^{-5}$ |
| dyn | g-cm/sec | 1 |
| dyn | poundals | $7.233 \times 10^{-5}$ |
| dyn | lb | $2.24809 \times 10^{-6}$ |
| dyn/cm | mg/in. | 2.5901 |
| $dyn/cm^2$ | atm | $9.8692 \times 10^{-7}$ |
| $dyn/cm^2$ | $lb/ft^2$ | $2.0886 \times 10^{-3}$ |
| $dyn/cm^2$ | $lb/in.^2$ | $1.4504 \times 10^{-5}$ |
| erg | J | $1 \times 10^{-7}$ |
| faraday | coulomb (absolute) | 96,500 |
| ft | m | 0.3048 |
| $ft^2$ | $m^2$ | 0.0929 |
| $ft^3$ | $cm^3$ | 28,317 |
| $ft^3$ | $m^3$ | 0.028317 |

**Conversion Tables — (continued)**

| To convert from | To | Multiply by |
|---|---|---|
| $ft^3$ | $yd^3$ | 0.03704 |
| $ft^3$ | gal | 7.481 |
| $ft^3$ | liters | 28.316 |
| $ft^3$ $H_2O$ (60°F) | lb | 62.37 |
| $ft^3$/min | $cm^3$/sec | 472.0 |
| $ft^3$/min | gal/sec | 0.1247 |
| $ft^3$/sec | gal/min | 448.8 |
| ft-poundals | Btu | $3.995 \times 10^{-5}$ |
| ft-poundals | J | 0.04214 |
| ft-lb | Btu | 0.0012856 |
| ft-lb | cal (g) | 0.3239 |
| ft-lb | ft-poundals | 32.174 |
| ft-lb | Horsepower-hr | $5.051 \times 10^{-7}$ |
| ft-lb | kW-hr | $3.766 \times 10^{-7}$ |
| ft-lb force | J | 1.3558 |
| ft-lb/sec | horsepower | 0.0018182 |
| ft-lb/sec | kW | 0.0013558 |
| g | grains | 15.432 |
| g | kg | 0.001 |
| g | lb (avoirdupois) | 0.0022046 |
| g | lb (troy) | 0.002679 |
| $g/cm^3$ | $lb/ft^3$ | 62.43 |
| $g/cm^3$ | lb/gal | 8.345 |
| g/liter | grains/gal | 58.42 |
| g/liter | $lb/ft^3$ | 0.0624 |
| $g/cm^2$ | $lb/ft^2$ | 2.0482 |
| $g/cm^2$ | lb/in.$^2$ | 0.014223 |
| gal (US liquid) | barrels (US liquid) | 0.03175 |
| gal | $m^3$ | 0.003785 |
| gal | $ft^3$ | 0.13368 |
| gal | gal (Imperial) | 0.8327 |
| gal | liters | 3.785 |
| gal | oz (US fluid) | 128 |
| grains | g | 0.06480 |
| grains | lb | $1.428 \times 10^{-4}$ |
| grains/$ft^3$ | $g/m^3$ | 2.2884 |
| grains/gal | ppm | 17.118 |
| horsepower (British) | Btu/min | 42.42 |
| horsepower (British) | Btu/hr | 2,545 |
| horsepower (British) | ft-lb/min | 33,000 |
| horsepower (British) | ft-lb/sec | 550 |
| horsepower | kW | 0.7457 |
| horsepower (British) | W | 745.7 |
| horsepower (British) | horsepower (metric) | 1.0139 |
| horsepower (metric) | ft-lb/sec | 542.47 |
| horsepower (metric) | kg-m/sec | 7.5 |
| in. | m | 0.0254 |

**Conversion Tables — (continued)**

| To convert from | To | Multiply by |
|---|---|---|
| in.$^2$ | cm$^2$ | 6.452 |
| in.$^2$ | m$^2$ | $6.452 \times 10^{-4}$ |
| in.$^3$ | m$^3$ | $1.6387 \times 10^{-5}$ |
| in. Hg (60°F) | N/m$^2$ | 3,376.9 |
| in. H$_2$O (60°F) | N/m$^2$ | 248.84 |
| J (absolute) | Btu (mean) | $9.480 \times 10^{-4}$ |
| J (absolute) | cal (g mean) | 0.2389 |
| J (absolute) | ft$^3$-atm | 0.3485 |
| J (absolute) | ft-lb | 0.7376 |
| J (absolute) | kW-hr | $2.7778 \times 10^{-7}$ |
| kcal | J | 4,186.8 |
| kg | lb (avoirdupois) | 2.2046 |
| kg (force) | N | 9.807 |
| kg/cm$^2$ | lb/in.$^2$ | 14.223 |
| kg/m$^2$ | lb/ft$^2$ | 0.2048 |
| km | miles | 0.6214 |
| kW-hr | Btu | 3,414 |
| kW-hr | ft-lb | $2.6552 \times 10^6$ |
| kW | horsepower | 1.3410 |
| knots (international) | m/sec | 0.5144 |
| knots (nautical miles per hour) | mph | 1.1516 |
| lb (avoirdupois) | grains | 7,000 |
| lb (avoirdupois) | kg | 0.45359 |
| lb (avoirdupois) | lb (troy) | 1.2153 |
| lb/ft$^3$ | g/cm$^3$ | 0.016018 |
| lb/ft$^3$ | kg/m$^3$ | 16.018 |
| lb/ft$^2$ | atm | $4.725 \times 10^{-4}$ |
| lb/ft$^2$ | kg/m$^2$ | 4.882 |
| lb/in$^2$ | atm | 0.06805 |
| lb/in.$^2$ | kg/cm$^2$ | 0.07031 |
| lb/in.$^2$ | N/m$^2$ | 6,894.8 |
| lb (force) | N | 4.4482 |
| lb (force)/ft$^2$ | N/m$^2$ | 47.88 |
| lb-centigrade units (p.c.u.) | Btu | 1.8 |
| liters | ft$^3$ | 0.03532 |
| liters | m$^3$ | 0.001 |
| liters | gal | 0.26418 |
| lumens | W | 0.001496 |
| m | ft | 3.2808 |
| micromicrons | microns | $1 \times 10^{-6}$ |
| microns | Å | $1 \times 10^4$ |
| microns | m | $1 \times 10^{-6}$ |
| miles (nautical) | ft | 6,080 |
| miles (nautical) | miles (US statute) | 1.1516 |
| miles | ft | 5,280 |

**Conversion Tables — (continued)**

| To convert from | To | Multiply by |
|---|---|---|
| miles | m | 1,609.3 |
| mils | in. | 0.001 |
| mils | m | $2.54 \times 10^{-5}$ |
| min (angle) | radians | $2.909 \times 10^{-4}$ |
| ml | $cm^3$ | 1 |
| mm | m | 0.001 |
| mm Hg ($0^{\circ}$C) | $N/m^2$ | 133.32 |
| millimicrons | microns | 0.001 |
| N | kg | 0.10197 |
| oz (avoirdupois) | kg | 0.02835 |
| oz (avoirdupois) | oz (troy) | 0.9115 |
| oz (US fluid) | $m^3$ | $2.957 \times 10^{-5}$ |
| oz (troy) | oz (apothecaries') | 1.000 |
| pints (US liquid) | $m^3$ | $4.732 \times 10^{-4}$ |
| poundals | N | 0.13826 |
| quarts (US liquid) | $m^3$ | $9.464 \times 10^{-4}$ |
| radians | degrees | 57.30 |
| rpm | radians/sec | 0.10472 |
| sec (angle) | radians | $4.848 \times 10^{-6}$ |
| slugs | kg | 14.594 |
| slugs | lb | 32.17 |
| tons (long) | kg | 1,016 |
| tons (long) | lb | 2,240 |
| tons (metric) | kg | 1,000 |
| tons (metric) | lb | 2,204.6 |
| tons (metric) | tons (short) | 1.1023 |
| tons (short) | kg | 907.18 |
| tons (short) | lb | 2,000 |
| tons (refrigeration) | Btu/hr | 12,000 |
| tons (British shipping) | $ft^3$ | 42.00 |
| tons (US shipping) | $ft^3$ | 40.00 |
| Torr (mm Hg, $0^{\circ}$C) | $N/m^2$ | 133.32 |
| W | Btu/hr | 3.413 |
| W | J/sec | 1 |
| W | kg-m/sec | 0.10197 |
| W-hr | J | 3,600 |
| yd | m | 0.9144 |
| $yd^2$ | $m^2$ | 0.8361 |
| $yd^3$ | $m^3$ | 0.76456 |

**Special Tables of Conversion Factors**

| To convert from | To | Multiply by |
|---|---|---|
| Heat Transfer | | |
| p.c.u./(hr-ft$^2$-$^\circ$C) | Btu/(hr-ft$^2$-$^\circ$F) | 1 |
| kg-cal/(hr-m$^2$-$^\circ$C) | Btu/(hr-ft$^2$-$^\circ$F) | 0.2048 |
| g-cal/(sec-cm$^2$-$^\circ$C) | Btu/(hr-ft$^2$-$^\circ$F) | 7380.0 |
| W/(cm$^2$-$^\circ$C) | Btu/(hr-ft$^2$-$^\circ$F) | 1760.0 |
| W/(in.$^2$-$^\circ$F) | Btu/(hr-ft$^2$-$^\circ$F) | 490.0 |
| Btu/(hr-ft$^2$-$^\circ$F) | p.c.u./(hr-ft$^2$-$^\circ$C) | 1 |
| Btu/(hr-ft$^2$-$^\circ$F) | kg-cal/(hr-m$^2$-$^\circ$C) | 4.88 |
| Btu/(hr-ft$^2$-$^\circ$F) | g-cal/(sec-cm$^2$-$^\circ$C) | 0.0001355 |
| Btu/(hr-ft$^2$-$^\circ$F) | W/(cm$^2$-$^\circ$C) | 0.000568 |
| Btu/(hr-ft$^2$-$^\circ$F) | W/(in.$^2$-$^\circ$F) | 0.00204 |
| Btu/(hr-ft$^2$-$^\circ$F) | hp/(ft$^2$-$^\circ$F) | 0.000394 |
| Btu/(hr-ft$^2$-$^\circ$F) | J/(sec-m$^2$-$^\circ$C) | 5.678 |
| kg-cal/(hr-m$^2$-$^\circ$C) | J/(sec-m$^2$-$^\circ$C) | 1.163 |
| W/(m$^2$-$^\circ$C) | J/(sec-m$^2$-$^\circ$C) | 1 |
| Thermal conductivity | | |
| g-cal/(sec-cm$^2$-$^\circ$C/cm) | Btu/(hr-ft$^2$-$^\circ$F/in.) | 2903.0 |
| W/(cm$^2$-$^\circ$C/cm) | Btu/(hr-ft$^2$-$^\circ$F/in.) | 694.0 |
| g-cal/(hr-cm$^2$-$^\circ$C/cm) | Btu/(hr-ft$^2$-$^\circ$F/in.) | 0.8064 |
| Btu/(hr-ft$^2$-$^\circ$F/ft) | J/(sec-m-$^\circ$C) | 1.731 |
| Btu/(hr-ft$^2$-$^\circ$F/in.) | J/(sec-m-$^\circ$C) | 0.1442 |

**Values of the Gas-Law Constant**

| Temperature scale | Pressure units | Volume units | Weight units | Energy units | $R$ |
|---|---|---|---|---|---|
| Kelvin | | | g-mol | cal | 1.9872 |
| | | | g-mol | J (absolute) | 8.3144 |
| | atm | $cm^3$ | g-mol | atm-$cm^3$ | 82.057 |
| | atm | liter | g-mol | atm-liter | 0.08205 |
| | mm Hg | liter | g-mol | mm Hg-liters | 62.361 |
| | bar | liter | g-mol | bar-liters | 0.08314 |
| | $kg/cm^2$ | liter | g-mol | liters-$kg/cm^2$ | 0.08478 |
| | atm | $ft^3$ | lb-mol | atm-$ft^3$ | 1.314 |
| | mm Hg | $ft^3$ | lb-mol | mm Hg-$ft^3$ | 998.9 |
| | psia | $ft^3$ | lb-mol | c.h.u. | 19.331 |
| | | | lb-mol | c.h.u. or p.c.u. | 1.9872 |
| Rankine | | | lb-mol | Btu | 1.9872 |
| | | | lb-mol | hp-hr | 0.0007805 |
| | | | lb-mol | kW-hr | 0.0005819 |
| | atm | $ft^3$ | lb-mol | atm-$ft^3$ | 0.7302 |
| | in. Hg | $ft^3$ | lb-mol | in. Hg-$ft^3$ | 21.85 |
| | mm Hg | $ft^3$ | lb-mol | mm Hg-$ft^3$ | 555.0 |
| | $lb/in.^2$ (absolute) | $ft^3$ | lb-mol | $(lb)(ft^3)/in.^2$ | 10.73 |
| | $lb/ft^2$ (absolute) | $ft^3$ | lb-mol | ft-lb | 1,545.0 |

# 1 Agitation and Mixing

## 1.1 Agitators

### PRELIMINARY TURBINE AGITATOR SIZING

1. Determine the dimensions of the agitator and the vessel. If these are unknown, use the section entitled Agitator Scale Up and Figure 1.1.
2. Calculate the Reynolds number [Equation (1.1)] and the power number [Equation (1.2)]. Use the section Agitator Scale Up to determine the agitator speed $N$ and Figure 1.2 to determine the power factor $N_p$.
3. Use Equation (1.3) to determine the motor horsepower.
4. Compare the result to the suggested horsepower requirements given below.
5. If the result is not within the range of the suggested values, repeat steps 1–4 with different agitator speed, and vessel and agitator dimensions.

### Suggested Horsepower Requirements (hp/1000 gal)

| Mild | 0.5–2 | Mixing, blending |
| Medium | 2–5 | Heat transfer, suspension, gas absorption |
| Violent | 5–10 | Reactions, emulsifications, suspension of fast settling slurries |

1. Higher viscosity fluids are more difficult to mix and, therefore, more horsepower is required.
2. Dual marine propellers or dual axial flow turbines provide the best means of circulating the entire contents of the vessel.

### AGITATOR SCALE UP

Assuming (1) geometrical similitude, (2) fully baffled tanks, and that (3) the Reynolds number $(N_{Re}) > 10^4$, then

$$\frac{N_1}{N_2} = \left(\frac{D_2}{D_1}\right)^x \quad \text{or} \quad ND^x = K$$

where $N$ is the rpm of the agitator, $D$ is the diameter of the agitator, and $K$ is a constant.

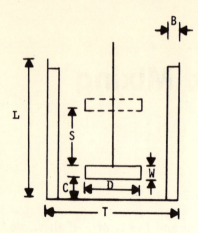

**Figure 1.1.**[2] Tank and agitator dimensions: $T$ is the tank diameter (in.), $D$ is the agitator diameter (in.), $C$ is the clearance (in.), $S$ is the spacing (in.), $B$ is the baffle width (in.), $W$ is the projected width (in.), and $L$ is the tank height (in.). Dimension ratios are as follows: $L/T = 1.0$; $B/T = 0.1$; $C/D = 0.25$–$0.3$; $S/D = 0.9$–$1.0$; $D/W = 8$; $D/T = 0.5$ to $0.6$. Where high shear is required, as in gas dispersions, use $D/T = 0.3$. For a given horsepower, the shear increases as $D/T$ decreases.

## GUIDELINES[1]

| $x$ | Criterion | Process |
|---|---|---|
| 1.0 | Constant tip speed | Suspension polymerization—equal shear |
| 0.75–0.85 | Maintain off-bottom suspension of solids | |
| 0.667 | Constant power/volume | Uniform solids suspension—complete gas–liquid homogeneity |
| 0.50 | Maintain same heat transfer and mass transfer coefficients | |
| 0 | Constant mix time, i.e., the average time to move a small mass of material around the tank is constant. | Very fast reactions |

As $x$ decreases, the ratio of power/volume increases with scale up.

## AGITATOR POWER

First calculate the Reynolds number,

$$N_{Re} = \frac{\rho N D^2}{\mu}$$

(1.1)

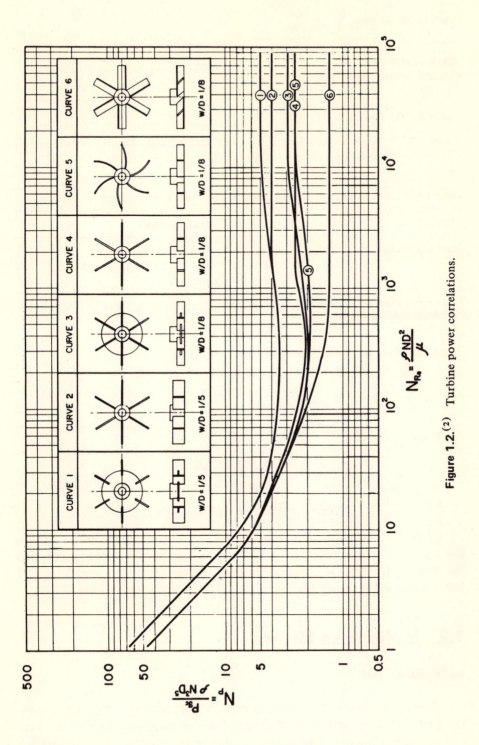

**Figure 1.2.**[2]  Turbine power correlations.

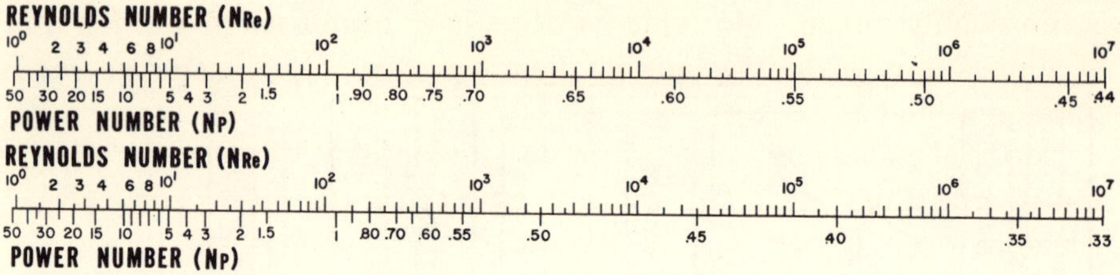

REYNOLDS NUMBER (N$_{Re}$)

POWER NUMBER (N$_P$)

REYNOLDS NUMBER (N$_{Re}$)

POWER NUMBER (N$_P$)

**Figure 1.3.**[3] Power correlations for glassed steel agitators (3-blade retreat curve). Upper scale: using 3-finger baffles; lower scale: using 1-finger baffle.

**Table 1.1.**[1]  Power Factors for Agitators in Turbulent Flow

| Single agitators | $N_p$ | Dual agitators | $N_p$ |
|---|---|---|---|
| Marine propeller | ~0.45 | Dual axial flow turbines[a] | 2.2 |
| | | Axial flow turbine above a flat blade turbine[b] | 3.2 |
| | | Dual marine propellers | 0.9 |

[a] Axial flow turbine (6 blade 45° pitch).
[b] Flat blade turbine (6 blade).

Then calculate the motor horsepower as follows:
For $N_{Re} > 10^4$ (turbulent conditions)

$$P = \frac{N_p N^3 \rho D^5}{550 g_c} \tag{1.2}$$

For $N_{Re} < 10^4$

$$P = \frac{N_p N_{Re} N^3 \rho D^5}{550 g_c}$$

$$\text{Motor horsepower} = (P \times 1.1) + 0.5 \tag{1.3}$$

$N_P$ is the power factor (from Figures 1.2 and 1.3, and Table 1.1); $P$ is the hydraulic horsepower; $N$ is the agitator speed (revolutions/sec); $\rho$ is the density (lb/ft$^3$); $D$ is the agitator diameter (ft); $g_c$ is equal to 32.2 ft/sec$^2$; $\mu$ is the viscosity (lb/ft-sec); $N_{Re}$ is the Reynolds number.

## 1.2  Motionless Mixers

### INTRODUCTION

Motionless mixers accomplish thorough fluid mixing without any moving parts. Various geometric configurations are used to split a stream into two or more segments. Further segmentation then follows until the stream is thoroughly mixed. Motionless mixers have been used in a wide variety of applications, including: mixing, heat transfer, mass transfer, gas–liquid contacting, and as reactors.

Motionless mixers such as those produced by Kenics Corporation and Charles Ross and Son Company are typical of what is commercially available.

## SCALE UP

Scale up for motionless mixers will be limited to pressure drop prediction.

While the general principles used in scaling up the mixers produced by various companies are the same, the actual calculations depend on the characteristics of the unit selected. Typical scale up examples of the units produced by Kenics and Charles Ross are presented here. For further information, the vendor should be contacted.

## KENICS STATIC MIXER, SINGLE PHASE, LIQUID FLOW

To design the Static Mixer for a blending application:

1. Derive the Reynolds number ($N_{Re}$) for the existing process line as shown below.

$$N_{Re} = 3{,}157 \frac{(Q)\,(SG)}{(\mu)\,(D)} = 50.6 \frac{(Q)\,(\rho)}{(\mu)\,(D)}$$

$$= 6.31 \frac{W}{(\mu)\,(D)} = 10 \frac{(\rho^*)\,(V)\,(D^*)}{\mu}$$

2. Using the table below, choose the approximate number of standard modules. One module consists of six mixer elements.

| $N_{Re}$ | Standard modules required | Flow characteristics |
|---|---|---|
| $<10$ | 4 | Laminar without ripples |
| 10 to 1,000 | 3 | Laminar with ripples |
| 1,000 to 2,000 | 2 | Laminar with ripples |
| $>2{,}000$ | 1 | Turbulent flow |

3. Obtain the length per module from Table 1.2. Calculate the overall length ($L$) of the Static Mixer unit.

**Table 1.2.**[(4)]   Static Mixer Unit Specification Table: Sample Table Only—Refer to Kenics Catalog for Specific Parameters

| NOM. PIPE SIZE | HOUSING SCHEDULE | OUTSIDE DIA. | | INSIDE DIA. | | MOD. LENGTH† | | $K_{OL}$ | $K'_{OL}$ | $K_{OT}$ |
|---|---|---|---|---|---|---|---|---|---|---|
| | | Inch | (mm) | Inch | (mm) | Feet | (m) | | | |
| 1/2 | 40 | 0.84 | 21.34 | 0.62 | 15.75 | .51 | .16 | 6.00 | 0.075 | 40.7 |
| 3/4 | 40 | 1.06 | 26.92 | 0.82 | 20.83 | .65 | .20 | 5.23 | 0.050 | 23.5 |
| 1 | 40 | 1.32 | 33.53 | 1.05 | 26.67 | .90 | .27 | 5.79 | 0.069 | 36.3 |
| 1 | 80 | 1.32 | 33.53 | 0.96 | 24.38 | .85 | .26 | 5.57 | 0.062 | 31.4 |
| 1½ | 40 | 1.90 | 48.26 | 1.61 | 40.89 | 1.27 | .39 | 5.72 | 0.071 | 36.8 |
| 1½ | 80 | 1.90 | 48.26 | 1.50 | 38.10 | 1.27 | .39 | 5.53 | 0.065 | 32.6 |
| 2 | 40 | 2.38 | 60.45 | 2.07 | 52.58 | 1.71 | .52 | 5.70 | 0.068 | 35.1 |
| 2 | 80 | 2.38 | 60.45 | 1.94 | 49.28 | 1.67 | .51 | 5.54 | 0.062 | 31.6 |
| 2½ | 40 | 2.88 | 73.15 | 2.47 | 62.74 | 2.30 | .70 | 5.04 | 0.053 | 24.3 |
| 2½ | 80 | 2.88 | 73.15 | 2.32 | 58.93 | 1.92 | .59 | 5.58 | 0.066 | 33.8 |
| 3 | 40 | 3.50 | 88.90 | 3.07 | 77.98 | 2.82 | .86 | 4.94 | 0.052 | 23.6 |
| 3 | 80 | 3.50 | 88.90 | 2.90 | 73.66 | 2.82 | .86 | 4.82 | 0.049 | 21.4 |
| 4 | 40 | 4.50 | 114.30 | 4.03 | 102.36 | 3.37 | 1.03 | 5.08 | 0.058 | 26.9 |
| 4 | 80 | 4.50 | 114.30 | 3.83 | 97.28 | 3.18 | .97 | 5.16 | 0.060 | 28.2 |
| 6 | 40 | 6.63 | 168.40 | 6.07 | 154.18 | 4.88 | 1.49 | 5.19 | 0.060 | 28.6 |
| 6 | 80 | 6.63 | 168.40 | 5.76 | 146.30 | 4.88 | 1.49 | 5.08 | 0.057 | 26.2 |
| 8 | 40 | 8.63 | 219.20 | 7.98 | 202.69 | 6.26 | 1.91 | 5.14 | 0.061 | 28.4 |
| 10 | 40 | 10.75 | 273.05 | 10.02 | 254.51 | 7.79 | 2.37 | 5.07 | 0.060 | 27.8 |
| 12 | 40 | 12.75 | 323.85 | 11.94 | 303.28 | 9.66 | 2.94 | 4.88 | 0.056 | 24.8 |

†Add $(T + \frac{1}{8})$ in. per flange up to a maximum of $\frac{3}{8}$ in. per flange for units with flanges ($T$ = wall thickness of housing; $\frac{1}{8}$ in. = 3.18 mm; $\frac{3}{8}$ in. = 9.53 mm).

4. Determine the pressure drop ($\Delta P$) for the fluid in an empty pipe of length $L$. See Section 6.1 for calculating $\Delta P$.

5. Obtain factor $K$:

$$N_{Re} < 10, \quad K = K_{OL}.$$

($K_{OL}$ can be obtained from Table 1.2)

$$10 < N_{Re} < 2000$$

$$K = (K'_{OL} \times A) + K_{OL} \text{ (See Figure 1.4 for A)}$$

$$N_{Re} > 2000$$

$$K = K_{OT} \times B \text{ (See Figure 1.5 for B)}$$

6. Calculate the Static Mixer pressure drop from

$$\Delta P_{SM} = (\Delta P)(K)$$

7. If the pressure drop is excessive, repeat the above procedure using a larger mixer diameter.

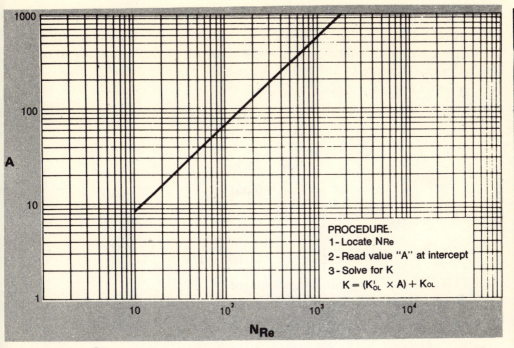

**Figure 1.4**[(4)] *A* factor vs. Reynolds number in the laminar flow region ($10 < N_{Re} < 2 \times 10^3$).

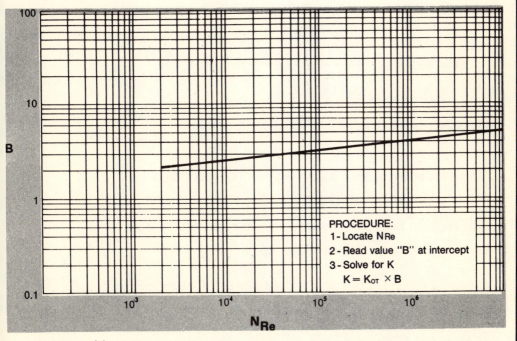

**Figure 1.5**[(4)] *B* factor vs. Reynolds number in the turbulent flow region ($N_{Re} > 2 \times 10^3$).

**Example 1.1***

How many Static Mixer modules of 2-in. Schedule 40 are required to process a Newtonian fluid with a viscosity of 100,000 cP, a density of 60 lb/ft³ and a flowrate of 500 lb/hr? What is the pressure drop?

From Table 1.2:

$$D = 2.07 \text{ in.}$$

$$K_{OL} = 5.70$$

$$K_{OT} = 35.1$$

$$K'_{OL} = 0.068$$

The Reynolds number is

$$N_{Re} = 6.31 \frac{W}{(\mu)(D)} = 6.31 \frac{500}{(10)^5 (2.07)} = 1.52 \times 10^{-2}$$

Since $N_{Re} < 10$, four modules are required.

From Figure 1.6

$$f = 4211 \quad \text{(note that } f = 64/N_{Re})$$

The length of the Static Mixer module is obtained from Table 1.2. Thus the length of four modules is

$$L = 4 \times 1.71 = 6.84$$

The pressure drop in the empty pipe is

$$\Delta P = 3.36 \times 10^{-6} \frac{(f)(L)(W^2)}{\rho(D^5)} = 3.36 \times 10^{-6} \frac{(4211)(6.84)(500)^2}{(60)(2.07)^5}$$

$$= 10.61 \text{ lb/in.}^2$$

Since $N_{Re}$ is less than 10, use Table 1.2 to obtain

$$K = K_{OL} = 5.70$$

Then, the pressure drop in four Static Mixer modules is

$$\Delta P_{SM} = (K)(\Delta P) = 5.70 \times 10.61 = 60.48 \text{ lb/in.}^2$$

The required theoretical horsepower can be calculated from

$$\text{Theoretical horsepower} = 0.262(\Delta P_{SM})(q)$$

$$= (0.262)(60.48) \frac{500}{(60)(3600)}$$

$$= 0.037 \text{ hp}$$

*After Ref. 4, with permission.

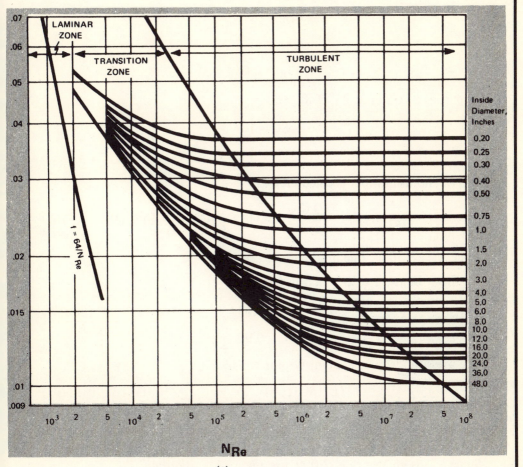

**Figure 1.6.**[4] Darcy's friction chart.

**Example 1.2\***

How many Static Mixer modules of 1-in. Schedule 40 are required to process water-like fluids at a flowrate of 5 gpm? What is the pressure drop?

From Table 1.2

$$D = 1.05 \text{ in.}$$

$$K_{OL} = 5.79$$

$$K_{OT} = 36.3$$

$$K'_{OL} = 0.069$$

Properties of water at 25°C are

$$SG = 1$$

$$\mu = 1 \text{ cP}$$

The Reynolds number is calculated as

$$N_{Re} = 3157 \frac{(Q)(SG)}{(\mu)(D)} = 3157 \frac{(5)(1)}{(1)(1.05)} = 15{,}033$$

Since $N_{Re} > 2000$, one module is required.

From Figure 1.6 the friction factor is

$$f = 0.031$$

$$L = 0.90$$

The pressure drop in the empty pipe is

$$\Delta P = 1.35 \times 10^{-2} \frac{(f)(L)(SG)(Q^2)}{D^5} = 1.35 \times 10^{-2} \frac{(0.031)(0.90)(1)(5)^2}{(1.05)^5}$$

$$= 7.38 \times 10^{-3} \text{ lb/in.}^2$$

Since $N_{Re}$ is greater than 2,000, use Figure 1.5 to obtain

$$B = 2.6; N_{Re} = 15{,}033$$

$$K = K_{OT} \times B = 36.3 \times 2.62 = 95.1$$

Thus, the pressure drop in the Static Mixer unit is

$$\Delta P_{SM} = K \, \Delta P = (95.1)(7.38 \times 10^{-3}) = 0.702 \text{ lb/in.}^2$$

\*After Ref. 4, with permission.

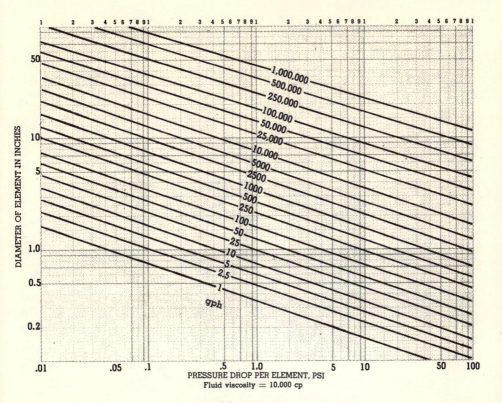

**Figure 1.7.**[5]  LPD laminar flow.

Required theoretical horsepower can be calculated from

$$\text{Theoretical horsepower} = 0.262 \, (\Delta P_{SM}) \, (q)$$

$$= (0.262) \, (0.702) \, \frac{(5) \, (0.134)}{60}$$

$$= 0.002 \text{ hp}$$

## ROSS LPD MIXER, LLPD MIXER, AND ISG MIXER, SINGLE-PHASE FLOW*

1. Compute the Reynolds number ($N_{Re}$) as outlined earlier.
2. If $N_{Re} < 500$, the flow is laminar.  Use Figure 1.7 for the LPD and Figure 1.8 for the ISG to obtain the pressure drop per element.  If

*After Ref. 5, with permission.

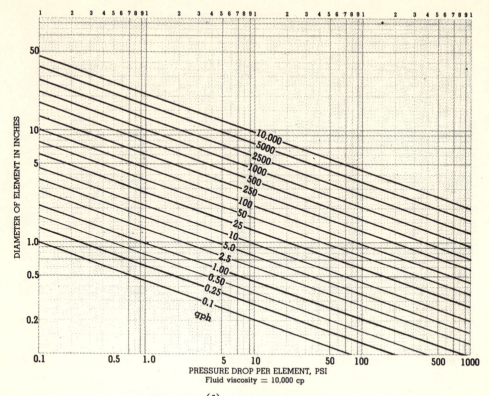

**Figure 1.8.**[5] ISG laminar flow.

$N_{Re} > 500$, the flow is turbulent. The pressure drop per element for an LPD is estimated from Figure 1.9; for an ISG, use Figure 1.10.

3. To determine the LLPD pressure drop, multiply the LPD pressure drop by 0.46.

4. Multiply the pressure drop per element by the number of elements to obtain the estimated pressure drop through the mixer. The number of elements can be estimated from Table 1.3.

5. The estimated pressure drop must be corrected for physical properties as follows. (a) For turbulent flow multiply the estimated pressure drop by the specific gravity and the correction factor $K'$ from Table 1.4 to obtain the actual pressure drop. (b) For laminar flow the pressure drop is proportional to the viscosity. Since Figures 1.7 and 1.8 are based on a fluid having a viscosity of 10,000 cP, the actual pressure drop is calculated as follows:

$$\text{actual pressure drop} = \frac{\text{fluid viscosity (cP)}}{10,000} \times \text{estimated pressure drop}$$

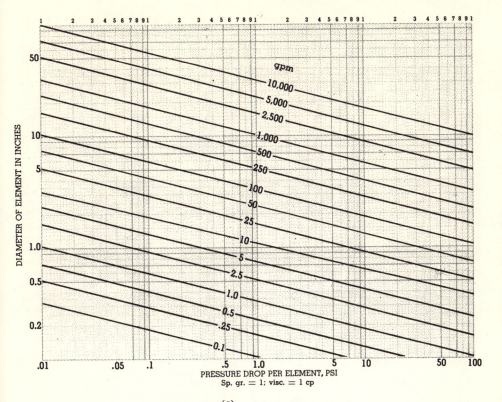

**Figure 1.9.**[5] LPD turbulent flow.

6. If the actual pressure drop exceeds the allowable value, repeat steps 1–5, using a larger-diameter mixer.

## ESTIMATION OF TWO-PHASE PRESSURE DROP (GAS–LIQUID)

A simplified version of the Lockhart and Martinelli correlation (Figure 1.11) is modified for use with the Static Mixer unit. To estimate the two-phase pressure drop, calculate the pressure drop for each phase assuming that each phase is flowing alone in the unit. The pressure drop for each phase is related to the $x$ factor as follows:

$$x = [\Delta P_L / \Delta P_G]^{1/2}$$

For flow systems with either turbulent gas phase and viscous liquid phase or vice versa, use the values midway between the TT and VV curves in Figure 1.11.

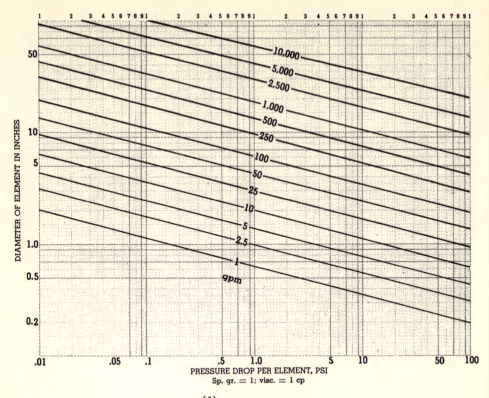

**Figure 1.10.**[5] ISG turbulent flow.

**Table 1.3.**[5] Estimating the Number of Mixer Elements

| Type of flow | Number of LPD modules | Number of ISG elements |
|---|---|---|
| Turbulent | 1, six elements | 4 |
| Laminar | | |
| Viscosity ratio[a] | | |
| ($\mu_p/\mu_m$) | | |
| 0.1 to 1,000 | 4, six elements | 10 |
| 1,000 to 10,000 | 6, six elements | 14 |
| 10,000 to 100,000 | 8, six elements | 20 |

[a]Viscosity ratio of primary stream ($\mu_p$) to minor stream ($\mu_m$).

**Table 1.4.**[5]  Viscosity Correction Factor $K'$

| Viscosity (cP) | $K'$ |
|---|---|
| 10 | 1.1 |
| 100 | 1.3 |
| 1000 | 1.5 |
| 5000 | 1.6 |

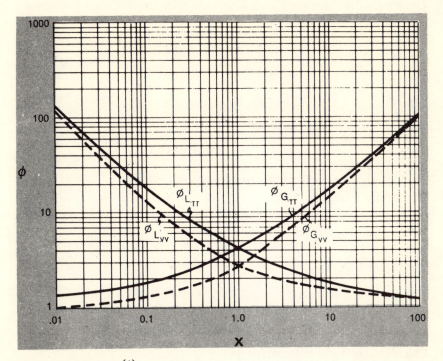

**Figure 1.11.**[4]  Parameters for pressure drop in liquid–gas flow.

## Example 1.3*

The following example illustrates the procedure: Estimate the pressure drop in a 4-in., Schedule 40 Static Mixer unit carrying 50,000 lb/hr water and 5,000 lb/hr air at 60°F. The inlet pressure is 100 psig.

The densities and viscosities for water and air are

$$\rho_{H_2O} = 62.3 \text{ lb/ft}^3 \qquad \rho_{air} = 0.56 \text{ lb/ft}^3$$

$$\mu_{H_2O} = 1 \text{ cP} \qquad\qquad \mu_{air} = 0.018 \text{ cP}$$

*After Ref. 4 with permission.

Using the methods described earlier, $\Delta P_L = 0.12$ psi/element and $\Delta P_G = 0.13$ psi/element.

1. Calculate $x$ factor

$$x = \left(\frac{\Delta P_L}{\Delta P_G}\right)^{1/2} = \left(\frac{0.12}{0.13}\right)^{1/2} = 0.96$$

2. Determine flow regime
   For water,

$$N_{Re} = 6.31 \frac{W}{(\mu)(D)} = 6.31 \frac{50,000}{(1)(4.026)} = 7.84 \times 10^4$$

For air,

$$N_{Re} = 6.31 \frac{W}{(\mu)(D)} = 6.31 \frac{5,000}{(0.018)(4.026)} = 4.35 \times 10^5$$

Thus, both phases are in the turbulent region.
3. Determine $\phi$ factors
   Since $x$ is 0.96 and both phases are turbulent, $\phi_{LTT}$ and $\phi_{GTT}$ are obtained from Figure 1.9:

$$\phi_{LTT} = 4.4$$

and

$$\phi_{GTT} = 4.2$$

4. Calculate the total pressure drop

$$\Delta P = \Delta P_L \cdot \phi_{LTT}^2 = (0.12)(4.4)^2 = 2.3 \text{ psi/element}$$

or

$$\Delta P = \Delta P_G \cdot \phi_{GTT}^2 = (0.13)(4.2)^2 = 2.1 \text{ psi/element}$$

Thus, the two-phase pressure drop is estimated to lie between 2.1 and 2.3 psi/element. Since the flow is turbulent, one Static Mixer module is required. The pressure drop in the module is

$$6\Delta P = 6 \times 2.3 = 13.8 \text{ psi}$$

or

$$6\Delta P = 6 \times 2.1 = 12.6 \text{ psi}$$

# NOMENCLATURE

### Agitators (Section 1.1)

| | |
|---|---|
| $B$ | Baffle width (in.) |
| $C$ | Clearance (in.) |
| $D$ | Agitator diameter (ft or in.) |
| $g_c$ | 32.2 ft/sec² |
| $K$ | A constant |
| $L$ | Vessel height (in.) |
| $N$ | Agitator speed (revolutions/sec) |
| $N_p$ | Power factor (from Figures 1.2 and 1.3 and Table 1.1) |
| $N_{Re}$ | Reynolds number |
| $P$ | Hydraulic horsepower |
| $S$ | Spacing (in.) |
| $T$ | Tank diameter (in.) |
| $W$ | Projected width (in.) |
| $\rho$ | Density (lb/ft³) |
| $\mu$ | Viscosity [lb/(ft-sec)] |

### Motionless Mixers (Section 1.2)

| | |
|---|---|
| $N_{Re}$ | Reynolds number (dimensionless) |
| $Q$ | Volume flowrate (gal/min) |
| $q$ | Volume flowrate (ft³/sec) |
| $W$ | Mass flowrate (lb/hr) |
| $\rho$ or $\rho^*$ | Density (lb/ft³) or (g/cm³)* |
| $\mu$ | Absolute viscosity (cP) |
| $\mu_p$ | Viscosity of primary stream |
| $\mu_m$ | Viscosity of minor stream |
| SG | Specific gravity (dimensionless) |
| $D$ or $D^*$ | ID of Static Mixer unit (in.) or (mm)* |
| $V$ | Fluid velocity (cm/sec) |
| $k'$ | Viscosity correction factor. See Table 1.4. |
| $x$ | A factor (dimensionless) |
| $\Delta P_L$ | Pressure drop of liquid phase only (psi) |
| $\Delta P_G$ | Pressure drop of gas phase only (psi) |

Two-phase pressure drop is obtained by multiplying either the liquid-phase pressure drop by $\phi_L^2$ or the gas-phase pressure drop by $\phi_G^2$. Figure 1.11 shows the correlations between $x$ and $\phi$'s.

| | |
|---|---|
| $\phi_{L_{TT}}$ | Liquid-phase pressure drop correction factor with both phases in turbulent flow region. |

$\phi_{L_{VV}}$  Liquid-phase pressure drop correction factor with both phases in viscous (or laminar) flow region.

$\phi_{G_{TT}}$  Gas-phase pressure drop correction factor with both phases in turbulent flow region.

$\phi_{G_{VV}}$  Gas-phase pressure drop correction factor with both phases in viscous (or laminar) flow region.

In most process piping applications, module diameter is identical to existing process line diameter. Power loss as a result of pressure drop across the Static Mixer module is best stated by the formula

$$K = \frac{\text{Pressure drop in the Static Mixer module}}{\text{Pressure drop in empty pipe of the same diameter and equal length}}$$

The $K$ factor for a specific process application is determined by Reynolds number as follows:

If $N_{Re}$ is less than 10, $K = K_{OL}$ (see Table 1.2)
If $N_{Re}$ is between 10 and 2,000, use Figure 1.4
If $N_{Re}$ is greater than 2,000, use Figure 1.5

Multiply $K$ by empty-pipe pressure drop to obtain the pressure drop caused by Static Mixer module installation.

## REFERENCES

1. D. Leng, Dow Chemical Co., USA, Midland, Michigan, 1973, unpublished data.
2. R. L. Bates, P. L. Fondy, and R. R. Corpstein, An examination of some geometric parameters of impeller power, *Ind. Eng. Chem. Process Des. Dev. 2* (4), 311 (1963).
3. Agitation Speed-Power Calculator, Bulletin 1018, Pfaudler Division, Ritter Pfaudler Corporation, Rochester, New York, 1961, p. 6.
4. Bulletin KTEK-2, Pressure Drop in the STATIC MIXER Unit, Kenics Corporation, N. Andover, Massachusetts, 1972, pp. K2-1–K2-9.
5. Bulletin M-376, Charles Ross and Son Company, Hauppauge, LI, New York, 1976, pp. 1–7.

## SELECTED READING

### Agitation

Agitation Speed-Power Calculator, Bulletin 1018, Pfaudler Division, Ritter Pfaudler Corporation, Rochester, N.Y., 1961.

R. L. Bates, P. L. Fondy, and R. R. Corpstein, An examination of some geometric parameters of impeller power, Industrial and Engineering Chemistry Process Design and Development, Vol. 2, No. 4, October (1963).

J. R. Connolly and R. L. Winter, Approaches to mixing operation scale-up, *Chem. Eng. Prog.*, Vol. 65, No. 8, August (1969).

D. S. Dickey and J. G. Fenic, Dimensional analysis for fluid agitation systems, *Chem. Eng.*, January 5 (1976).

D. S. Dickey and R. W. Hicks, Fundamentals of agitation, *Chem. Eng.*, February 2 (1976).

L. E. Gates, T. L. Henley, and J. G. Fenic, How to select the optimum turbine agitator, *Chem. Eng.*, December 8 (1976).

L. E. Gates, R. W. Hicks, and D. S. Dickey, Application guidelines for turbine agitators, *Chem. Eng.*, December 6 (1976).

L. E. Gates, J. R. Morton, and P. L. Fondy, Selecting agitator systems to suspend solids in liquids, *Chem. Eng.*, May 24 (1976).

R. W. Hicks and D. S. Dickey, Applications analysis for turbine agitators, *Chem. Eng.*, November 8 (1976).

R. W. Hicks and L. E. Gates, How to select turbine agitators for dispersing gas into liquids, *Chem. Eng.*, July 19 (1976).

R. W. Hicks, J. R. Morton, and J. G. Fenic, How to design agitators for desired process response, *Chem. Eng.*, April 26 (1976).

R. S. Hill and D. L. Kime, How to specify drive trains for turbine agitators, *Chem. Eng.*, August 2 (1976).

F. A. Holland and L. S. Chapman, *Liquid Mixing and Process in Stirred Tanks*, Reinhold Publishing Corp., New York, 1966.

W. S. Meyer and D. L. Kime, Cost estimation for turbine agitators, *Chem. Eng.*, September 27 (1976).

N. H. Parker, Mixing—Modern theory and practice, *Chem. Eng.*, June 8 (1964).

W. R. Penney, Recent trends in mixing equipment, *Chem. Eng.*, March 22 (1971).

W. D. Ramsey and G. C. Zoller, How the design of shafts, seals and impellers affects agitator performance, *Chem. Eng.*, August 30 (1976).

R. R. Rautzen, R. R. Corpstein, and D. S. Dickey, How to use scale-up methods for turbine agitators, *Chem. Eng.*, October 25 (1976).

J. H. Rushton and J. Y. Oldshue, Mixing—Present theory and practice, *Chem. Eng. Prog.*, Vol. 49, No. 4, April (1953) and Vol. 49, No. 5, May (1953).

V. W. Uhl and J. B. Gray, *Mixing—Theory and practice*, Vols. I and II, Academic Press, New York, 1966.

V. W. Uhl and H. P. Voznick, The anchor agitator, *Chem. Eng. Prog.*, Vol. 56, No. 3, March (1960).

A. P. Weber, Selecting turbine agitators, *Chem. Eng.*, December 7 (1964).

## Motionless Mixers

S. J. Chen, Bulletin, KTEK-1, The Static Mixer® Unit and Principles of Operation; KTEK-2, Pressure Drop in the Static Mixer® Unit; KTEK-3, Heat Transfer and Thermal Homogenization of Viscous Flow in the Static Mixer® Unit; KTEK-4, Radial Mixing and Residence Time Distribution in the Static Mixer® Unit; KTEK-5, Drop Formation of Low-Viscosity Fluids in the Static Mixer® Unit; KTEK-6, Interphase Heat and Mass Transfer Operations in the Static Mixer® Unit; KTEK-7, Dry Solids Mixing and Handling in the Static Mixer® Unit; KTEK-8, Comparative Costs of the Static Mixer® Unit and Several Conventional Mixing Devices, Kenics Corporation, N. Andover, Massachusetts, 1972.

S. J. Chen and A. R. MacDonald, Motionless mixers for viscous polymers, *Chem. Eng.*, March 19 (1973).

S. J. Chen, Static mixing of polymers, *Chem. Eng. Prog.*, Vol. 71, No. 8, August (1975).

R. Devellion, Motionless mixers, *Automation*, February (1972).

N. R. Shott, B. Weinstein, and D. LaBombard, Motionless mixers in plastic processing, *Chem. Eng. Prog.*, Vol. 71, No. 1, January (1975).

# 2  Cooling Towers[1]

This chapter outlines a procedure that may be used to select an induced draft cooling tower as normally used in process plants.

## VARIABLES IN COOLING TOWER SELECTION

The variables in selecting a cooling tower are as follows:

1. *Cooling range*  The difference in temperature between the hot water entering the tower ($T_1$) and the cold water leaving the tower ($T_2$) is the cooling range of the tower.
2. *Approach*  The difference between the temperature of the cold water leaving the tower ($T_2$) and the wet-bulb temperature of the air ($T_{WB}$) is known as the approach; establishment of the approach fixes the operating temperature of the tower and is the single most important parameter in determining both tower size and cost.
3. *Wet bulb*  That temperature ($T_{WB}$) to which air can be cooled adiabatically to saturation by the additon of water vapor.  More practically, wet-bulb temperature is the temperature indicated by a thermometer, the bulb of which is kept moist by a wick, and over which air is circulated.
4. *Dry bulb*  The temperature ($T_{DB}$) indicated by a dry-bulb thermometer.
5. *Heat load*  This is the amount of heat to be dissipated by the tower, usually expressed in Btu/hr; heat load is a function of the water circulation rate and the cooling range.  Heat load is also an important parameter in determining tower size and cost.

## PRELIMINARY DESIGN

With the above variables known, enter Figure 2.1 at *range* on the left margin and proceed horizontally to the point of intersection with the *approach* line.  At this point proceed downward to intersect with the *wet-*

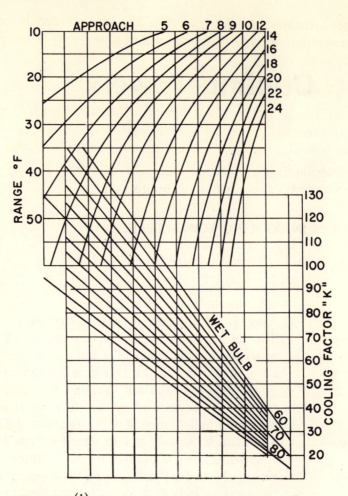

**Figure 2.1.**[1]  Cooling tower performance curves. Curves show the relationship of approach, range, wet-bulb temperature, and cooling factor $K$.

*bulb* line.  From this point proceed horizontally to the right-hand margin and read $K$ factor:

$$K \text{ factor} \times \text{gpm} \times 10^{-6} = B \text{ factor}$$

With $B$ factor known, enter Figure 2.2 and proceed vertically to horsepower and read $C$ fan horsepower factor at right hand margin, continue vertically to basin area and read basin area $C$ factor at right-hand margin.

Fan horsepower = (Fan horsepower factor $C$) $\times$ 100
Basin size = (Basin area factor $C$) $\times$ 1000 ft$^2$
Pump horsepower = 0.012 (gpm)

When estimates are made from flow diagrams or if details are not available, *concrete cold water basins* for cooling towers may be estimated from the following.

### Cooling Tower Concrete Cold Water Basins

Using Figure 2.2 the cold water basin area can be determined. For estimating purposes, assume a basin length to width ratio of $3:1$.

The following relationship assumes an 8-in. uniform thickness of basin slab and walls and allows for typical pump pit and piers for cooling tower supports.

$$yd^3 = 0.18 (L + W) + 0.033 \, WL$$

where $L$ is the length of the basin and $W$ is its width. Assume 100 lb of reinforcing steel per cubic yard of concrete.

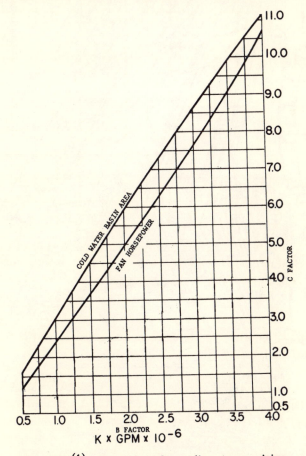

**Figure 2.2.**[1] Induced-draft cooling tower sizing curve. Basin area = $C \times 1000$ ft$^2$; pump horsepower = gpm $\times$ .012; fan horsepower = $C \times 100$.

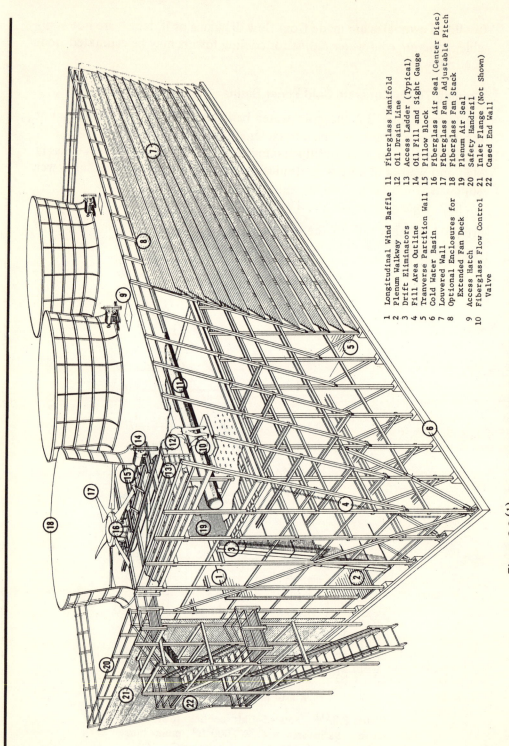

| | | | |
|---|---|---|---|
| 1 | Longitudinal Wind Baffle | 11 | Fiberglass Manifold |
| 2 | Plenum Walkway | 12 | Oil Drain Line |
| 3 | Drift Eliminators | 13 | Access Ladder (Typical) |
| 4 | Fill Area Outline | 14 | Oil Fill and Sight Gauge |
| 5 | Tranverse Partition Wall | 15 | Pillow Block |
| 6 | Cold Water Basin | 16 | Fiberglass Air Seal (Center Disc) |
| 7 | Louvered Wall | 17 | Fiberglass Fan, Adjustable Pitch |
| 8 | Optional Enclosures for | 18 | Fiberglass Fan Stack |
| | Extended Fan Deck | 19 | Plenum Air Seal |
| 9 | Access Hatch | 20 | Safety Handrail |
| 10 | Fiberglass Flow Control | 21 | Inlet Flange (Not Shown) |
| | Valve | 22 | Cased End Wall |

**Figure 2.3.**[1] Typical parts and framing for a crossflow cooling tower.

### Cooling Tower Height

Should the time of contact be insufficient, no amount of increase in the ratio of air to water will produce the desired cooling. It is therefore necessary to maintain a certain minimum height of cooling tower. Where a wide approach of 15–20°F to the wet-bulb temperature and a 25–35°F cooling range is required, a relatively low cooling tower will suffice. A tower in which the water travels 15–20 ft from the distributing system to the basin is sufficient. Where a moderate approach of 8–15°F and a cooling range of 25–35°F is required, a tower in which the water travels 25–30 ft is adequate. Where a close approach of 4–8°F with a 25–35°F cooling range is required, a tower in which the water travels from 35–40 ft is required. It is usually not economical to design a cooling tower with an approach of less than 5°F, but it can be accomplished satisfactorily with a tower in which the water travels 35–40 ft.

## REFERENCES

1. Ecodyne Corporation, private communication, 1976.

## SELECTED READING

C. A. Baird and J. E. Behen, Jr., Automation solves winter cooling tower problems, *Power Eng*. May (1975).

J. R. Buss, How to control fog from cooling towers, *Power* January (1968).

J. C. Campbell, How to prevent cooling tower fog, *Hydrocarbon Processing* December (1976).

F. Caplan, Quick calculation of cooling tower blowdown and makeup, *Chem. Eng.* July 7 (1975).

*Cooling Tower Fundamentals and Application Principles*, Marley Company, Kansas City, Mo., 1967.

*Cooling Tower Performance Curves*, Cooling Tower Institute, Houston, Texas, 1967.

D. R. DeHarpporte, Cooling tower site considerations, *Power Eng.* August (1970).

F. Friar, Cooling-tower basin design, *Chem. Eng.* July 22 (1974).

T. H. Hamilton, Estimating cooling tower evaporation rates, *Power Eng.* March (1977).

R. W. Maze, Practical tips on cooling tower sizing, *Hydrocarbon Processing* Vol. 46, No. 2, February (1967).

A. M. Rubin and P. S. Klanian, Visible plume abatement with the wet/dry cooling tower, *Power Eng.* March (1975).

E. C. Smith and M. W. Larinoff, Alternative arrangements and designs for wet/dry cooling towers, *Power Eng.* May (1976).

S. D. Strauss, Guide to evaluating cooling-tower performance, *Power* October (1975).

T. Uchiyama, Cooling tower estimates made easy, *Hydrocarbon Processing* December (1976).

# 3    Decanters[5]

The following can be used for sizing horizontal decanters for the separation of immiscible liquids. Do not use for emulsions.

## SIZING VESSEL[1]

1. Calculate holdup time with the formula:[1]

$$T = 0.1[\mu/(S_H - S_L)] \tag{3.1}$$

where $T$ is the holdup time (hr), $\mu$ is the viscosity of the continuous phase (cP), $S_H$ is the specific gravity of the bottom phase, and $S_L$ is the specific gravity of the top phase.
2. Assign a length-to-diameter ratio of 3–5:1 and size the tank to accommodate the required holdup time.
3. Provide inlet nozzles at one end at the calculated interface level. Provide outlet nozzles at other end of tank as shown in Figure 3.1.

## SIZING LOOP SEAL[2,5]

1. Use Equation (3.2) to calculate loop seal as shown in Figure 3.1.

$$Z_2 = \frac{(h_L + Z_1 - Z_3)S_L}{S_H} + Z_3 - h_H \tag{3.2}$$

where $Z_1$, $Z_2$, and $Z_3$ are the heights shown in Figure 3.1 (in. or ft); $S_L$ and $S_H$ are the specific gravities of light and heavy phases, respectively; $h_L$ and $h_H$ are the head losses in light and heavy liquid discharge piping (in. or ft). These friction losses should be minimized for proper operation.
2. Use Figure 3.2 for sizing discharge piping. To get actual pressure drop refer to Section 6.1. Figure 3.2 is based on the pipe size to carry a given volume of fluid without being completely full. This would give a minimum pressure drop in the discharge piping.
3. Consider the sensitivity of the interface[4] $Z_3$ to a change in height of the overflow $Z_2$ during flow conditions. Variations in flow can typically cause a 1-in. to 2-in. change in $Z_2$. However, in a typical system (xylene–water), this can result in a 7-in. to 15-in. change in $Z_3$. If head

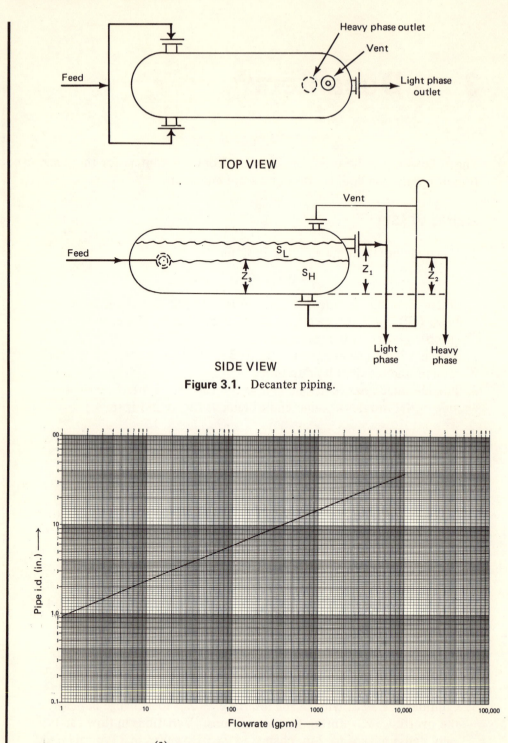

TOP VIEW

SIDE VIEW

**Figure 3.1.** Decanter piping.

**Figure 3.2.**[3] Sizing discharge piping from gravity decanters.

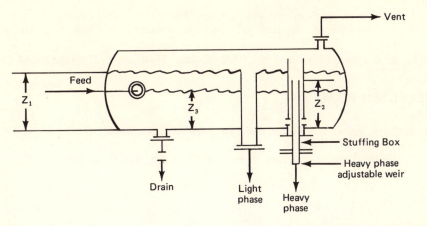

**Figure 3.3.** Liquid–liquid gravity decanter with circular overflow weirs and adjustable interface position.

losses are ignored,

$$Z_3 = Z_2 - \frac{(Z_1 - Z_3)S_L}{S_H}$$

If $Z_3$ is greater than or equal to $Z_1$, the decanter becomes unstable. The instability can be corrected by:

a. Using a larger-diameter vessel, or;
b. Designing the overflow weir such that the crest height is no more than $\frac{1}{8}$ in. at maximum flow.

In decanters a convenient weir is a circular weir—i.e., a pipe. The weir can then be adjusted with a simple stuffing box on the pipe. Figure 3.3 illustrates such a decanter.

Turbulence at the decanter inlet can be minimized by installing a tee on the inlet line.[4] A hole drilled in the tee directly opposite the incoming flow is needed. Performance can be further improved by installing several tees, mounted such that the flow from the tees opposes each other.

## REFERENCES

1. R. L. Barton, Sizing liquid–liquid phase separators empirically, *Chem. Eng.* July 8, 11 (1974).
2. R. G. Perry, C. H. Chilton, and S. D. Kirkpatrick, *Chemical Engineers' Handbook*, 4th ed., McGraw-Hill, New York, 1963 pp. 18–21.

3. L. L. Simpson, Sizing process piping, *Chem. Eng.* June 17, 204 (1968).
4. W. B. Hopper, Predicting flow patterns in plant equipment, *Chem. Eng.* August 4 (1975).
5. L. A. Robbins, Dow Chemical, USA, Midland, Michigan, 1977, unpublished data.

## SELECTED READING

R. H. Perry, and C. H. Chilton, *Chemical Engineers' Handbook*, 5th ed., McGraw-Hill, New York, 1973.
R. E. Treybal, *Liquid Extraction*, 2nd ed., McGraw-Hill, New York, 1963.

# 4     Distillation

## 4.1   Basic Laws

**Raoult's Law**

$$p_i = P_i \chi_i \tag{4.1}$$

where $p_i$ is the partial pressure of component $i$, $P_i$ is the vapor pressure of pure component $i$, and $\chi_i$ is the mole fraction of $i$ in the liquid. Holds only for *ideal* solutions; does not apply for the solute in a dilute solution.

**Henry's Law**

$$p_i = H\chi_i \tag{4.2}$$

where $H$ is Henry's law constant. Applies to the solute in a dilute solution.

**Dalton's Law**

$$p_i + p_j + p_k + \cdots = \pi \tag{4.3}$$

or

$$p_i = \pi Y_i \tag{4.4}$$

where $y_i$ is the mole fraction of $i$ in vapor and $\pi$ is the total pressure of the system. Applies only to *perfect* gas mixtures.

Dalton's law and Raoult's law can be combined to give

$$Y_i \pi = \chi_i P_i \tag{4.5}$$

## 4.2   Shortcut Method[1]—Optimum Trays and Optimum Reflux Ratio

**Minimum Plates—Fenske**

$$N_M = \frac{\log\left[(\chi_{LK}/\chi_{HK})_D \, (\chi_{HK}/\chi_{LK})_B\right]}{\log(\alpha_{LK/HK})_{avg}} \tag{4.6}$$

### Minimum Reflux—Underwood

For $\theta$,

$$\sum_{i=1}^{n} \frac{X_{iF}\alpha_{ir}}{\alpha_{ir} - \theta} = 1 - q = \frac{X_{1F}\alpha_{1r}}{\alpha_{1r} - \theta} + \frac{X_{2F}\alpha_{2r}}{\alpha_{2r} - \theta} + \cdots \qquad (4.7)$$

For $(L_0/D)_M$,

$$(L_0/D)_M + 1 = \sum_{i=1}^{n} \frac{\alpha_{ir} X_{iD}}{\alpha_{ir} - \theta} \qquad (4.8)$$

where $\alpha_{LK} > \theta > \alpha_{HK}$.

### Feed-Plate Location—Kirkbride

$$\frac{N_u}{N_L} = \left[ \left( \frac{X_{HK}}{X_{LK}} \right)_F \left( \frac{X_{LK,B}}{X_{HK,D}} \right)^2 \frac{B}{D} \right]^{0.206} \qquad (4.9)$$

### Component Distribution—Geddes

$$\log (i_D/i_B) = C + M \log \alpha_{ir} \qquad (4.10)$$

The constants $C$ and $M$ are found by solving Equation (4.10) for the split of the heavy key $(i_D/i_B)_{HK}$ and the light key $(i_D/i_B)_{LK}$ in the distillate and the bottoms. After determining $C$ and $M$, the distribution of other components may also be calculated.

## PROCEDURE*

The stepwise procedure for using the correlations presented here is as follows.

1. From the feedrate, composition and condition, and the desired recovery and/or composition of the products, calculate the distribution of the components in the distillate and bottoms.
2. From the most desirable pressure and condenser system (total or partial), calculate the dewpoint temperature of the vapor from the top plate, and the bubble-point temperature of the bottoms. If the condensing-medium temperature is the controlling factor, the corresponding dewpoint pressure at the top is determined.
3. Calculate $\alpha_{LK/HK}$ at the average column temperature, or at the feed-plate temperature if it is a "normal" column, i.e., not a stripping or rectifying column.
4. Calculate the value of $\log [(X_{LK}/X_{HK})_D (X_{HK}/X_{LK})_B]$; then, using $(\alpha_{LK/HK})_{avg}$, determine $N_M$ from Figure 4.1.

*After Ref. 1, with permission.

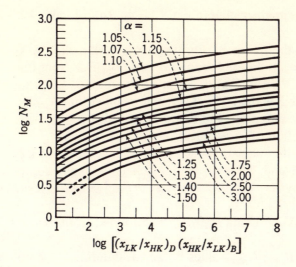

**Figure 4.1.**[1] Fenske equation for minimum plates expressed in graph form.

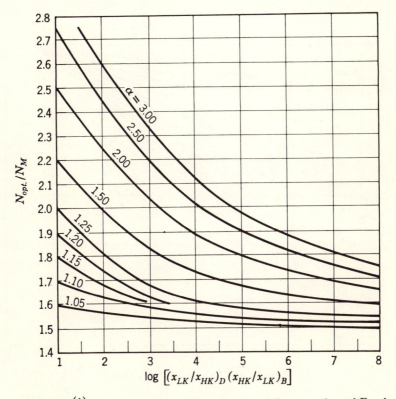

**Figure 4.2.**[1] Relation between optimum-to-minimum ratio and Fenske separation factor of $\alpha_{avg}$ values.

5. In a similar manner, determine $N_{opt}/N_M$ from Figure 4.2 and calculate $N_{opt}$.

6. Calculate the value of log $[(\chi_{LK}/\chi_{HK})_D (\chi_{HK}/\chi_{LK})_B (\chi_{LK}/\chi_{HK})_F^{0.55\alpha}]$ and using $\alpha_{avg}$, determine the ratio of $R_{opt}/R_M$ from Figure 4.3.

7. Determine $\theta$ from Figure 4.4 using the ratio of $(\chi_{LK}/\chi_{HK})_F$ and $\alpha$. If the ratio is less than 1.5, it is recommended that $N_M$ be calculated by the Fenske equation, and $\theta$ be checked by $\Sigma\alpha\chi_F/(\alpha - \theta) = 1 - q$. For an explanation of $q$, see Figure 4.8.

8. Determine the value of $(\alpha - \theta)/\alpha$ for each component from Figures 4.5 and/or 4.6 and 4.7 and divide each component composition in the distillate by its corresponding $(\alpha - \theta)/\alpha$. Sum the resulting values to get $R_M + 1$.

9. Calculate $R_M$. Using $R_{opt}/R_M$ determined from Figure 4.2, calculate $R_{opt}$.

10. From the $R_{opt}$, $N_{opt}$ values obtained, determine the column size and accessories. Figure 4.7, for instance, includes negative values of $(\alpha - \theta)/\alpha$ for the case of distributed and heavier-than-heavy keys.

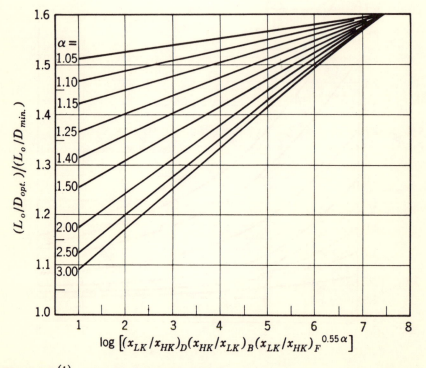

**Figure 4.3.**[(1)] Optimum–minimum reflux ratio relationship to the column's feed, distillate, and bottoms composition.

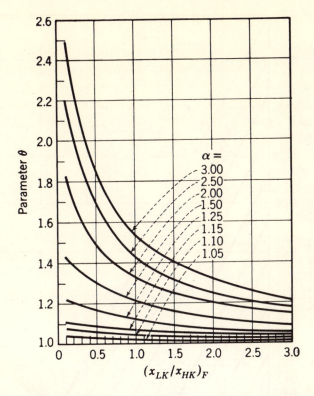

**Figure 4.4.**[1]   Underwood's $\theta$ vs. key ratios in feed.

## SAMPLE DETERMINATION*

Separation of propane (LK) and butadiene (HK) in the presence of propylene, butane, and pentane.  Desired recovery: 99% of both keys at 400 psia.  Feed is liquid at its bubble-point temperature.  Determine $N_M$, $R_M$, $N_{theor}$, and $R_{theor}$.

1. Stream compositions

| Component | Feed (%) | Distillate (%) | Bottoms (%) |
|---|---|---|---|
| Propylene (LK – 1) | 21.58 | 54.05 | 0.11 |
| Propane (LK) | 18.17 | 45.20 | 0.30 |
| Butadiene (HK) | 20.10 | 0.51 | 33.05 |
| Butane (HK + 1) | 23.12 | 0.25 | 38.24 |
| Pentane (HK + 2) | 17.03 | 0.00 | 28.29 |

*After Ref. 1, with permission.

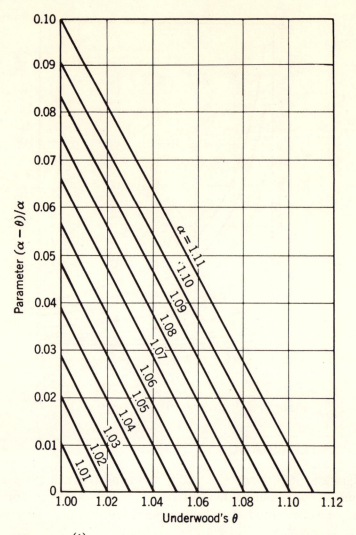

**Figure 4.5.**[1]  Underwood's $\theta$ vs. $(\alpha-\theta)/\alpha$ for $\alpha$ in range 1.01 to 1.11.

2. Temperatures

$$t_{top} = 155.44°F \qquad t_{bot} = 290.12°F \qquad t_{avg} = 222.78°F$$

3. Relative volatilities

$$\alpha_{LK-1} = 2.245 \qquad \alpha_{HK} = 1.000 \qquad \alpha_{HK+2} = 0.4333$$
$$\alpha_{LK} = 2.052 \qquad \alpha_{HK+1} = 0.935$$

4. $\log [(\chi_{LK}/\chi_{HK})_D (\chi_{HK}/\chi_{LK})_B] = 3.989$. From Figure 4.1, $\log (N_M) = 1.11$ and $N_M = 12.9$.

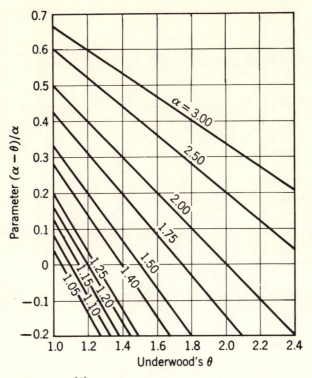

**Figure 4.6.**[1] Underwood's $\theta$ vs. $(\alpha-\theta)/\alpha$ for $\alpha$ in range of 1.05 to 3.00.

5. From Figure 4.2, $N_{opt}/N_M$ = 1.90 and $N_{opt}$ = 24.5.
6. log $[(\chi_{LK}/\chi_{HK})_D(\chi_{HK}/\chi_{LK})_B(\chi_{LK}/\chi_{HK})_F^{0.55\alpha LK}]$ = 3.94. From Figure 4.3, $R_{opt}/R_M$ = 1.37.
7. From Figure 4.4, $\theta$ = 1.35; for $\chi_{LK}/\chi_{HK}$ = (18.17/20.10) = 0.904 and $\alpha_{LK}$ = 2.052.
8. Using Figures 4.5–4.7,

| Component | Distillate (%) | $\alpha_i$ | $\dfrac{\alpha_i - \theta}{\alpha_i}$ | $\dfrac{\alpha_i \chi_{iD}}{\alpha_i - \theta}$ |
|---|---|---|---|---|
| Propylene | 54.05 | 2.245 | 0.38 | 1.4224 |
| Propane | 45.20 | 2.052 | 0.32 | 1.4125 |
| Butadiene | 0.51 | 1.000 | −0.35 | −0.0156 |
| Butane | 0.25 | 0.935 | −0.45 | −0.0055 |
| Pentane | 0.00 | 0.433 | — | — |

$$\Sigma = 2.8138$$

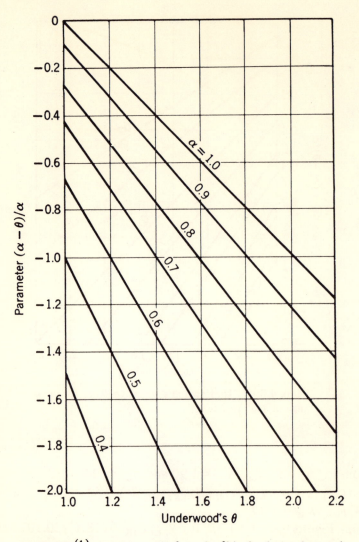

**Figure 4.7.**[1]  Underwood's $\theta$ vs. $(\alpha-\theta)/\alpha$ for heavy key and heavier components.

Therefore

$$R_M + 1 = \sum \frac{\alpha_i X_{iD}}{\alpha_i - \theta} = 2.80$$

and

$$R_M = 1.80$$

9. $R_{opt} = (1.80)\,(1.37) = 2.47$

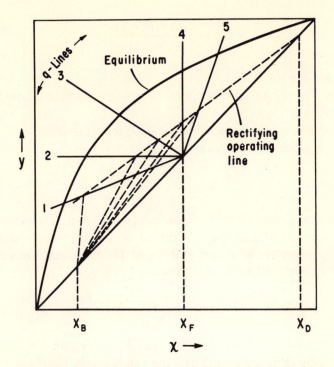

**Figure 4.8.** Effect of thermal condition of feed on feed tray location. $q$ lines: (1) $q < 0$, superheated vapor feed; (2) $q = 0$, vapor feed; (3) $0 < q < 1$, partially vaporized feed; (4) $q = 1$, liquid feed at b.p.; (5) $1 < q$, cold feed. $q$-line equation—$y = [q/(q-1)x] - [x_F/(q-1)]$.

Summary of calculations:

| Parameter | Graphical method | By equations |
|---|---|---|
| $N_M$ | 12.9 | 12.8 |
| $R_M$ | 1.80 | 1.75 |
| $N_{theor}$ | 24.5 | 24.0 |
| $R_{theor}$ | 2.47 | 2.45 |

## 4.3  Flash Vaporization

$$Y_{iD} = \frac{Z_{iF}(B/D+1)}{1 + B/(DK_i)} \qquad (4.11)$$

$$X_{iB} = \frac{Z_{iF}(B/D + 1)}{K_i + B/D} \tag{4.12}$$

1. Assume a value for $B/D$.
2. Compute $Y_{iD}$ for each component. For solutions that follow Raoult's law $K_i = P_i/\pi$.
3. If $\Sigma Y_{iD} \neq 1.0$, assume new value for $B/D$; repeat until $\Sigma Y_{iD} = 1.0$.
4. Calculate $X_{iB}$ for each component.

### Illustration*

A liquid containing 50 mol % benzene (A), 25 mol % toluene (B), and 25 mol % o-xylene (C) is flash vaporized at 1-atm pressure and 100°C. Compute the amounts of liquid and vapor products and their compositions. The system follows Raoult's law.

### Solution

Since the system follows Raoult's law, $K_i = P_i/\pi$ and $\pi = 760$ mm Hg. In the following table, column 2 lists the vapor pressures $(P_i)$ at 100°C for each component, and column 3 gives the value of $K_i$. Feed composition $(Z_{iF})$ is given in column 4. Assume $B/D = 3.0$. $Y_{iD}$ is calculated for each component. As $\Sigma Y_{iD} \neq 1.0$, a new $B/D$ is assumed until a value of $B/D = 2.08$ is found to be correct. $X_{iB}$ is then calculated for each component.

| Substance | Vapor pressure $(P,$ mm Hg$)$ | $K_i = \dfrac{P_i}{760}$ | $Z_F$ | $B/D = 3.0$ $Y_{iD} = \dfrac{Z_F(B/D + 1)}{1 + B/(DK_i)}$ | $B/D = 2.08$ $Y_{iD}$ | $X_{iB} = \dfrac{FZ_{iF} - DY_{iD}}{B}$ $= \dfrac{Y_{iD}}{K_i}$ |
|---|---|---|---|---|---|---|
| A | 1,370 | 1.803 | 0.50 | $\dfrac{0.5(3+1)}{1 + 3/1.803} = 0.750$ | 0.715 | 0.397 |
| B | 550 | 0.724 | 0.25 | | 0.1940 | 0.1983 | 0.274 |
| C | 200 | 0.263 | 0.25 | | 0.0805 | 0.0865 | 0.329 |
| | | | | $\Sigma = 1.0245$ | 0.9998 | 1.000 |

*After Ref. 2, with permission.

# 4.4   Selection of Internals[3]

**Table 4.1.**[3]   Relative Performance Ratings of Contacting Devices for Distillation Columns

|  | Bubble-cap trays | Sieve trays | Valve trays | Counterflow trays | Packings (high void)[a] | Packings (normal)[a] |
|---|---|---|---|---|---|---|
| Vapor capacity | 3 | 4 | 4 | 4 | 5 | 2 |
| Liquid capacity | 4 | 4 | 4 | 5 | 5 | 3 |
| Efficiency | 3 | 4 | 4 | 4 | 5 | 2 |
| Flexibility | 5 | 3 | 5 | 1 | 2 | 2 |
| Pressure drop | 3 | 4 | 4 | 4 | 5 | 2 |
| Cost | 3 | 5 | 4 | 5 | 1 | 3 |
| Design reliability[b] | 4 | 4 | 3 | 2 | 2 | 3 |

[a] Examples of high-void packings: Pall rings, Glitschgrid; of normal packings: Raschig rings, Berl saddles.
[b] Based on published literature.
[c] Key to ratings: (5) excellent; (4) very good; (3) good; (2) fair; (1) poor.

# 4.5   Tray Column Diameter[3-5]

Usually the vapor-flow criterion is satisfied on the basis of a *limiting* (incipient flooding) velocity, $U_F$.

For sieve or bubble-cap plates

$$U_F = Q_V / A_N = C_{SB} [(\rho_L - \rho_V)/\rho_V]^{1/2} \qquad (4.13)$$

where $U_F$ is the limiting vapor velocity (ft/sec); $Q_V$ is the vapor rate at flooding (ft$^3$/sec); $A_N$ is the net cross sectional area (ft$^2$); $C_{SB}$ is the capacity parameter (ft/sec) (see Figure 4.9); $\rho_L$ is the liquid density (lb/ft$^3$); $\rho_V$ is the vapor density (lb/ft$^3$). Since $U_F$ is the vapor velocity at incipient flooding, it should be multiplied by an appropriate factor (0.75 to 0.8) to obtain the column design velocity.

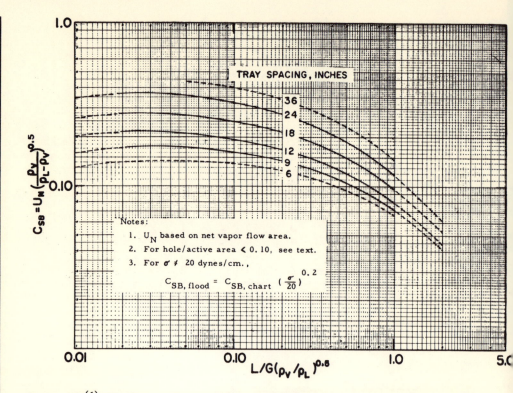

**Figure 4.9.**[6] Capacity parameter for column diameter. $\sigma$ is the surface tension (dyn/cm); $L$ is the liquid rate (lb-mol/hr); $V$ is the vapor rate (lb-mol/hr); $U_N$ is the velocity (ft/sec).

Limitations on Figure 4.9: (1) Do not use for foaming systems; (2) downcomer area $\leqslant 15\%$ of tray area; (3) use the following correction factors for $C_{SB}$ from Figure 4.9:

| $A_h/A_a$ | Correction factor |
|---|---|
| $\geqslant 10\%$ | 1.0 |
| 8% | 0.9 |
| 6% | 0.8 |

where $A_h$ is the hole area and $A_a$ is the active area of the tray.

## 4.6 Tray Overall Efficiency[6]

Figure 4.10 will usually suffice for *preliminary* evaluations. More rigorous methods are needed for final design.

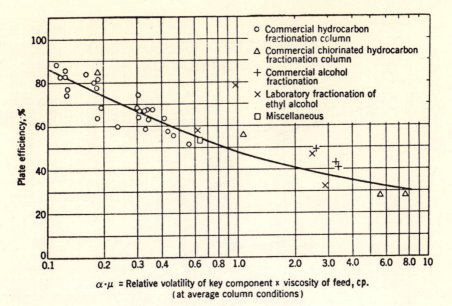

$\alpha \cdot \mu$ = Relative volatility of key component x viscosity of feed, cp.
(at average column conditions)

**Figure 4.10.**[7]   Tray overall efficiency.

## 4.7  Packed Column Design

### HEIGHT OF PACKING PER THEORETICAL STAGE*

For distillation systems where there is no chemical reaction in the liquid phase, no unusually high values of liquid viscosity or surface tension, and where there is good vapor/liquid distribution, the designer can expect to achieve a theoretical plate of separation[7] for each:

1. $2\frac{1}{4}$–$2\frac{1}{2}$ ft of packed depth for all 2-in. packings
2. $1\frac{3}{4}$–2 ft of packed depth of all $1\frac{1}{2}$-in. packings
3. $1\frac{1}{4}$–$1\frac{1}{2}$ ft of packed depth for all 1-in. packings.

These efficiencies will be realized when the flowrates, expressed as pressure drop per foot (inches of water per foot of packing) of packed depth, are within the following ranges:

1. 0.4–0.75 for Raschig rings
2. 0.2–0.75 for ceramic Intalox saddles
3. 0.1–0.75 for Pall rings and plastic Intalox saddles.

*After Ref. 7, with permission.

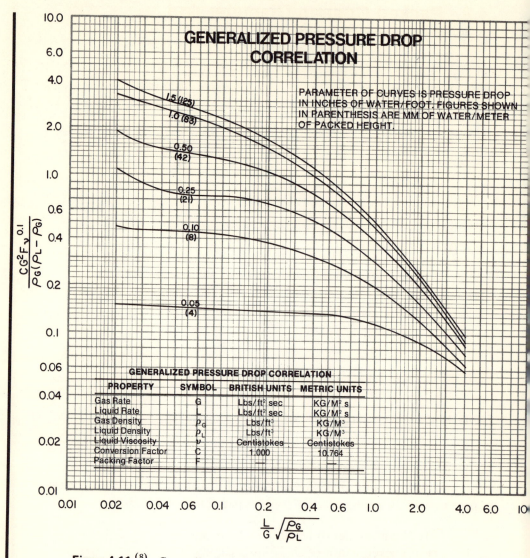

**Figure 4.11.**[8]  Generalized pressure drop correlation in packed towers.

Most packed distillation towers are designed with a safety factor of 10% to 35%, not because of variations in the behavior of the packing itself, but rather, because of imperfect distribution of the liquid,[7] improper vapor liquid equilibrium data, improper loading of the bed, or any combination of these factors.  Thus, while a 2-in. Pall ring will usually deliver a theoretical plate for each $2$–$2\frac{1}{4}$ ft of packed depth, distillation towers using them are always designed for a theoretical plate height of $2\frac{1}{4}$–$2\frac{3}{4}$ ft.

# 4.8 Packed Column Diameter and Pressure Drop[8]

For a given (or assumed) diameter and packed depth, calculate $\Delta P$. If unacceptable, repeat with new diameter and/or depth (Table 4.2, Figure 4.11).

## NOMENCLATURE

| | |
|---|---|
| $A_N$ | Net cross-sectional area (ft$^2$) |
| $B$ | Bottoms product (lb-mol/hr) |
| $C_{SB}$ | Capacity parameter (ft/sec) |
| $D$ | Distillate product (lb-mol/hr) |
| $F$ | Feed (lb-mol/hr) |
| $i$ | Any component |
| $K$ | Equilibrium vaporization ratio |
| $L_0$ | Liquid (lb-mol/hr) of external reflux returned to column |
| $H$ | Henry's law constant |
| HK | Heavy key component |
| $L_0/D$ | Reflux ratio |
| $(L_0/D)_M$ | Minimum reflux ratio |
| $(L_0/D)_{opt}$ | Optimum reflux ratio |
| LK | Light key |
| $N$ | Number of theoretical stages |
| $n$ | Number of components |
| $N_F$ | Feed-plate location, trays numbered from top to bottom |
| $N_M$ | Minimum number of theoretical stages |
| $N_L$ | Number of theoretical stages below feed |
| $N_{opt}$ | Optimum number of theoretical stages below feed |
| $N_u$ | Number of theoretical stages above feed |
| $P$ | Vapor pressure |
| $p$ | Partial pressure |
| $Q_V$ | Vapor rate at flooding (ft$^3$/sec) |
| $q$ | Ratio of moles of saturated liquid in the feed to total moles of feed. See Figure 4.8 |
| $R$ | Reflux ratio |
| $R_M$ | Minimum reflux ratio |
| $R_{opt}$ | Optimum reflux ratio |
| $t$ | Temperature (°F) |
| $U_F$ | Limiting vapor velocity (ft/sec) |
| $\chi$ | Mole fraction component in liquid |
| $Y$ | Mole fraction component in vapor |
| $Z$ | Mole fraction component in feed |

**Table 4.2.** Packing Factors for Tower Packing: Wet and Dump Packed Nominal Packing Sizes (in.)

| Type of packing | Material | $\frac{1}{4}$ | $\frac{3}{8}$ | $\frac{1}{2}$ | $\frac{5}{8}$ | $\frac{3}{4}$ | 1 | $1\frac{1}{4}$ | $1\frac{1}{2}$ | 2 | 3 | $3\frac{1}{2}$ |
|---|---|---|---|---|---|---|---|---|---|---|---|---|
| Intalox saddles | Ceramic | 725 | 330 | 200 | – | 145 | 92 | – | 52 | 40 | 22 | – |
| Intalox saddles | Plastic | – | – | – | – | – | 33 | – | – | 21 | 16 | – |
| Super Intalox | Ceramic | – | – | – | – | – | 60 | – | – | 30 | – | – |
| Super Intalox | Plastic | – | – | – | – | – | 33 | – | – | 21 | 16 | – |
| Berl saddles | Ceramic | 900[a] | – | 240[a] | – | 170[b,i] | 110[b,i] | – | 65[b,i] | 45[a] | – | – |
| Pall rings | Plastic | – | – | – | 97 | – | 52 | – | 40 | 24 | – | 16 |
| Pall rings | Metal | – | – | – | 70 | – | 48 | – | 33 | 20 | – | 16 |
| Raschig rings | Ceramic | 1,600[a,b] | 1,000[a,b] | 580[c] | 380[c] | 255[c] | 155[d] | 125[a,e] | 95[e] | 65[f] | 37[a,h] | – |
| Raschig rings $\frac{1}{32}$-in. wall | Metal | 700[a] | 390[a] | 300[a] | 170 | 155 | 115[a] | – | – | – | – | – |
| Raschig rings $\frac{1}{16}$-in. wall | Metal | – | – | 410 | 290 | 220 | 137 | 110[a] | 83 | 57 | 32[a] | – |
| Hy-Pak rings | Metal | – | – | – | – | – | 43 | – | – | 18 | 15 | – |
| Tellerettes | Plastic | – | – | – | – | – | 40 | – | – | 20 | – | – |
| Maspak | Plastic | – | – | – | – | – | – | – | – | 32 | 20 | – |
| Lessing exp. | Metal | – | – | – | – | – | – | – | 30 | – | – | – |
| Cross partition | Ceramic | – | – | – | – | – | – | – | – | – | 70 | – |

[a] Extrapolated.
[b] $\frac{1}{16}$-in. wall.
[c] $\frac{3}{32}$-in. wall.
[d] $\frac{1}{8}$-in. wall.
[e] $\frac{3}{16}$-in. wall.
[f] $\frac{1}{4}$-in. wall.
[g] $\frac{3}{8}$-in. wall.
[h] Data by Leva.

*Greek letters*

$\alpha$    Relative volatility (see below)

$\theta$    Underwood's parameter

$\rho_L, \rho_V$    Liquid or vapor density (lb/ft$^3$)

$\Sigma$    Summation

$\pi$    Total system pressure

*Subscripts*

$B$    Bottoms

$D$    Distillate

$F$    Feed

HK    Heavy key component

$i, j, k$    Any component

$ir$    As used in $\alpha_{ir}$, to indicate the $\alpha$ of component $i$ compared to a reference component $r$

LK    Light key component

LK/HK    Relative, LK to HK

M, min    Minimum

Relative volatility—for ideal solutions: $\alpha_{AB} = \dfrac{\text{vapor pressure of A}}{\text{vapor pressure of B}} = \dfrac{P_A}{P_B}$

## REFERENCES

1. M. VanWinkle and W. G. Todd, Optimum fractionation design by simple graphical methods, *Chem. Eng.* Sept. 20, 136–148 (1971).
2. R. E. Treybal, *Mass Transfer Operations*, 2nd ed., McGraw-Hill, New York, 1968, pp. 301–305.
3. J. R. Fair and W. L. Bolles, Modern design of distillation columns, *Chem. Eng.*, April 22, 156–178 (1968).
4. B. Smith, *Design of Equilibrium Stage Processes*, McGraw-Hill, New York, 1963, p. 545.
5. M. VanWinkle and J. R. Fair, Distillation in Practice, American Institute of Chemical Engineering Today Series, 1971, p. B-7.
6. H. E. O'Connell, *Trans. Am. Inst. Chem. Eng. 42*, 741 (1946).
7. J. Eckert, *Chem. Eng. Prog. 66*, 39–44 (1970).
8. Bulletin DC-11, Design Information for Packed Towers, Norton Chemical Process Products Division, Norton Company, Akron, Ohio, 1976, p. 4.

## SELECTED READING

R. K. Badhwar, Quick sizing of distillation columns, *Chem. Eng. Prog.* Vol. 66, No. 3, March (1970).

R. Billet, Recent investigations of metal pall rings, *Chem. Eng. Prog.* Vol. 63, No. 9, September (1967).

J. D. Chase, Sieve tray design (Parts I and II), *Chem. Eng.* July 31 (1967) and August 28 (1967).

H. A. Clay, J. W. Clark, and B. L. Munro, Which packing for which job, *Chem. Eng. Prog.* Vol. 62, No. 1, January (1966).

J. S. Eckert, Tower packings—Comparative performance, *Chem. Eng. Prog.* Vol. 59, No. 5, May (1963).

J. S. Eckert, Selecting the proper distillation column packing, *Chem. Eng. Prog.* Vol. 66, No. 3, March (1970).

J. S. Eckert, How tower packings behave, *Chem. Eng.* April 14 (1975).

F. G. Eichel, Capacity of packed columns in vacuum distillations, *Chem. Eng.* September 12 (1966).

R. W. Ellerbe, Batch distillation basics, *Chem. Eng.* May 28 (1973).

R. W. Ellerbe, Steam distillation basics, *Chem. Eng.* March 4 (1974).

J. R. Fair, Comparing trays and packings, *Chem. Eng. Prog.* Vol. 66, No. 3, March (1970).

J. R. Fair and W. L. Bolles, Modern design of distillation columns, *Chem. Eng.* April 22 (1968).

J. A. Gerster, Azeotropic and extractive distillation, *Chem. Eng. Prog.* Vol. 65, No. 9, September (1969).

G. C. Gester, Design and operation of a light hydrocarbon distillation drier, *Chem. Eng. Prog.* Vol. 43, No. 3, March (1947).

C. J. Haung and J. R. Hodson, Perforated trays—Designed this way, *Petroleum Refiner*, Vol. 37, No. 2, February (1958).

C. D. Holland, *Multicomponent Distillation*, Prentice-Hall, Englewood Cliffs, New Jersey, 1963.

K. K. Mahajan, Analyze tower vibration quicker, *Hydrocarbon Processing* May (1977).

A. Osborne, How to calculate three-phase flash vaporization, *Chem. Eng.* December 21 (1964).

B. D. Smith, *Design of Equilibrium Stage Processes*, McGraw-Hill, New York, 1963.

R. E. Treybal, *Mass Transfer Operations*, 2nd ed. McGraw-Hill, New York, 1968.

M. VanWinkle and J. R. Fair, Distillation in practice, *A.I.Ch.E. Today Series*, 1971.

M. VanWinkle and W. G. Todd, Optimum fractionation design by simple graphical methods, *Chem. Eng.* September 20 (1971).

# 5  Economic Evaluation

Two separate evaluation methods will be looked at: (1) Return on investment (ROI) and (2) net present value (NPV). ROI is a relatively simple measure to calculate and understand. However, it has several limitations, the most serious of which is its failure to account for the time value of money.

By time value of money, we mean the future worth of a sum of money $F$ is greater than its present worth $P$. The amount by which $F$ exceeds $P$ depends on the rate of interest used.

NPV, by contrast, is more difficult to calculate and understand. It does, however, take into account the time value of money. Therefore, it becomes a more reliable measure of a project's economic desirability.

## ROI (RETURN ON INVESTMENT)[1]

ROI can be calculated from the following:

$$ROI = \frac{sales - costs}{investment}$$

The terms in Equation (5.1) include:
1. Sales = billings – distribution costs
2. Costs

| | |
|---|---|
| Raw materials | XXXX |
| Labor | XXXX |
| Maintenance | XXXX |
| Utilities | XXXX |
| Overhead | XXXX |
| Depreciation | XXXX |
| Insurance and taxes | XXXX |
| Miscellaneous (quality control, waste disposal, etc.) | XXXX |
| Total bulk cost .................................. | XXXX |
| Packaging | XXXX |
| Total plant cost  .................................. | XXXX |
| Selling | XXXX |

|  |  |
|---|---|
| Administration | XXXX |
| Research | XXXX |
| Total cost for sale .............................. | XXXX |

3.  Investment (capital)

|  |  |
|---|---|
| Direct fixed capital | XXXX |
| Raw materials manufactured capital | XXXX |
| Utilities manufactured capital | XXXX |
| Overhead capital | XXXX |
| Materials and supplies | XXXX |
| Raw material inventory | XXXX |
| In process inventory | XXXX |
| Total bulk manufacturing capital . ................... | XXXX |
| Inventory for sale | XXXX |
| Selling | XXXX |
| Administration | XXXX |
| Research | XXXX |
| Cash and accounts receivable | XXXX |
| Total capital for sale .............................. | XXXX |

## ROI Ground Rules

1.  Plant is operating at capacity.
2.  Evaluation is made after the plant has achieved smooth operation following start up.
3.  Straight line depreciation.
4.  Total capital is used in the calculations.

## ROI Limitations

1.  Time value of money is ignored.
2.  Cannot handle irregular costs well.
3.  Measures the operation only at capacity.
4.  Does not allow for depreciation as it occurs (e.g., accelerated depreciation schedules).
5.  Does not allow for cost and price changes with time.
6.  Cannot be used in special cases such as (a) where salvage value is appreciable and (b) lease vs. purchase.

## ROI Example

| | |
|---|---|
| Plant capacity ............................... | 20,000,000 lb/yr |
| Selling price .......................................... | 32¢/lb |
| Freight to customers (average) ...................... | $40.00/ton |
| Raw material costs .......................... | 10¢/lb of product |

Conversion costs ................................... $700M/yr
Packaging costs ............................... 1¢/lb of product
Additional costs for sale ........................... $600 M/yr
Capital required to build plant ........................ $3000 M
Allocated fixed capital ............................. $1500 M
Working capital
   A.  Average days of raw material on hand .............. 36.5 days
   B.  Supplies and spare parts ............. 2% of direct fixed capital
   C.  Others ................................... $240 M

1. Prepare a Cost Summary showing cost per year.
2. Prepare a Capital Summary showing capital requirements.
3. Calculate ROI.

**Solution**

1. Cost Summary
   Freight
      20,000,000 lb = 10,000 tons
      10,000 × $40. ................................. 0.4 MM
   Raw material ..................................... 2.0 MM
   Conversion costs ................................. 0.7 MM
   Packaging ....................................... 0.2 MM
   Additional costs for sale ........................ 0.6 MM
   Cost/yr ...................................... $3.9 MM
2. Capital Summary
   Direct fixed capital ............................. 3.0 MM
   Allocated fixed capital .......................... 1.5 MM
   Working capital
      36.5 days = 2 MM lb raw material
      2 MM × 0.1 .................................. 0.2 MM
   Supplies
      3 MM × 0.02 ................................. 0.06 MM
   Others ......................................... 0.24 MM
   Capital ....................................... $5.0  MM
3. ROI

$$ROI = \frac{sales - costs}{investment}$$

$$ROI = \frac{6.4 \text{ MM} - 3.9 \text{ MM}}{5.0 \text{ MM}} = 50\%$$

## TIME VALUE OF MONEY

The future value $F$ of a sum of money is related to its present value $P$ by the interest rate $i$. This relationship is known as the time value of money. The interest can be either simple or compound.

### Example: Simple Interest

$$F = P(1 + ni)$$

where $P$ is the present value; $i$ is the interest rate per period of time; $n$ is the number of time periods; $F$ is the value in the future.

For example, $1,000 is deposited for 6 months at an annual interest rate of 6%. What is the value of the deposit after 6 months?

$P = \$1,000.$
$i = 0.06$ (rate per year)
$n = 0.5$ (years)
$F = 1,000(1 + 0.5 \times 0.06) = \$1,030.00$

### Examples: Compound Interest

1. Future value of an amount $P$, with compound interest:

$$F = P(1 + i)^n = PC_F$$

where

$$C_F = (1 + i)^n$$

For example, $1,000 is deposited for 18 months at an annual interest rate of 6%. Interest is compounded and paid quarterly. What is the value of the deposit after 18 months?

$P = \$1,000$
$i = 0.06/4 = 0.015$ (rate per quarter)
$n = 18/3 = 6$ quarters
$F = 1,000 (1 + 0.015)^6 = \$1,093.44$

2. Present value of a periodic payment $R$, with compound interest:

$$P = \frac{R}{i} \left[ \frac{(1 + i)^n - 1}{(1 + i)^n} \right] = \frac{R}{i} \left( \frac{C_F - 1}{C_F} \right)$$

For example: A 20-year mortgage is taken out on a house for $20,000 with an annual interest rate of 9%. Interest is compounded monthly on the unpaid balance. What will be the monthly payment?

$P = 20,000$

$i = 0.09/12 = 0.0075$ (rate per month)

$n = 20$ years $\times$ 12 months/year $= 240$ months

$R = ?$

$$20,000 = R \left[ \frac{(1 + 0.0075)^{240} - 1}{0.0075(1 + 0.0075)^{240}} \right]$$

$R = \$179.95$

3. Future worth of a periodic payment with compound interest:

$$F = R \left[ \frac{(1 + i)^n - 1}{i} \right] = \frac{R}{i}(C_F - 1)$$

For example, suppose the $179.95 of the above example is deposited each month in an account that pays 9% annual interest, compounded and paid monthly. After 20 years, how much will be in the account?

$R = 179.95$

$i = 0.0075$ (rate per month)

$n = 240$ months

$$F = \frac{179.95}{0.0075} [(1 + 0.0075)^{240} - 1] = \$120,186.24$$

## Example: Continuous Compounding

$$F = Pe^{in}$$

For example, $1,000 is deposited at 6% annual interest, compounded continuously. How much will be in the account after 18 months? (Compare result to example 1.)

$P = \$1,000$

$i = 0.06/12 = 0.005$ (rate per month)

$n = 18$ months

$F = 1,000e^{0.005 \times 18} = \$1,094.17$

## NPV (NET PRESENT VALUE)*

If $P$ represents a net annual dollar profit after taxes and depreciation, we may define a venture profit as

$$V = P - i(I + I_w) \tag{5.2}$$

where $V$ is the venture profit (dollars/yr); $I$ is a capital investment for facilities, considered here, for simplicity as a single lump sum (in dollars);

*After Ref. 2, with permission.

$I_w$ is the working capital (in dollars); $i$ is a corporate base minimum acceptable return rate (as a fraction); $P$ is the net annual dollar profit after taxes and depreciation.

Equation (5.2) defines $V$ as an annual net dollar profit after taxes and depreciation, reduced by an amount of money which represents the minimum acceptable charge to a specific project under consideration for all the capital it requires. The rate $i$ may be considered to be the project borrowing rate or the charge which the project must bear for its capital requirements. This quantity is usually, but not necessarily, fixed somewhat higher than the company cost of capital.

If a project operates over a period of $n$ years the attractiveness may be judged by summing up a series of annual venture profits $V$, each discounted to its present worth at a discount rate $i$. Thus, we may define the net present value (NPV) as

$$\text{NPV} = \sum_{k=1}^{k=n} \frac{V_k}{(1+i)} k \qquad (5.3)$$

where $V_k$ is the venture profit for any year $k$ and $i$ is a discount rate, representing the rate at which future earnings are to be discounted to arrive at a present value.

If the different cash flows which contribute to the numerical value of $V_k$ are separately identified, various expanded algebraic expressions are possible. One such alternate expression for NPV may be written as:

$$\text{NPV} = \sum_{k=1}^{k=n} \frac{(1-t)R_k}{(1+i)^k} + \sum_{k=1}^{k=r} \frac{d_k tI}{(1+i)^k}$$

| Net present value | = | Summation of present worth of gross untaxed income | + | Summation of present worth of discounted tax credit |
|---|---|---|---|---|

$$- \quad I \quad - \left[\frac{(1+i)^n - 1}{i(1+i)^n}\right] iI_w + \frac{(1-t)S_a}{(1+i)^n} \qquad (5.4)$$

| − Initial investment | − | Present worth of cost of working capital | + | Present worth of salvage |
|---|---|---|---|---|

In more complicated cases, tabulated year by year, cash flows discounted to their present worth will be more convenient than using the analytical expressions. The algebraic expressions do show certain relations more clearly, however.

In Equation (5.4), NPV is seen to be the present worth of the total cash flows into and out of the project. $R_k$ is a net profit in any year $k$ before deductions for income taxes and depreciation; $t$ the levelized effective income tax rate as a fraction; $d_k$ a depreciation rate over $r$ years;

and $S_a$ the estimated salvage value after $n$ years of operation. $I$ and $I_w$ are the investment and working capital, respectively. Note that the annual summations are discounted basically at the rate $i$, which may be viewed as the company cost of capital.

The NPV measures the direct incentive in dollars to invest in a given proposal as a bonus or premium over the amount an investor would otherwise earn by investing the same money in some presumably safe alternative, that would give him a return calculated at the rate $i_m$. The more positive NPV is, the more attractive the proposition. If NPV is 0 the project is marginal, and if it is negative, the proposal is unattractive.

## REFERENCES

1. J. L. Schick, Dow Chemical Co., USA, Midland, Michigan, 1976, unpublished data.
2. W. H. Kapfer, Appraising rate of return methods, *Chem. Eng. Prog.*, *65*, 55–60 (1969).

## SELECTED READING

J. Happel, *Chemical Process Economics*, John Wiley & Sons, New York, 1958.

F. A. Holland, F. A. Watson, and J. K. Wilkinson, Part 1, Engineering economics for chemical engineers, June 25, (1973); Part 2, Capital costs and depreciation, July 23, (1973); Part 3, Profitability of invested capital, August 20 (1973); Part 4, Time value of money, September 17 (1973); Part 5, Methods of estimating project profitability, October 1 (1973); Part 6, Sensitivity analysis of project profitabilities, October 29 (1973); Part 7, Time, capital and interest affect choice of project, November 26 (1973); Part 8, Statistical techniques improve decision-making, December 24 (1973); Part 9, Probability techniques for estimates of profitability, January 7 (1974); Part 10, Estimating profitabilities when uncertainties exist, February 4 (1974); Part 11, Numerical measures of risk, March 4 (1974); Part 12, How to estimate capital costs, April 1 (1974); Part 13, Manufacturing costs and how to estimate them, April 15 (1974); Part 14, How to budget and control manufacturing costs, May 13 (1974); Part 15, How to allocate overhead cost & appraise inventory, June 10 (1974); Part 16, Principles of accounting, July 8 (1974); Part 17, How to evaluate working capital for a company, August 5 (1974); Part 18, Financing assets by equity and debt, September 2 (1974); Part 19, How to assess your company's progress, September 16 (1974); Part 20, Inflation and its impact on costs and prices, October 28 (1974), *Chem. Eng.*

W. H. Kapfer, Appraising rate of return methods, *Chem. Eng. Prog.* Vol. 65, No. 11, November (1969).

M. Souders, Engineering economy, *Chem. Eng. Prog.* Vol. 62, No. 3, March (1966).

K. D. Timmerhaus and M. S. Peters, *Plant Design and Economics for Chemical Engineers*, 2nd Ed., McGraw-Hill, New York, 1968.

# 6 Fluid Flow

## 6.1 Fluid Flow—Single Phase

### FRICTION LOSS CALCULATION

1. Choose a trial line size by knowing the flow capacity and suggested velocity in Table 6.1.
2. Estimate or measure total linear feet of pipe.
3. Estimate equivalent length of all fittings, valves, expansions, contractors, entrances, and exits using Tables 6.2–6.6. Use Table 6.7 for estimating pressure drop for Teflon-lined valves.
4. Use Table 6.8 for estimating $\Delta P/100$ ft of piping. Use Figure 6.1 to correct values of Table 6.8 for fluids other than water.
5. For plastic-lined pipe use Figure 6.2 for estimating pressure drop.
6. Since Table 6.8 and Figure 6.2 are only good for turbulent flow, use Figure 6.3 to check for turbulent flow. If laminar, either reduce line size to get turbulent flow or use Figure 6.7 to calculate pressure drop in laminar flow.

Table 6.9 gives the flow of air through Schedule 40 pipe. Figures 6.4 and 6.5 are used for estimating the pressure drop of gases in pipelines where the gas density does not change by more than 10% between inlet and outlet. To use Figure 6.4 or 6.5 (1) read factor from gas flow and pipe size and (2) divide factor by gas density $\rho_V$ to get

$$\Delta P = \frac{\text{psi}}{100 \text{ ft}}$$

$$\rho_V = \frac{(\text{psia}) (144) (\text{mol. wt.})}{(1545) (460 + {}^\circ\text{F})}$$

For a high-pressure drop in gas piping the reader should consult the section on compressible flow (see Figures 6.9 and 6.10).

**Table 6.1.**[1]  Suggested Fluid Velocities in Pipe and Tubing [a]

| Fluid | Suggested Trial Velocity | Pipe Material |
|---|---|---|
| Acetylene (Observe pressure limitations) | 4000 fpm | Steel |
| Air, 0 to 30 psig | 4000 fpm | Steel |
| Ammonia | | |
| Liquid | 6 fps | Steel |
| Gas | 6000 fpm | Steel |
| Benzene | 6 fps | Steel |
| Bromine | | |
| Liquid | 4 fps | Glass |
| Gas | 2000 fpm | Glass |
| Calcium Chloride | 4 fps | Steel |
| Carbon Tetrachloride | 6 fps | Steel |
| Chlorine (Dry) | | |
| Liquid | 5 fps | Steel, Sch. 80 |
| Gas | 2000—5000 fpm | Steel, Sch. 80 |
| Chloroform | | |
| Liquid | 6 fps | Copper & Steel |
| Gas | 2000 fpm | Copper & Steel |
| Ethylene Gas | 6000 fpm | Steel |
| Ethylene Dibromide | 4 fps | Glass |
| Ethylene Dichloride | 6 fps | Steel |
| Ethylene Glycol | 6 fps | Steel |
| Hydrogen | 4000 fpm | Steel |
| Hydrochloric Acid | | |
| Liquid | 5 fps | Rubber Lined R. L., Saran, Haveg |
| Gas | 4000 fpm | |
| Methyl Chloride | | |
| Liquid | 6 fps | Steel |
| Gas | 4000 fpm | Steel |
| Natural Gas | 6000 fpm | Steel |
| Oils, lubricating | 6 fps | Steel |
| Oxygen (ambient temp.) | 1800 fpm Max. | Steel (300 psig Max.) |
| (Low temp.) | 4000 fpm | Type 304 SS |
| Propylene Glycol | 5 fps | Steel |

| Fluid | Suggested Trial Velocity | Pipe Material |
|---|---|---|
| Sodium Hydroxide | | |
| 0—30 Percent | 6 fps | Steel |
| 30—50 Percent | 5 fps | and |
| 50—73 Percent | 4 | Nickel |
| Sodium Chloride Sol'n. | | |
| No Solids | 5 fps | Steel |
| With Solids | (6 Min.—15 Max.) | Monel or nickel |
| Perchlorethylene | 7.5 fps | Steel |
| Steam | | |
| 0—30 psi Saturated* | 6 fps | Steel |
| 30—150 spi Saturated or superheated* | 4000—6000 fpm | Steel |
| 150 psi up superheated | 6000—10000 fpm | |
| *Short lines | 6500—15000 fpm 15,000 fpm (max.) | |
| Sulfuric Acid | | |
| 88—93 Percent | 4 fps | S. S.—316, Lead |
| 93—100 Percent | 4 fps | Cast Iron & Steel, Sch. 80 |
| Sulfur Dioxide | 4000 fpm | Steel |
| Styrene | 6 fps | Steel |
| Trichlorethylene | 6 fps | Steel |
| Vinyl Chloride | 6 fps | Steel |
| Vinylidene Chloride | 6 fps | Steel |
| Water | | |
| Average service | 3–8 (avg. 6) fps | Steel |
| Pump suction lines | 3–8 fps | Steel |
| Maximum economical (usual) | 7–10 fps | Steel |
| Sea and brackish water, lined pipe | 5-8 fps}3 | R. L., concrete, asphalt-line, saran-lined pipe |
| Concrete | 5-12 fps (Min.) | transite |

[a]The velocities are suggestive only and and are to be used to approximate line size as a starting point for pressure drop calculations.  The final line size should be such as to give an economical balance between pressure drop and reasonable velocity.  R. L. is rubber-lined steel.

**Table 6.2.**[2]   Resistance of Flanged Elbows, Tees, and Bends in Equivalent Pipe Length (ft)

| Pipe size (inch) | 90° Elbows* | | 90° Bends* | | Tee | | Pipe size (inch) |
| | Short radius R=1D | Long radius R=1.5D | R=5D | R=10D | Flow through branch | Flow through | |
|---|---|---|---|---|---|---|---|
| 1½ | 4.5 | 3 | 2.5 | 4 | 8 | 3 | 1½ |
| 2 | 5.25 | 3.5 | 3 | 5 | 11 | 3.5 | 2 |
| 2½ | 6 | 4 | 3.5 | 6 | 13 | 4 | 2½ |
| 3 | 7.5 | 5 | 4 | 7.5 | 16 | 5 | 3 |
| 4 | 10.5 | 7 | 5.5 | 10 | 20 | 7 | 4 |
| 6 | 15 | 10 | 8.5 | 15 | 30 | 10 | 6 |
| 8 | 21 | 14 | 11 | 20 | 40 | 14 | 8 |
| 10 | 24 | 16 | 14 | 25 | 50 | 16 | 10 |
| 12 | 32 | 21 | 16 | 30 | 60 | 21 | 12 |
| 14 | 33 | 22 | 19 | 33 | 65 | 22 | 14 |
| 16 | 39 | 26 | 21 | 38 | 75 | 26 | 16 |
| 18 | 44 | 29 | 24 | 42 | 86 | 29 | 18 |
| 20 | 48 | 32 | 27 | 50 | 100 | 32 | 20 |
| 24 | 57 | 38 | 32 | 60 | 120 | 38 | 24 |

* Estimate 50 percent of tabulated values for 45° elbows and bends. Double tabulated values for 180° returns.

**Table 6.3.**[3]   Resistance of Screwed Elbows, Tees, and Bends in Equivalent Pipe Length (ft)

| | 1 in. | 1½ in. | 2 in. | 2½ in. | 3 in. | 4 in. |
|---|---|---|---|---|---|---|
| Elbow—45° | 1.3 | 2.1 | 2.7 | 3.2 | 4.0 | 5.5 |
| Elbow—90° | 5.2 | 7.4 | 8.5 | 9.3 | 11 | 13 |
| Elbow—90° long radius | 2.7 | 3.4 | 3.6 | 3.6 | 4.0 | 4.6 |
| Tee—run thru | 3.2 | 5.6 | 7.7 | 9.3 | 12 | 17 |
| Tee—thru side | 6.6 | 9.9 | 12 | 13 | 17 | 21 |
| 180° return bend | 5.2 | 7.4 | 8.5 | 9.3 | 11 | 13 |
| Gate valve | 0.84 | 1.2 | 1.5 | 1.7 | 1.9 | 2.5 |
| Globe valve | 29 | 42 | 54 | 62 | 79 | 110 |
| Swing check valve | 11 | 15 | 19 | 22 | 27 | 38 |
| Angle valve | 17 | 18 | 18 | 18 | 18 | 18 |

**Table 6.4.(2)** Resistance of Valves in Equivalent Pipe Length (ft)

| Nominal pipe size, in. | Gate valves, fully open | Globe valves, fully open, bevel or plug type seat* 90° | Globe valves, fully open, bevel or plug type seat* 60° | Globe valves, fully open, bevel or plug type seat* 45° | Check valves Swing | Check valves Ball | Straight through cock port area open Port area = pipe area | Three way cock Port area equals 80% of pipe area Straight through Flow | Three way cock Port area equals 80% of pipe area Flow through branch | Butterfly valve fully open | Nominal pipe size, in. |
|---|---|---|---|---|---|---|---|---|---|---|---|
| 1½ | 1.75 | 46 | 23 | 18 | 17 | 20 | 2.5 | 6 | 20 | 6 | 1½ |
| 2 | 2.25 | 60 | 30 | 24 | 22 | 25 | 3.5 | 7.5 | 24 | 8 | 2 |
| 2½ | 2.75 | 70 | 38 | 30 | 27 | 30 | 4 | 9 | 30 | 10 | 2½ |
| 3 | 3.5 | 90 | 45 | 38 | 35 | 38 | 5 | 12 | 36 | 12 | 3 |
| 4 | 4.5 | 120 | 60 | 48 | 45 | 50 | 6.5 | 15 | 48 | 15 | 4 |
| 6 | 6.5 | 175 | 88 | 72 | 65 | 75 | 10 | 22 | 70 | 23 | 6 |
| 8 | 9 | 230 | 120 | 95 | 90 | 100 | 13 | 30 | 95 | 27 | 8 |
| 10 | 12 | 280 | 150 | 130 | 120 | 130 | 16 | 38 | 120 | 35 | 10 |
| 12 | 14 | 320 | 170 | 145 | 140 | 150 | 19 | ... | ... | 40 | 12 |
| 14 | 15 | 380 | 190 | 160 | 150 | 170 | 20 | ... | ... | 45 | 14 |
| 16 | 17 | 420 | 220 | 180 | 170 | 190 | 22 | ... | ... | 50 | 16 |
| 18 | 18 | 480 | 250 | 205 | 180 | 210 | 24 | ... | ... | 58 | 18 |
| 20 | 20 | 530 | 290 | 240 | 200 | 240 | 27 | ... | ... | 64 | 20 |
| 24 | 32 | 630 | 330 | 270 | 250 | 290 | 33 | ... | ... | 78 | 24 |

* For partially closed globe valves multiply tabulated values as follows;
Three-quarter open   x3
One-half open   x12
One-quarter open   x70

**Table 6.5.**[(2)] Resistance of Eccentric and Concentric Reducers and Sudden Line Size Changes in Equivalent Length (ft)

| Nominal sizes (inch) | | | | Nominal sizes (inch) | | | |
|---|---|---|---|---|---|---|---|
| $d_1$ | $d_2$ | | | $d_1$ | $d_2$ | | |
| ¾ | ½ | .6 | .5 | | | | |
| 1 | ½ | 1.2 | .7 | 14 | 6 | 22 | 13 |
|   | ¾ | .6 | .6 | | 8 | 22 | 14 |
| 1½ | ¾ | 1.6 | 1.0 | | 10 | 15 | 13 |
|   | 1 | 1.2 | .9 | | 12 | 6 | 6 |
| 2 | 1 | 2.2 | 1.3 | 16 | 8 | 27 | 17 |
|   | 1½ | 1.3 | 1.3 | | 10 | 23 | 17 |
| 3 | 1½ | 3.8 | 2.4 | | 12 | 15 | 15 |
|   | 2 | 2.7 | 2.3 | | 14 | 7 | 7 |
| 4 | 2 | 5 | 3.2 | 18 | 10 | 30 | 18 |
|   | 3 | 3 | 3 | | 12 | 23 | 19 |
| 6 | 3 | 8 | 5 | | 14 | 15 | 15 |
|   | 4 | 4 | 4 | | 16 | 4 | 4 |
| 8 | 4 | 12 | 7 | 20 | 12 | 30 | 23 |
|   | 6 | 7 | 7 | | 14 | 21 | 19 |
| 10 | 4 | 15 | 8 | | 16 | 13 | 13 |
|   | 6 | 14 | 9.5 | | 18 | 5 | 5 |
|   | 8 | 6 | 6 | | | | |
| 12 | 6 | 19 | 12 | 24 | 16 | 30 | 25 |
|   | 8 | 14 | 12 | | 18 | 25 | 25 |
|   | 10 | 6.5 | 6.5 | | 20 | 12 | 12 |

Note: Add these lengths to the smaller pipe equivalent length.

**Table 6.6.**[2]   Resistance of Horizontal and Vertical Inlets
and Outlets in Equivalent Pipe Length (ft)

| Resistance coeff.: | K = 1.0 | K = .78 | K = .5 | K = .23 |
|---|---|---|---|---|
| Nominal pipe size (inch) | | | | |
| ½ | 2 | 1.5 | 1 | 0.5 |
| ¾ | 3 | 2.5 | 1.5 | 0.75 |
| 1 | 4 | 3 | 2 | 1 |
| 1½ | 7 | 5.5 | 3.5 | 1.75 |
| 2 | 9 | 7 | 4.5 | 2.25 |
| 3 | 15 | 12 | 7.5 | 3.75 |
| 4 | 20 | 16 | 10 | 5 |
| 6 | 36 | 29 | 18 | 9 |
| 8 | 48 | 38 | 24 | 12 |
| 10 | 62 | 49 | 31 | 15 |
| 12 | 78 | 60 | 39 | 19 |
| 14 | 88 | 70 | 44 | 22 |
| 16 | 100 | 78 | 50 | 25 |
| 18 | 120 | 95 | 60 | 30 |
| 20 | 136 | 107 | 68 | 34 |
| 24 | 170 | 135 | 85 | 42 |

**Table 6.7A.**[4]   $C_v$ Factors for Tufline Valves (Teflon-Lined Valves and Sleeved Valves)

| | $\frac{1}{2}$ in. | $\frac{3}{4}$ in. | 1 in. | $1\frac{1}{2}$ in. | 2 in. | 3 in. | 4 in. | 6 in. | 8 in. | 10 in. | 12 in. |
|---|---|---|---|---|---|---|---|---|---|---|---|
| 2 Way | 9 | 9 | 25 | 55 | 144 | 254 | 433 | 900 | 1400 | 2100 | 3100 |
| 2 Way butterfly | – | – | – | – | – | – | 740 | 2000 | 3850 | 5020 | 8550 |
| 3 Way (A, AX, C) | 7 | 7 | 20 | 40 | 70 | 100 | 175 | 350 | 475 | 650 | – |
| 3 Way (D, position 1 & 3) | 4 | 4 | 13 | 23 | 43 | 60 | 100 | 230 | 475 | – | – |
| 3 Way (D, position 2) | 5 | 5 | 24 | 42 | 55 | 95 | 174 | 400 | 600 | – | – |
| 4 Way | 4 | 4 | 10 | 30 | 75 | 110 | 180 | – | – | – | – |
| 5 Way | 6 | 6 | 27 | 42 | 69 | 120 | 200 | – | – | – | – |

**Table 6.7B.**[4]  Plug Position for 3-Way Tufline Valves

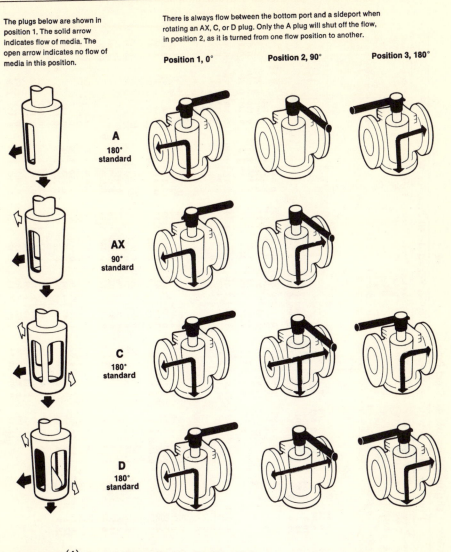

The plugs below are shown in position 1. The solid arrow indicates flow of media. The open arrow indicates no flow of media in this position.

There is always flow between the bottom port and a sideport when rotating an AX, C, or D plug. Only the A plug will shut off the flow, in position 2, as it is turned from one flow position to another.

Position 1, 0°     Position 2, 90°     Position 3, 180°

A
180°
standard

AX
90°
standard

C
180°
standard

D
180°
standard

**Table 6.7C.**[4]  Sizing Formulas for Tufline Valves

| $\Delta P$ | Liquids | Key: |
|---|---|---|
| Noncritical | $Q = C_v \sqrt{\dfrac{\Delta P}{SG}}$ | $Q$ = flow in (gpm)<br>$\Delta P = (P_1 - P_2)$ pressure drop (psi)<br>$P_1$ = inlet pressure (psia)<br>$P_2$ = outlet pressure (psia)<br>psia = psig + 14.7<br>$SG$ = specific gravity |
| Critical | Same as above | |

**Table 6.8.[5]**  Flow of Water through Schedule 40 Steel Pipe [a]

Pressure Drop per 100 feet and Velocity in Schedule 40 Pipe for Water at 60 F.

*Each Veloc./Press. Drop column pair serves two pipe sizes: the smaller size at low flow, the larger size at high flow. Velocity is in Feet per Second; Press. Drop is in Lbs. per Sq. In.*

| Gallons per Minute | Cubic Ft. per Second | 1/8" & 2" Veloc. | Press. Drop | 1/4" & 2 1/2" Veloc. | Press. Drop | 3/8" & 3" Veloc. | Press. Drop | 1/2" & 3 1/2" Veloc. | Press. Drop | 3/4" & 4" Veloc. | Press. Drop | 1" & 5" Veloc. | Press. Drop | 1 1/4" & 6" Veloc. | Press. Drop | 1 1/2" & 8" Veloc. | Press. Drop |
|---|---|---|---|---|---|---|---|---|---|---|---|---|---|---|---|---|---|
| .2 | 0.000446 | 1.13 | 1.86 | 0.616 | 0.359 |  |  |  |  |  |  |  |  |  |  |  |  |
| .3 | 0.000668 | 1.69 | 4.22 | 0.924 | 0.903 | 0.504 | 0.159 | 0.317 | 0.061 |  |  |  |  |  |  |  |  |
| .4 | 0.000891 | 2.26 | 6.98 | 1.23 | 1.61 | 0.672 | 0.345 | 0.422 | 0.086 |  |  |  |  |  |  |  |  |
| .5 | 0.00111 | 2.82 | 10.5 | 1.54 | 2.39 | 0.840 | 0.539 | 0.528 | 0.167 | 0.301 | 0.033 |  |  |  |  |  |  |
| .6 | 0.00134 | 3.39 | 14.7 | 1.85 | 3.29 | 1.01 | 0.751 | 0.633 | 0.240 | 0.361 | 0.041 |  |  |  |  |  |  |
| .8 | 0.00178 | 4.52 | 25.0 | 2.46 | 5.44 | 1.34 | 1.25 | 0.844 | 0.408 | 0.481 | 0.102 |  |  |  |  |  |  |
| 1 | 0.00223 | 5.65 | 37.2 | 3.08 | 8.28 | 1.68 | 1.85 | 1.06 | 0.600 | 0.602 | 0.155 | 0.371 | 0.048 |  |  |  |  |
| 2 | 0.00446 | 11.29 | 134.4 | 6.16 | 30.1 | 3.36 | 6.58 | 2.11 | 2.10 | 1.20 | 0.526 | 0.743 | 0.164 | 0.429 | 0.044 |  |  |
| 3 | 0.00668 |  |  | 9.25 | 64.1 | 5.04 | 13.9 | 3.17 | 4.33 | 1.81 | 1.09 | 1.114 | 0.336 | 0.644 | 0.090 | 0.473 | 0.043 |
| 4 | 0.00891 |  |  | 12.33 | 111.2 | 6.72 |  | 4.22 | 7.42 | 2.41 | 1.83 | 1.49 | 0.565 | 0.858 | 0.150 | 0.630 | 0.071 |
| 5 | 0.01114 |  |  |  |  | 8.40 | 36.7 | 5.28 | 11.2 | 3.01 | 2.75 | 1.86 | 0.835 | 1.073 | 0.223 | 0.788 | 0.104 |
| 6 | 0.01337 | 0.574 | 0.044 |  |  | 10.08 | 51.9 | 6.33 | 15.8 | 3.61 | 3.84 | 2.23 | 1.17 | 1.29 | 0.309 | 0.946 | 0.145 |
| 8 | 0.01782 | 0.765 | 0.073 |  |  | 13.44 | 91.1 | 8.45 | 27.7 | 4.81 | 6.60 | 2.97 | 1.99 | 1.72 | 0.518 | 1.26 | 0.241 |
| 10 | 0.02228 | 0.956 | 0.108 | 0.670 | 0.046 |  |  | 10.56 | 42.4 | 6.02 | 9.99 | 3.71 | 2.99 | 2.15 | 0.774 | 1.58 | 0.361 |
| 15 | 0.03342 | 1.43 | 0.224 | 1.01 | 0.094 |  |  |  |  | 9.03 | 21.6 | 5.57 | 6.36 | 3.22 | 1.63 | 2.37 | 0.755 |
| 20 | 0.04456 | 1.91 | 0.375 | 1.34 | 0.158 | 0.868 | 0.056 |  |  | 12.03 | 37.8 | 7.43 | 10.9 | 4.29 | 2.78 | 3.16 | 1.28 |
| 25 | 0.05570 | 2.39 | 0.561 | 1.68 | 0.234 | 1.09 | 0.083 | 0.812 | 0.041 |  |  | 9.28 | 16.7 | 5.37 | 4.22 | 3.94 | 1.93 |
| 30 | 0.06684 | 2.87 | 0.786 | 2.01 | 0.327 | 1.30 | 0.114 | 0.974 | 0.056 |  |  | 11.14 | 23.8 | 6.44 | 5.92 | 4.73 | 2.72 |
| 35 | 0.07798 | 3.35 | 1.05 | 2.35 | 0.436 | 1.52 | 0.151 | 1.14 | 0.074 | 0.882 | 0.041 | 12.99 | 32.2 | 7.51 | 7.90 | 5.52 | 3.64 |
| 40 | 0.08912 | 3.83 | 1.35 | 2.68 | 0.556 | 1.74 | 0.192 | 1.30 | 0.095 | 1.01 | 0.052 | 14.85 | 41.5 | 8.59 | 10.24 | 6.30 | 4.65 |
| 45 | 0.1003 | 4.30 | 1.67 | 3.02 | 0.668 | 1.95 | 0.239 | 1.46 | 0.117 | 1.13 | 0.064 |  |  | 9.67 | 12.80 | 7.09 | 5.85 |
| 50 | 0.1114 | 4.78 | 2.03 | 3.35 | 0.839 | 2.17 | 0.288 | 1.62 | 0.142 | 1.26 | 0.076 |  |  |  |  |  |  |
| 60 | 0.1337 | 5.74 | 2.87 | 4.02 | 1.18 | 2.60 | 0.406 | 1.95 | 0.204 | 1.51 | 0.107 |  |  |  |  |  |  |
| 70 | 0.1560 | 6.70 | 3.84 | 4.69 | 1.59 | 3.04 | 0.540 | 2.27 | 0.261 | 1.76 | 0.143 | 1.12 | 0.047 |  |  |  |  |
| 80 | 0.1782 | 7.65 | 4.97 | 5.36 | 2.03 | 3.47 | 0.687 | 2.60 | 0.334 | 2.02 | 0.180 | 1.28 | 0.060 |  |  |  |  |
| 90 | 0.2005 | 8.60 | 6.20 | 6.03 | 2.53 | 3.91 | 0.861 | 2.92 | 0.416 | 2.27 | 0.224 | 1.44 | 0.074 |  |  |  |  |
| 100 | 0.2228 | 9.56 | 7.59 | 6.70 | 3.09 | 4.34 | 1.05 | 3.25 | 0.509 | 2.52 | 0.272 | 1.60 | 0.090 | 1.11 | 0.036 |  |  |
| 125 | 0.2785 | 11.97 | 11.76 | 8.38 | 4.71 | 5.43 | 1.61 | 4.06 | 0.769 | 3.15 | 0.415 | 2.01 | 0.135 | 1.39 | 0.055 |  |  |
| 150 | 0.3342 | 14.36 | 16.70 | 10.05 | 6.69 | 6.51 | 2.24 | 4.87 | 1.08 | 3.78 | 0.580 | 2.41 | 0.190 | 1.67 | 0.077 |  |  |
| 175 | 0.3899 | 16.75 | 22.3 | 11.73 | 8.97 | 7.60 | 3.00 | 5.68 | 1.44 | 4.41 | 0.774 | 2.81 | 0.253 | 1.94 | 0.102 |  |  |
| 200 | 0.4456 | 19.14 | 28.8 | 13.42 | 11.68 | 8.68 | 3.87 | 6.49 | 1.85 | 5.04 | 0.985 | 3.21 | 0.323 | 2.22 | 0.130 |  |  |
| 225 | 0.5013 |  |  | 15.09 | 14.63 | 9.77 | 4.83 | 7.30 | 2.32 | 5.67 | 1.23 | 3.61 | 0.401 | 2.50 | 0.162 | 1.44 | 0.043 |
| 250 | 0.557 |  |  |  |  | 10.85 | 5.93 | 8.12 | 2.84 | 6.30 | 1.46 | 4.01 | 0.495 | 2.78 | 0.195 | 1.60 | 0.051 |
| 275 | 0.6127 |  |  |  |  | 11.94 | 7.14 | 8.93 | 3.40 | 6.93 | 1.79 | 4.41 | 0.583 | 3.05 | 0.234 | 1.76 | 0.061 |
| 300 | 0.6684 |  |  |  |  | 13.00 | 8.36 | 9.74 | 4.02 | 7.56 | 2.11 | 4.81 | 0.683 | 3.33 | 0.275 | 1.92 | 0.072 |
| 325 | 0.7241 |  |  |  |  | 14.12 | 9.89 | 10.53 | 4.09 | 8.19 | 2.47 | 5.21 | 0.797 | 3.61 | 0.320 | 2.08 | 0.083 |

Flow of Water Through Schedule 40 Steel Pipe — Velocity (ft/s) and Pressure Drop (psi per 100 ft) [a]

| Q (gpm) | Q (ft³/s) | 3½″ Vel | 3½″ ΔP | 4″ Vel | 4″ ΔP | 5″ Vel | 5″ ΔP | 6″ Vel | 6″ ΔP | 8″ Vel | 8″ ΔP | 10″ Vel | 10″ ΔP | 12″ Vel | 12″ ΔP | 14″ Vel | 14″ ΔP | 16″ Vel | 16″ ΔP | 18″ Vel | 18″ ΔP | 20″ Vel | 20″ ΔP | 24″ Vel | 24″ ΔP |
|---|---|---|---|---|---|---|---|---|---|---|---|---|---|---|---|---|---|---|---|---|---|---|---|---|---|
| 350 | 0.7798 | 11.36 | 5.41 | 8.82 | 2.84 | 5.62 | 0.919 | 3.89 | 0.367 | 2.24 | 0.095 | | | | | | | | | | | | | | |
| 375 | 0.8355 | 12.17 | 6.18 | 9.45 | 3.25 | 6.02 | 1.05 | 4.16 | 0.416 | 2.40 | 0.108 | | | | | | | | | | | | | | |
| 400 | 0.8912 | 12.98 | 7.03 | 10.08 | 3.68 | 6.42 | 1.19 | 4.44 | 0.471 | 2.56 | 0.121 | | | | | | | | | | | | | | |
| 425 | 0.9469 | 13.80 | 7.89 | 10.71 | 4.12 | 6.82 | 1.33 | 4.72 | 0.529 | 2.73 | 0.136 | | | | | | | | | | | | | | |
| 450 | 1.003 | 14.61 | 8.80 | 11.34 | 4.60 | 7.22 | 1.48 | 5.00 | 0.590 | 2.89 | 0.151 | | | | | | | | | | | | | | |
| 475 | 1.059 | | | 11.97 | 5.12 | 7.62 | 1.64 | 5.27 | 0.653 | 3.04 | 0.166 | 1.93 | 0.054 | | | | | | | | | | | | |
| 500 | 1.114 | | | 12.60 | 5.65 | 8.02 | 1.81 | 5.55 | 0.720 | 3.21 | 0.182 | 2.03 | 0.059 | | | | | | | | | | | | |
| 550 | 1.225 | | | 13.86 | 6.79 | 8.82 | 2.17 | 6.11 | 0.861 | 3.53 | 0.219 | 2.24 | 0.071 | | | | | | | | | | | | |
| 600 | 1.337 | | | 15.12 | 8.04 | 9.63 | 2.55 | 6.66 | 1.02 | 3.85 | 0.258 | 2.44 | 0.083 | | | | | | | | | | | | |
| 650 | 1.448 | | | | | 10.43 | 2.98 | 7.22 | 1.18 | 4.17 | 0.301 | 2.64 | 0.097 | | | | | | | | | | | | |
| 700 | 1.560 | | | | | 11.23 | 3.43 | 7.78 | 1.35 | 4.49 | 0.343 | 2.85 | 0.112 | 2.01 | 0.047 | | | | | | | | | | |
| 750 | 1.671 | | | | | 12.03 | 3.92 | 8.33 | 1.55 | 4.81 | 0.392 | 3.05 | 0.127 | 2.15 | 0.054 | | | | | | | | | | |
| 800 | 1.782 | | | | | 12.83 | 4.43 | 8.88 | 1.75 | 5.13 | 0.443 | 3.25 | 0.143 | 2.29 | 0.061 | | | | | | | | | | |
| 850 | 1.894 | | | | | 13.64 | 5.00 | 9.44 | 1.96 | 5.45 | 0.497 | 3.46 | 0.160 | 2.44 | 0.068 | 2.02 | 0.042 | | | | | | | | |
| 900 | 2.005 | | | | | 14.44 | 5.58 | 9.99 | 2.18 | 5.77 | 0.554 | 3.66 | 0.179 | 2.58 | 0.075 | 2.13 | 0.047 | | | | | | | | |
| 950 | 2.117 | | | | | 15.24 | 6.21 | 10.55 | 2.42 | 6.09 | 0.613 | 3.86 | 0.198 | 2.72 | 0.083 | 2.25 | 0.052 | | | | | | | | |
| 1 000 | 2.228 | | | | | 16.04 | 6.84 | 11.10 | 2.68 | 6.41 | 0.675 | 4.07 | 0.218 | 2.87 | 0.091 | 2.37 | 0.057 | | | | | | | | |
| 1 100 | 2.451 | | | | | 17.65 | 8.23 | 12.22 | 3.22 | 7.05 | 0.807 | 4.48 | 0.260 | 3.15 | 0.110 | 2.61 | 0.068 | | | | | | | | |
| 1 200 | 2.674 | | | | | | | 13.33 | 3.81 | 7.70 | 0.948 | 4.88 | 0.306 | 3.44 | 0.128 | 2.85 | 0.080 | 2.18 | 0.042 | | | | | | |
| 1 300 | 2.896 | | | | | | | 14.43 | 4.45 | 8.33 | 1.11 | 5.29 | 0.355 | 3.73 | 0.150 | 3.08 | 0.093 | 2.36 | 0.047 | | | | | | |
| 1 400 | 3.119 | | | | | | | 15.55 | 5.13 | 8.98 | 1.28 | 5.70 | 0.409 | 4.01 | 0.171 | 3.32 | 0.107 | 2.54 | 0.055 | | | | | | |
| 1 500 | 3.342 | | | | | | | 16.66 | 5.85 | 9.62 | 1.46 | 6.10 | 0.466 | 4.30 | 0.195 | 3.56 | 0.122 | 2.72 | 0.063 | | | | | | |
| 1 600 | 3.565 | | | | | | | 17.77 | 6.61 | 10.26 | 1.65 | 6.51 | 0.527 | 4.59 | 0.219 | 3.79 | 0.138 | 2.90 | 0.071 | | | | | | |
| 1 800 | 4.010 | | | | | | | 19.99 | 8.37 | 11.54 | 2.08 | 7.32 | 0.663 | 5.16 | 0.276 | 4.27 | 0.172 | 3.27 | 0.088 | 2.58 | 0.050 | | | | |
| 2 000 | 4.456 | | | | | | | 22.21 | 10.3 | 12.82 | 2.55 | 8.14 | 0.808 | 5.73 | 0.339 | 4.74 | 0.209 | 3.63 | 0.107 | 2.87 | 0.060 | | | | |
| 2 500 | 5.570 | | | | | | | | | 16.03 | 3.94 | 10.17 | 1.24 | 7.17 | 0.515 | 5.93 | 0.321 | 4.54 | 0.163 | 3.59 | 0.091 | | | | |
| 3 000 | 6.684 | | | | | | | | | 19.24 | 5.59 | 12.20 | 1.76 | 8.60 | 0.731 | 7.11 | 0.451 | 5.45 | 0.232 | 4.30 | 0.129 | 3.46 | 0.075 | | |
| 3 500 | 7.798 | | | | | | | | | 22.44 | 7.56 | 14.24 | 2.38 | 10.03 | 0.982 | 8.30 | 0.607 | 6.35 | 0.312 | 5.02 | 0.173 | 4.04 | 0.101 | | |
| 4 000 | 8.912 | | | | | | | | | 25.65 | 9.80 | 16.27 | 3.08 | 11.47 | 1.27 | 9.48 | 0.787 | 7.26 | 0.401 | 5.74 | 0.222 | 4.62 | 0.129 | 3.19 | 0.052 |
| 4 500 | 10.03 | | | | | | | | | 28.87 | 12.2 | 18.31 | 3.87 | 12.90 | 1.60 | 10.67 | 0.990 | 8.17 | 0.503 | 6.46 | 0.280 | 5.20 | 0.162 | 3.59 | 0.065 |
| 5 000 | 11.14 | | | | | | | | | | | 20.35 | 4.71 | 14.33 | 1.95 | 11.85 | 1.21 | 9.08 | 0.617 | 7.17 | 0.340 | 5.77 | 0.199 | 3.99 | 0.079 |
| 6 000 | 13.37 | | | | | | | | | | | 24.41 | 6.74 | 17.20 | 2.77 | 14.23 | 1.71 | 10.89 | 0.877 | 8.61 | 0.483 | 6.93 | 0.280 | 4.79 | 0.111 |
| 7 000 | 15.60 | | | | | | | | | | | 28.49 | 9.11 | 20.07 | 3.74 | 16.60 | 2.31 | 12.71 | 1.18 | 10.04 | 0.652 | 8.08 | 0.376 | 5.59 | 0.150 |
| 8 000 | 17.82 | | | | | | | | | | | | | 22.93 | 4.84 | 18.96 | 2.99 | 14.52 | 1.51 | 11.47 | 0.839 | 9.23 | 0.488 | 6.38 | 0.192 |
| 9 000 | 20.05 | | | | | | | | | | | | | 25.79 | 6.09 | 21.34 | 3.76 | 16.34 | 1.90 | 12.91 | 1.05 | 10.39 | 0.608 | 7.18 | 0.242 |
| 10 000 | 22.28 | | | | | | | | | | | | | 28.66 | 7.46 | 23.71 | 4.61 | 18.15 | 2.34 | 14.34 | 1.28 | 11.54 | 0.739 | 7.98 | 0.294 |
| 12 000 | 26.74 | | | | | | | | | | | | | 34.40 | 10.7 | 28.45 | 6.59 | 21.79 | 3.33 | 17.21 | 1.83 | 13.85 | 1.06 | 9.58 | 0.416 |
| 14 000 | 31.19 | | | | | | | | | | | | | | | 33.19 | 8.89 | 25.42 | 4.49 | 20.08 | 2.45 | 16.16 | 1.43 | 11.17 | 0.562 |
| 16 000 | 35.65 | | | | | | | | | | | | | | | | | 29.05 | 5.83 | 22.95 | 3.18 | 18.47 | 1.85 | 12.77 | 0.723 |
| 18 000 | 40.10 | | | | | | | | | | | | | | | | | 32.68 | 7.31 | 25.82 | 4.03 | 20.77 | 2.32 | 14.36 | 0.907 |
| 20 000 | 44.56 | | | | | | | | | | | | | | | | | 36.31 | 9.03 | 28.69 | 4.93 | 23.08 | 2.86 | 15.96 | 1.12 |

[a]For pipe lengths other than 100 feet, the pressure drop is proportional to the length. Thus, for 50 feet of pipe, the pressure drop is approximately one-half the value given in the table . . . for 300 feet, three times the given value, etc. Velocity is a function of the cross sectional flow area; thus, it is constant for a given flow rate and is independent of pipe length. For pipe other than schedule 40: $V = V_{40}\,(d_{40}/d)^2$ and $\Delta P = \Delta P_{40}\,(d_{40}/d)^5$.

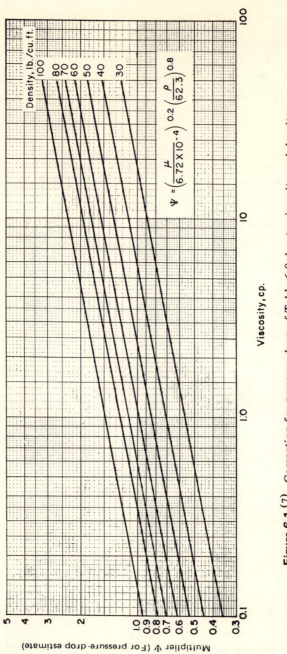

$$\Psi = \left(\frac{\mu}{6.72 \times 10^{-4}}\right)^{0.2} \left(\frac{\rho}{62.3}\right)^{0.8}$$

**Figure 6.1.**[7] Correction for pressure drop of Table 6.8 due to viscosity and density.

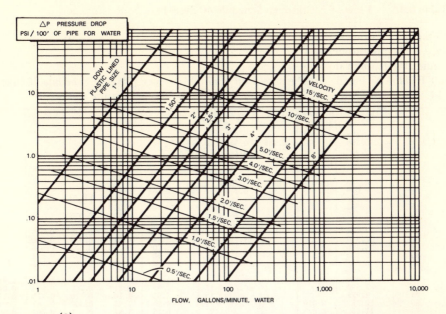

**Figure 6.2.**[8] Pressure drop and flow velocity of water in DOW plastic-lined pipe. Values are based on $P = 0.0286/D^3 \ (G/D)^{1.97}$; where $P$ is the pressure loss in psi per 100 ft of pipe; $G$ is the flow of water in gallons per minute; $D$ is the inside diameter of pipe in inches.

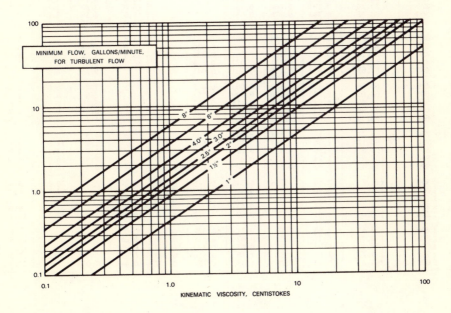

**Figure 6.3.**[8] Viscosity vs. minimum flow to produce turbulent flow (cSt = cP/density).

**Table 6.9.**[5] Flow of Air through Schedule 40 Steel Pipe[a]

| Free air $q'_m$, ft$^3$/min at 60°F and 14.7 psia | Compressed air ft$^3$/min at 60°F and 100 psig | Pressure drop of air, lb/in.$^2$ per 100 ft of Schedule 40 pipe for air at 100 lb/in.$^2$ gauge pressure and 60°F | | | | | | | | |
|---|---|---|---|---|---|---|---|---|---|---|
| | | 1/8" | 1/4" | 3/8" | 1/2" | | | | | |
| 1 | 0.128 | 0.361 | 0.083 | 0.018 | | | | | | |
| 2 | 0.256 | 1.31 | 0.285 | 0.064 | 0.020 | | | | | |
| 3 | 0.384 | 3.06 | 0.605 | 0.133 | 0.042 | 3/4" | | | | |
| 4 | 0.513 | 4.83 | 1.04 | 0.226 | 0.071 | | | | | |
| 5 | 0.641 | 7.45 | 1.58 | 0.343 | 0.106 | 0.027 | 1" | | | |
| 6 | 0.769 | 10.6 | 2.23 | 0.408 | 0.148 | 0.037 | | | | |
| 8 | 1.025 | 18.6 | 3.89 | 0.848 | 0.255 | 0.062 | 0.019 | | | |
| 10 | 1.282 | 28.7 | 5.96 | 1.26 | 0.356 | 0.094 | 0.029 | 1¼" | 1½" | |
| 15 | 1.922 | ... | 13.0 | 2.73 | 0.834 | 0.201 | 0.062 | | | |
| 20 | 2.563 | ... | 22.8 | 4.76 | 1.43 | 0.345 | 0.102 | 0.026 | | |
| 25 | 3.204 | ... | 35.6 | 7.34 | 2.21 | 0.526 | 0.156 | 0.039 | 0.019 | |
| 30 | 3.845 | ... | ... | 10.5 | 3.15 | 0.748 | 0.219 | 0.055 | 0.026 | |
| 35 | 4.486 | ... | ... | 14.2 | 4.24 | 1.00 | 0.293 | 0.073 | 0.035 | |
| 40 | 5.126 | ... | ... | 18.4 | 5.49 | 1.30 | 0.379 | 0.095 | 0.044 | |
| 45 | 5.767 | ... | ... | 23.1 | 6.90 | 1.62 | 0.474 | 0.116 | 0.055 | 2" |
| 50 | 6.408 | | | 28.5 | 8.49 | 1.99 | 0.578 | 0.149 | 0.067 | 0.019 |
| 60 | 7.690 | 2½" | | 40.7 | 12.2 | 2.85 | 0.819 | 0.200 | 0.094 | 0.027 |
| 70 | 8.971 | | | ... | 16.5 | 3.83 | 1.10 | 0.270 | 0.126 | 0.036 |
| 80 | 10.25 | 0.019 | | ... | 21.4 | 4.96 | 1.43 | 0.350 | 0.162 | 0.046 |
| 90 | 11.53 | 0.023 | | ... | 27.0 | 6.25 | 1.80 | 0.437 | 0.203 | 0.058 |
| 100 | 12.82 | 0.029 | 3" | | 33.2 | 7.69 | 2.21 | 0.534 | 0.247 | 0.070 |
| 125 | 16.02 | 0.044 | | | ... | 11.9 | 3.39 | 0.825 | 0.380 | 0.107 |
| 150 | 19.22 | 0.062 | 0.021 | | ... | 17.0 | 4.87 | 1.17 | 0.537 | 0.151 |
| 175 | 22.43 | 0.083 | 0.028 | | ... | 23.1 | 6.60 | 1.58 | 0.727 | 0.205 |
| 200 | 25.63 | 0.107 | 0.036 | 3½" | ... | 30.0 | 8.54 | 2.05 | 0.937 | 0.264 |
| 225 | 28.84 | 0.134 | 0.045 | 0.022 | | 37.9 | 10.8 | 2.59 | 1.19 | 0.331 |
| 250 | 32.04 | 0.164 | 0.055 | 0.027 | | ... | 13.3 | 3.18 | 1.45 | 0.404 |
| 275 | 35.24 | 0.191 | 0.066 | 0.032 | | ... | 16.0 | 3.83 | 1.75 | 0.484 |
| 300 | 38.45 | 0.232 | 0.078 | 0.037 | | ... | 19.0 | 4.56 | 2.07 | 0.573 |
| 325 | 41.65 | 0.270 | 0.090 | 0.043 | 4" | ... | 22.3 | 5.32 | 2.42 | 0.673 |
| 350 | 44.87 | 0.313 | 0.104 | 0.050 | | ... | 25.8 | 6.17 | 2.80 | 0.776 |
| 375 | 48.06 | 0.356 | 0.119 | 0.057 | 0.030 | ... | 29.6 | 7.05 | 3.20 | 0.887 |
| 400 | 51.26 | 0.402 | 0.134 | 0.064 | 0.034 | ... | 33.6 | 8.02 | 3.64 | 1.00 |
| 425 | 54.47 | 0.452 | 0.151 | 0.072 | 0.038 | ... | 37.9 | 9.01 | 4.09 | 1.13 |
| 450 | 57.67 | 0.507 | 0.168 | 0.081 | 0.042 | ... | ... | 10.2 | 4.59 | 1.26 |
| 475 | 60.88 | 0.562 | 0.187 | 0.089 | 0.047 | | ... | 11.3 | 5.09 | 1.40 |
| 500 | 64.08 | 0.623 | 0.206 | 0.099 | 0.052 | | ... | 12.5 | 5.61 | 1.55 |
| 550 | 70.49 | 0.749 | 0.248 | 0.118 | 0.062 | | ... | 15.1 | 6.79 | 1.87 |
| 600 | 76.90 | 0.887 | 0.293 | 0.139 | 0.073 | | ... | 18.0 | 8.04 | 2.21 |
| 650 | 83.30 | 1.04 | 0.342 | 0.163 | 0.086 | 5" | ... | 21.1 | 9.43 | 2.60 |
| 700 | 89.71 | 1.19 | 0.395 | 0.188 | 0.099 | 0.032 | | 24.3 | 10.9 | 3.00 |
| 750 | 96.12 | 1.36 | 0.451 | 0.214 | 0.113 | 0.036 | | 27.9 | 12.6 | 3.44 |
| 800 | 102.5 | 1.55 | 0.513 | 0.244 | 0.127 | 0.041 | | 31.8 | 14.2 | 3.90 |
| 850 | 108.9 | 1.74 | 0.576 | 0.274 | 0.144 | 0.046 | | 35.9 | 16.0 | 4.40 |
| 900 | 115.3 | 1.95 | 0.642 | 0.305 | 0.160 | 0.051 | 6" | 40.2 | 18.0 | 4.91 |
| 950 | 121.8 | 2.18 | 0.715 | 0.340 | 0.178 | 0.057 | 0.023 | ... | 20.0 | 5.47 |
| 1 000 | 128.2 | 2.40 | 0.788 | 0.375 | 0.197 | 0.063 | 0.025 | ... | 22.1 | 6.06 |
| 1 100 | 141.0 | 2.89 | 0.948 | 0.451 | 0.236 | 0.075 | 0.030 | ... | 26.7 | 7.29 |
| 1 200 | 153.8 | 3.44 | 1.13 | 0.533 | 0.279 | 0.089 | 0.035 | ... | 31.8 | 8.63 |
| 1 300 | 166.6 | 4.01 | 1.32 | 0.626 | 0.327 | 0.103 | 0.041 | ... | 37.3 | 10.1 |
| 1 400 | 179.4 | 4.65 | 1.52 | 0.718 | 0.377 | 0.119 | 0.047 | | | 11.8 |
| 1 500 | 192.2 | 5.31 | 1.74 | 0.824 | 0.431 | 0.136 | 0.054 | | | 13.5 |
| 1 600 | 205.1 | 6.04 | 1.97 | 0.932 | 0.490 | 0.154 | 0.061 | 8" | | 15.3 |
| 1 800 | 230.7 | 7.65 | 2.50 | 1.18 | 0.616 | 0.193 | 0.075 | | | 19.3 |
| 2 000 | 256.3 | 9.44 | 3.06 | 1.45 | 0.757 | 0.237 | 0.094 | 0.023 | | 23.9 |

**Table 6.9.**[s]  *Cont'd.*

| Free air $q'_m$, ft$^3$/min at 60°F and 14.7 psia | Compressed air ft$^3$/min at 60°F and 100 psig | Pressure drop of air, lb/in.$^2$ per 100 ft of Schedule 40 pipe for air at 100 lb/in.$^2$ gauge pressure and 60°F | | | | | | | 10" | 12" |
|---|---|---|---|---|---|---|---|---|---|---|
| 2 500 | 320.4 | 14.7 | 4.76 | 2.25 | 1.17 | 0.366 | 0.143 | 0.035 |  | 37.3 |
| 3 000 | 384.5 | 21.1 | 6.82 | 3.20 | 1.67 | 0.524 | 0.204 | 0.051 | 0.016 |  |
| 3 500 | 448.6 | 28.8 | 9.23 | 4.33 | 2.26 | 0.709 | 0.276 | 0.068 | 0.022 |  |
| 4 000 | 512.6 | 37.6 | 12.1 | 5.66 | 2.94 | 0.919 | 0.358 | 0.088 | 0.028 |  |
| 4 500 | 576.7 | 47.6 | 15.3 | 7.16 | 3.69 | 1.16 | 0.450 | 0.111 | 0.035 | 12" |
| 5 000 | 640.8 | ... | 18.8 | 8.85 | 4.56 | 1.42 | 0.552 | 0.136 | 0.043 | 0.018 |
| 6 000 | 769.0 | ... | 27.1 | 12.7 | 6.57 | 2.03 | 0.794 | 0.195 | 0.061 | 0.025 |
| 7 000 | 897.1 | ... | 36.9 | 17.2 | 8.94 | 2.76 | 1.07 | 0.262 | 0.082 | 0.034 |
| 8 000 | 1025 | ... | ... | 22.5 | 11.7 | 3.59 | 1.39 | 0.339 | 0.107 | 0.044 |
| 9 000 | 1153 | ... | ... | 28.5 | 14.9 | 4.54 | 1.76 | 0.427 | 0.134 | 0.055 |
| 10 000 | 1282 | ... | ... | 35.2 | 18.4 | 5.60 | 2.16 | 0.526 | 0.164 | 0.067 |
| 11 000 | 1410 | ... | ... | ... | 22.2 | 6.78 | 2.62 | 0.633 | 0.197 | 0.081 |
| 12 000 | 1538 | ... | ... | ... | 26.4 | 8.07 | 3.09 | 0.753 | 0.234 | 0.096 |
| 13 000 | 1666 | ... | ... | ... | 31.0 | 9.47 | 3.63 | 0.884 | 0.273 | 0.112 |
| 14 000 | 1794 | ... | ... | ... | 36.0 | 11.0 | 4.21 | 1.02 | 0.316 | 0.129 |
| 15 000 | 1922 | ... | ... | ... | ... | 12.6 | 4.84 | 1.17 | 0.364 | 0.148 |
| 16 000 | 2051 | ... | ... | ... | ... | 14.3 | 5.50 | 1.33 | 0.411 | 0.167 |
| 18 000 | 2307 | ... | ... | ... | ... | 18.2 | 6.96 | 1.68 | 0.520 | 0.213 |
| 20 000 | 2563 | ... | ... | ... | ... | 22.4 | 8.60 | 2.01 | 0.642 | 0.260 |
| 22 000 | 2820 | ... | ... | ... | ... | 27.1 | 10.4 | 2.50 | 0.771 | 0.314 |
| 24 000 | 3076 | ... | ... | ... | ... | 32.3 | 12.4 | 2.97 | 0.918 | 0.371 |
| 26 000 | 3332 | ... | ... | ... | ... | 37.9 | 14.5 | 3.49 | 1.12 | 0.435 |
| 28 000 | 3588 | ... | ... | ... | ... | ... | 16.9 | 4.04 | 1.25 | 0.505 |
| 30 000 | 3845 | ... | ... | ... | ... | ... | 19.3 | 4.64 | 1.42 | 0.520 |

[a] For lengths of pipe other than 100 ft, the pressure drop is proportional to the length. Thus, for 50 ft of pipe, the pressure drop is approximately one-half the value given in the table ... for 300 ft, three times the given value, etc.

The pressure drop is also inversely proportional to the absolute pressure and directly proportional to the absolute temperature. Therefore, to determine the pressure drop for inlet or average pressures other than 100 psi and at temperatures other than 60°F, multiply the values given in the table by the ratio

$$\left(\frac{100 + 14.7}{P + 14.7}\right) \left(\frac{460 + t}{520}\right)$$

where $P$ is the inlet or average gauge pressure in pounds per square inch, and $t$ is the temperature in degrees Fahrenheit under consideration.

The cubic feet per minute of compressed air at any pressure is inversely proportional to the absolute pressure and directly proportional to the absolute temperature. To determine the cubic feet per minute of compressed air at any temperature and pressure other than standard conditions, multiply the value of cubic feet per minute of free air by the ratio

$$\left(\frac{14.7}{14.7 + P}\right) \left(\frac{460 + t}{520}\right)$$

For pipe other than Schedule 40:

$$V = V_{40}(d_{40}/d)^2 \quad \text{and} \quad \Delta P = \Delta P_{40}(d_{40}/d)^5$$

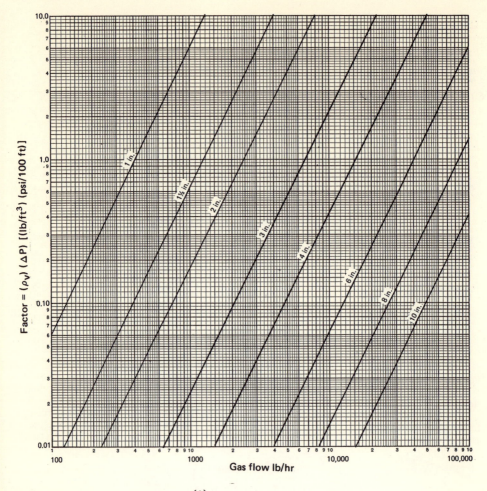

**Figure 6.4.**[9]   Pressure drop for gas flow.

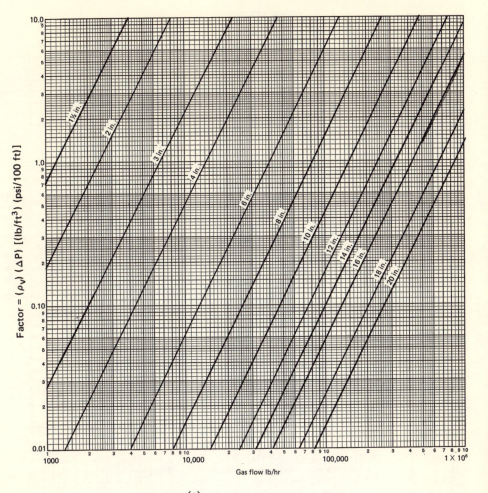

**Figure 6.5.**[9]  Pressure drop for gas flow.

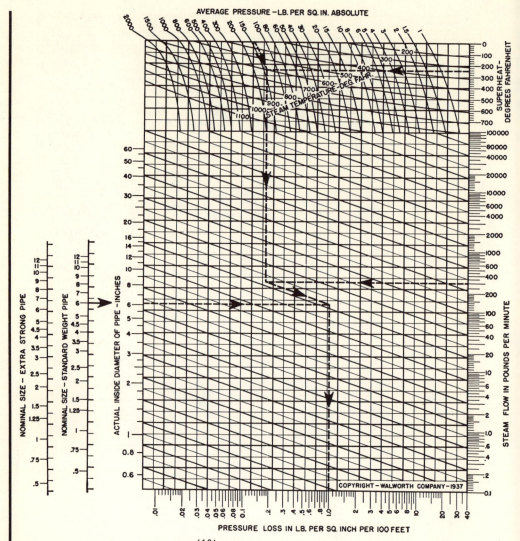

**Figure 6.6.**[10]   Steam flow chart: based on Babcock's formula:
$$P = 0.000131 \cdot [1 + (3.6/d) (w^2 L/\rho d^5)].$$

Figure 6.6 can be used for estimating pressure drop in steam piping. Figure 6.7 can be used for more accurate determination of pressure drop of gases and liquids. The following equations are useful when using Figure 6.7.

| Reynolds number | Frictional pressure loss |
|---|---|
| $N_{Re} = \dfrac{124dV\rho}{\mu'}$ | $(-\Delta P_{100})' = \dfrac{0.518f\rho V^2}{d}$ |
| $N_{Re} = \dfrac{50.65\rho Q_a}{\mu'd}$ | $(-\Delta P_{100})' = \dfrac{0.0864f\rho(Q_a)^2}{d^5}$ |
| $N_{Re} = \dfrac{379\rho q_a}{\mu'd}$ | $(-\Delta P_{100})' = \dfrac{4.84f\rho(q_a)^2}{d^5}$ |
| $N_{Re} = \dfrac{6.32W}{\mu'd}$ | $(-\Delta P_{100})' = \dfrac{0.00134fW^2}{\rho d^5}$ |

Where $N_{Re}$ is the Reynolds number; $\mu'$ is the viscosity (cP); $d$ is the internal pipe diameter (in.); $(-\Delta P_{100})'$ is the frictional head loss (psi/100-ft. pipe); $f$ is the fanning friction factor; $\rho$ is the density (lb/ft$^3$); $V$ is the pipe velocity (ft/sec); $Q_a$ is the flowrate (gpm); $q_a$ is the flowrate (ft$^3$/min); $W$ is the flowrate (lb/hr).

Figure 6.8 can be used for sizing vertical piping, for example, discharge lines from gravity decanters, or return lines from condensers.

Table 6.10 can be used for estimating flows from triangular notch weirs with end contractions. Table 6.11 can be used for estimating flows from rectangular weirs.

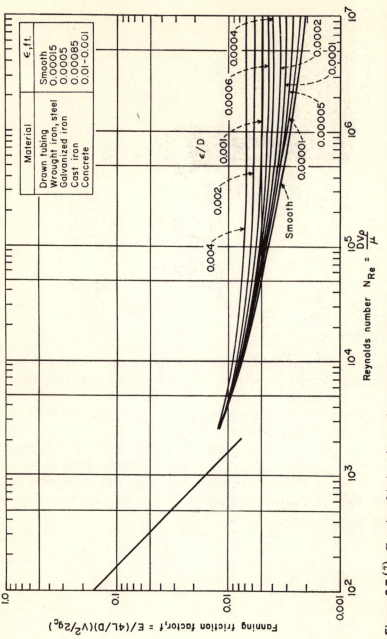

**Figure 6.7.**[7] Fanning friction factor for pipe flow, where $E$ is the internal head loss (ft-lb$_f$/lb); $D$ is the internal head diameter (ft); $g_c = 32.174$ ft-lb/(lb$_f$) (sec$^2$); $L$ is the pipe length (ft); $\epsilon$ is the pipe roughness (ft); $\mu$ is the viscosity [lb/(sec) (ft)], $f$ is the Fanning friction number; $N_{Re}$ is the Reynold's number; $V$ is the pipe velocity (ft/sec); and $\rho$ is the density (lb/ft$^3$).

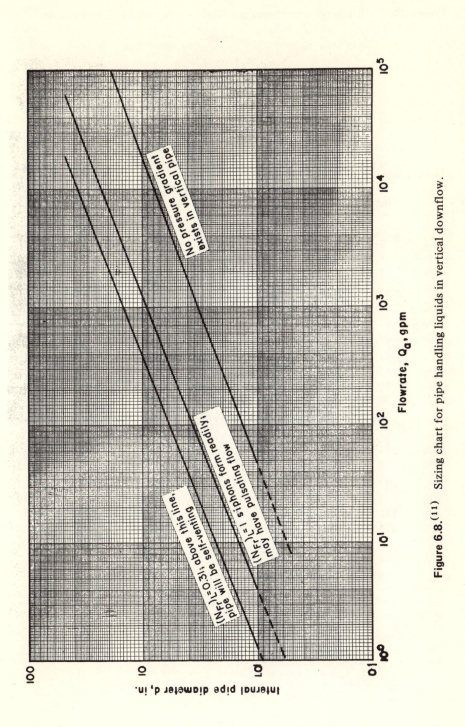

**Figure 6.8.**[11] Sizing chart for pipe handling liquids in vertical downflow.

**Table 6.10.**[6] Discharge from Triangular Notch Weirs with End Contractions [a]

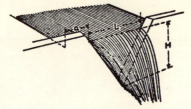

| Head (H) in inches | Flow (gpm) 90° Notch | Flow (gpm) 60° Notch | Head (H) in inches | Flow (gpm) 90° Notch | Flow (gpm) 60° Notch | Head (H) in inches | Flow (gpm) 90° Notch | Flow (gpm) 60° Notch |
|---|---|---|---|---|---|---|---|---|
| 1    | 2.19 | 1.27 | 6¾   | 260  | 150  | 15   | 1912 | 1104 |
| 1¼   | 3.83 | 2.21 | 7    | 284  | 164  | 15½  | 2073 | 1197 |
| 1½   | 6.05 | 3.49 | 7¼   | 310  | 179  | 16   | 2246 | 1297 |
| 1¾   | 8.89 | 5.13 | 7½   | 338  | 195  | 16½  | 2426 | 1401 |
| 2    | 12.4 | 7.16 | 7¾   | 367  | 212  | 17   | 2614 | 1509 |
| 2¼   | 16.7 | 9.62 | 8    | 397  | 229  | 17½  | 2810 | 1623 |
| 2½   | 21.7 | 12.5 | 8¼   | 429  | 248  | 18   | 3016 | 1741 |
| 2¾   | 27.5 | 15.9 | 8½   | 462  | 267  | 18½  | 3229 | 1864 |
| 3    | 34.2 | 19.7 | 8¾   | 498  | 287  | 19   | 3452 | 1993 |
| 3¼   | 41.8 | 24.1 | 9    | 533  | 308  | 19½  | 3684 | 2127 |
| 3½   | 50.3 | 29.0 | 9¼   | 571  | 330  | 20   | 3924 | 2266 |
| 3¾   | 59.7 | 34.5 | 9½   | 610  | 352  | 20½  | 4174 | 2410 |
| 4    | 70.2 | 40.5 | 9¾   | 651  | 376  | 21   | 4433 | 2560 |
| 4¼   | 81.7 | 47.2 | 10   | 694  | 401  | 21½  | 4702 | 2715 |
| 4½   | 94.2 | 54.4 | 10½  | 784  | 452  | 22   | 4980 | 2875 |
| 4¾   | 108  | 62.3 | 11   | 880  | 508  | 22½  | 5268 | 3041 |
| 5    | 123  | 70.8 | 11½  | 984  | 568  | 23   | 5565 | 3213 |
| 5¼   | 139  | 80.0 | 12   | 1094 | 632  | 23½  | 5873 | 3391 |
| 5½   | 156  | 89.9 | 12½  | 1212 | 700  | 24   | 6190 | 3574 |
| 5¾   | 174  | 100  | 13   | 1337 | 772  | 24½  | 6518 | 3763 |
| 6    | 193  | 112  | 13½  | 1469 | 848  | 25   | 6855 | 3958 |
| 6¼   | 214  | 124  | 14   | 1609 | 929  |      |      |      |
| 6½   | 236  | 136  | 14½  | 1756 | 1014 |      |      |      |

[a]Based on formula: $Q = (C) (4/15) (L) (H) (2gH)^{1/2}$, in which $Q$ is the flow of water in cu ft per sec; $L$ is the width of notch in ft at $H$ distance above apex; $H$ is the head of water above apex of notch in ft; $C$ is the constant varying with conditions, .57 being used for this table; $a$ should be not less than ¾$L$.
For 90° notch the formula becomes $Q = 2.4381 H^{5/2}$.
For 60° notch the formula becomes $Q = 1.4076 H^{5/2}$.

**Table 6.11.**[6]  Discharge from Rectangular Weir and End Contractions [a]

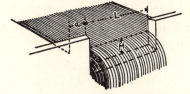

| Head (H) in inches | Length (L) of weir | | | Additional gpm for each ft over 5 ft | Head (H) in inches | Length (L) of weir | | Additional gpm for each ft over 5 ft |
|---|---|---|---|---|---|---|---|---|
| | 1 ft | 3 ft | 5 ft | | | 3 ft | 5 ft | |
| 1 | 35.4 | 107.5 | 179.8 | 36.05 | 8 | 2338 | 3956 | 814 |
| 1¼ | 49.5 | 150.4 | 250.4 | 50.4 | 8¼ | 2442 | 4140 | 850 |
| 1½ | 64.9 | 197 | 329.5 | 66.2 | 8½ | 2540 | 4312 | 890 |
| 1¾ | 81 | 248 | 415 | 83.5 | 8¾ | 2656 | 4511 | 929 |
| 2 | 98.5 | 302 | 506 | 102 | 9 | 2765 | 4699 | 970 |
| 2¼ | 117 | 361 | 605 | 122 | 9¼ | 2876 | 4899 | 1011 |
| 2½ | 136.2 | 422 | 706 | 143 | 9½ | 2985 | 5098 | 1051 |
| 2¾ | 157 | 485 | 815 | 165 | 9¾ | 3101 | 5288 | 1091 |
| 3 | 177.8 | 552 | 926 | 187 | 10 | 3216 | 5490 | 1136 |
| 3¼ | 199.8 | 624 | 1047 | 211 | 10½ | 3480 | 5940 | 1230 |
| 3½ | 222 | 695 | 1167 | 236 | 11 | 3716 | 6355 | 1320 |
| 3¾ | 245 | 769 | 1292 | 261 | 11½ | 3960 | 6780 | 1410 |
| 4 | 269 | 846 | 1424 | 288 | 12 | 4185 | 7165 | 1495 |
| 4¼ | 293.6 | 925 | 1559 | 316 | 12½ | 4430 | 7595 | 1575 |
| 4½ | 318 | 1006 | 1696 | 345 | 13 | 4660 | 8010 | 1660 |
| 4¾ | 344 | 1091 | 1835 | 374 | 13½ | 4950 | 8510 | 1780 |
| 5 | 370 | 1175 | 1985 | 405 | 14 | 5215 | 8980 | 1885 |
| 5¼ | 395.5 | 1262 | 2130 | 434 | 14½ | 5475 | 9440 | 1985 |
| 5½ | 421.6 | 1352 | 2282 | 465 | 15 | 5740 | 9920 | 2090 |
| 5¾ | 449 | 1442 | 2440 | 495 | 15½ | 6015 | 10400 | 2165 |
| 6 | 476.5 | 1535 | 2600 | 528 | 16 | 6290 | 10900 | 2300 |
| 6¼ | | 1632 | 2760 | 560 | 16½ | 6565 | 11380 | 2410 |
| 6½ | | 1742 | 2920 | 596 | 17 | 6925 | 11970 | 2520 |
| 6¾ | | 1826 | 3094 | 630 | 17½ | 7140 | 12410 | 2640 |
| 7 | | 1928 | 3260 | 668 | 18 | 7410 | 12900 | 2745 |
| 7¼ | | 2029 | 3436 | 701.5 | 18½ | 7695 | 13410 | 2855 |
| 7½ | | 2130 | 3609 | 736 | 19 | 7980 | 13940 | 2970 |
| 7¾ | | 2238 | 3785 | 774 | 19½ | 8280 | 14460 | 3090 |

[a] This table is based on Francis formula: $Q = 3.33 (L - 0.2 H) H^{1.5}$, in which $Q$ is cu ft of water flowing per second; $L$ is the length of weir opening in feet (should be 4–8 times $H$); $H$ is the head on weir in feet (to be measured at least 6 ft back of weir opening); $a$ should be at least 3 $H$.

## COMPRESSIBLE FLOW CALCULATIONS

Compressible flow calculations using Lapple[13] charts are carried out using Figures 6.9 and 6.10. To calculate a pipe size for a given flowrate the following are required:

1. Knowing $P_0$ and $P_3$, calculate the ratio $P_3/P_0$.
2. Using $f = 0.0035$, calculate $N = 4fL/D$ assuming a pipe size $D$. $L$ is equivalent feet of pipe.
3. Using Table 11.5 or other sources, get the specific heat ratio of gas $\gamma$.
4. Using Figure 6.10, check to see if $P_3/P_0$ is below the dashed line. If below dashed line, flow will be at sonic velocity.* In this case, for a given value for $N$, use the value of $P_2/P_0$ and $G/G_{cni}$ at the dashed line. If the ratio $P_3/P_0$ is above the dashed line use this value as $P_2/P_0$ and read $G/G_{cni}$ as before.
5. After calculating the maximum gas density, calculate $G_{cni} = (g_c P_0 \rho_0 / e)^{1/2}$, where $e = 2.718$.
6. Calculate $G = G_{cni}(G/G_{cni})$.
7. Calculate required flow area $A = w/G$.
8. Calculate pipe size $D = (A4/\pi)^{1/2}$.
9. Knowing pipe size use Figure 6.7 to calculate $N$. Repeat steps 2–8 until the assumed value of $N$ is close to the calculated value of $N$.

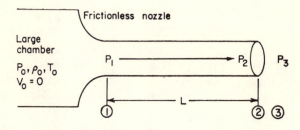

**Figure 6.9.**[13]   Basis for Lapple charts.

## THE PARSHALL FLUME[14]

This type of flow-measuring device is widely used for gauging flows in open channels. A complete description of the construction and operation of the flume is given in references 14 and 15.

Figure 6.11 shows the capacity of the Parshall flume as a function of size and head height. It should be noted that this figure is only good for free flow where the ratio of downstream head to upstream head is less than 0.6 for sizes less than 1 ft and 0.7 for flume sizes greater than 1 ft. Figures 6.12 and 6.13 and Tables 6.12 and 6.13 give some construction details of the Parshall flume.

*Sonic velocity is the maximum velocity that can be attained in a pipe of constant cross sectional area. It is equal to the velocity of sound in the fluid. Further increase in upstream pressure will not result any further increase in velocity.

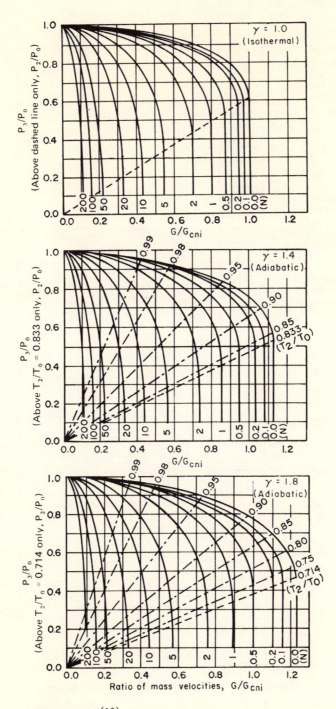

**Figure 6.10.**[13] Lapple charts for compressible flow. (Above dashed line $P_2 = P_3$.)

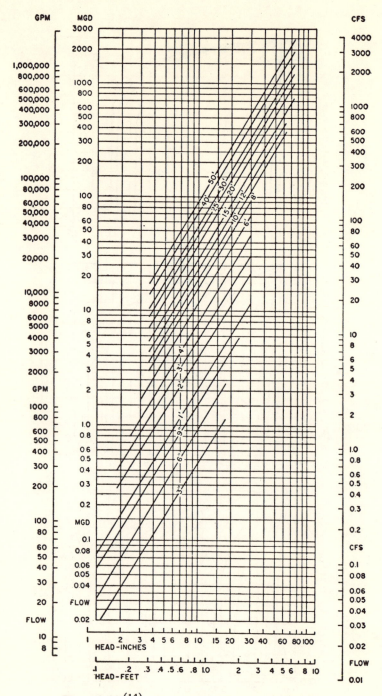

**Figure 6.11.**[14]   Flow curves for Parshall flumes.

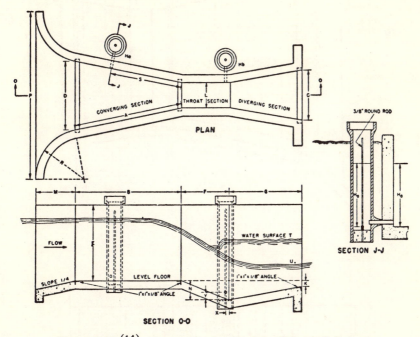

**Figure 6.12.**[14]   Composite sketch of small Parshall flumes.

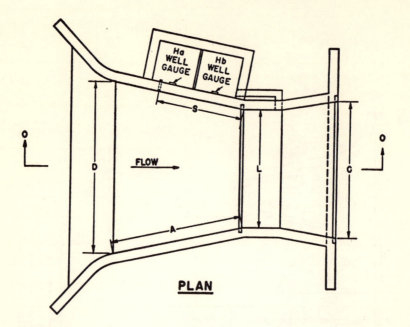

**PLAN**

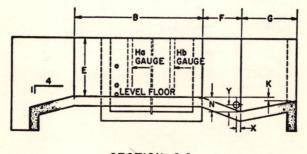

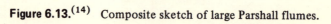

**SECTION O-O**

**Figure 6.13.**[(14)]  Composite sketch of large Parshall flumes.

**Table 6.12.**[14] Dimensions and Capacities of Small Parshall Measuring Flumes

| L | A | S | B | C | D | E | F | G | K | N | R | M | P | X | Y | Free-flow Capacity | |
|---|---|---|---|---|---|---|---|---|---|---|---|---|---|---|---|---|---|
| | | | | | | | | | | | | | | | | Min. | Max. |
| Ft. In. | Ft. In. | Ft. In. | Ft. In. | Ft. In. | Ft. In. | Ft. In. | Ft. In. | Ft. In. | In. | In. | Ft. In. | Ft. In. | Ft. In. | In. | In. | Sec.-Ft. | Sec.-Ft. |
| 0–3 | 1–6 3/8 | 1–1/4 | 1–6 | 0–7 | 0–10 3/16 | 2–0 | 0–6 | 1–0 | 1 | 2 1/4 | 1–4 | 1–0 | 2–6 1/4 | 1 | 1 1/2 | .03 | 1.9 |
| 0–6 | 2–0 7/16 | 1–4 5/16 | 2–0 | 1–3 5/8 | 1–3 5/8 | 2–0 | 1–0 | 2–0 | 3 | 4 1/2 | 1–4 | 1–0 | 2–11 1/2 | 2 | 3 | .05 | 3.9 |
| 0–9 | 2–10 5/8 | 1–11 1/8 | 2–10 | 1–3 | 1–10 5/8 | 2–6 | 1–0 | 1–6 | 3 | 4 1/2 | 1–4 | 1–0 | 3–6 1/2 | 2 | 3 | .09 | 8.9 |
| 1–0 | 4–6 | 3–0 | 4–4 7/8 | 2–0 | 2–9 1/4 | 3–0 | 2–0 | 3–0 | 3 | 9 | 1–8 | 1–3 | 4–10 3/4 | 2 | 3 | .11 | 16.1 |
| 1–6 | 4–9 | 3–2 | 4–7 7/8 | 2–6 | 3–4 3/8 | 3–0 | 2–0 | 3–0 | 3 | 9 | 1–8 | 1–3 | 5–6 | 2 | 3 | .15 | 24.6 |
| 2–0 | 5–0 | 3–4 | 4–10 7/8 | 3–0 | 3–11 1/2 | 3–0 | 2–0 | 3–0 | 3 | 9 | 1–8 | 1–3 | 6–1 | 2 | 3 | .42 | 33.1 |
| 3–0 | 5–6 | 3–8 | 5–4 3/4 | 4–0 | 5–1 7/8 | 3–0 | 2–0 | 3–0 | 3 | 9 | 1–8 | 1–3 | 7–3 1/2 | 2 | 3 | .61 | 50.4 |
| 4–0 | 6–0 | 4–0 | 5–10 5/8 | 5–0 | 6–4 1/4 | 3–0 | 2–0 | 3–0 | 3 | 9 | 2–0 | 1–6 | 8–10 3/4 | 2 | 3 | 1.3 | 67.9 |
| 5–0 | 6–6 | 4–4 | 6–4 1/2 | 6–0 | 7–6 5/8 | 3–0 | 2–0 | 3–0 | 3 | 9 | 2–0 | 1–6 | 10–1 1/4 | 2 | 3 | 1.6 | 85.6 |
| 6–0 | 7–0 | 4–8 | 6–10 3/8 | 7–0 | 8–9 | 3–0 | 2–0 | 3–0 | 3 | 9 | 2–0 | 1–6 | 11–3 1/2 | 2 | 3 | 2.6 | 103.5 |
| 7–0 | 7–6 | 5–0 | 7–4 1/4 | 8–0 | 9–11 3/8 | 3–0 | 2–0 | 3–0 | 3 | 9 | 2–0 | 1–6 | 12–6 | 2 | 3 | 3.0 | 121.4 |
| 8–0 | 8–0 | 5–4 | 7–10 1/8 | 9–0 | 11–1 3/4 | 3–0 | 2–0 | 3–0 | 3 | 9 | 2–0 | 1–6 | 13–8 1/4 | 2 | 3 | 3.5 | 139.5 |

**Table 6.13.**[14] Dimensions of Large Parshall Measuring Flumes

| L | A | S | B | C | D | E | F | G | K | N | X | Y | Free-flow Capacity | |
|---|---|---|---|---|---|---|---|---|---|---|---|---|---|---|
| | | | | | | | | | | | | | Max. | Min. |
| Ft. In. | Ft. In. | Ft. In. | Ft. In. | Ft. In. | Ft. In. | Ft. In. | Ft. | Ft. | In. | Ft. In. | In. | In. | Sec.-Ft. | Sec.-Ft. |
| 10–0 | 14–3 3/8 | 6–0 | 14–0 | 14–8 | 15–7 1/4 | 4–0 | 3 | 6 | 6 | 1–1 1/2 | 12 | 9 | 200 | 6 |
| 12–0 | 16–3 3/4 | 6–8 | 16–0 | 18–4 | 18–4 3/4 | 5–0 | 3 | 8 | 6 | 1–1 1/2 | 12 | 9 | 350 | 8 |
| 15–0 | 25–6 | 7–8 | 25–0 | 24–0 | 25–0 | 6–0 | 4 | 10 | 9 | 1–6 | 12 | 9 | 600 | 8 |
| 20–0 | 25–6 | 9–4 | 25–0 | 29–4 | 30–0 | 7–0 | 6 | 12 | 12 | 2–3 | 12 | 9 | 1,000 | 10 |
| 25–0 | 25–6 | 11–0 | 25–0 | 34–8 | 35–0 | 7–0 | 6 | 13 | 12 | 2–3 | 12 | 9 | 1,200 | 15 |
| 30–0 | 26–6 | 12–8 | 26–0 | 45–4 | 40–4 3/4 | 7–0 | 6 | 14 | 12 | 2–3 | 12 | 9 | 1,500 | 15 |
| 40–0 | 27–6 | 16–0 | 27–0 | 56–8 | 50–9 1/2 | 7–0 | 6 | 16 | 12 | 2–3 | 12 | 9 | 2,000 | 20 |
| 50–0 | 27–6 | 19–4 | 27–0 | | 60–9 1/2 | 7–0 | 6 | 20 | 12 | 2–3 | 12 | 9 | 3,000 | 25 |

## 6.2 Fluid Flow—Two Phase

### INTRODUCTION

The prediction of pressure drop for two-phase fluid flow can be very complex. In fact, under some conditions, even the best present methods give answers that are off by 50%—or more! The methods presented here are greatly simplified, and should be used only within the limitations stated below.

### CALCULATIONAL PROCEDURE*

The pressure drop for a very small increment of pipe can be regarded as the sum of three factors: pressure drop due to friction, pressure drop due to gravity, and pressure drop due to momentum changes. This can be expressed as

$$\Delta P_{tp} = \left[ \left( \frac{\Delta P}{\Delta L} \right)_{tpf} + \left( \frac{\Delta P}{\Delta L} \right)_{grav} \right] \Delta L + \Delta P_m \qquad (6.1)$$

**Friction Pressure Gradient $(\Delta P/\Delta L)_{tpf}$**

1. Calculate the friction gradient for each phase as if it were flowing alone through the pipe, $(\Delta P/\Delta L)_l$ and $(\Delta P/\Delta L)_g$. See section 6.1 for the procedure to do this. Use average fluid properties over $\Delta L$.
2. From the vapor Reynolds number, $N_{reg}$, and the liquid Reynolds number, $N_{rel}$, determine the combination of flow mechanics involved. Use Table 6.14.
3. Using Equation (6.2), calculate $x$.

$$x = \left[ \frac{(\Delta P/\Delta L)_l}{(\Delta P/\Delta L)_g} \right]^{1/2} \qquad (6.2)$$

4. From Figure 1.11, Figure 6.14, or Table 6.15, determine $\emptyset_l$ and/or $\emptyset_g$.
5. Calculate the two phase friction gradient $(\Delta P/\Delta L)_{tpf}$ from Equation 6.3.

$$\left( \frac{\Delta P}{\Delta L} \right)_{tpf} = \emptyset_l^2 \left( \frac{\Delta P}{\Delta L} \right)_l = \emptyset_g^2 \left( \frac{\Delta P}{\Delta L} \right)_g \qquad (6.3)$$

**Gravity Pressure Gradient $(\Delta P/\Delta L)_{grav}$**

The pressure gradient due to gravity is found by Equation (6.4). $R_l$ and $R_g$ can be read from Figure 6.14 or Table 6.15.

$$\left( \frac{\Delta P}{\Delta L} \right)_{grav} = (\rho_l R_l + \rho_g R_g) \left( \frac{g}{g_0} \right) (\sin \theta) \qquad (6.4)$$

*All of the pressure losses in this section are expressed in lb/ft$^2$. These results must be divided by 144 to obtain lb/in$^2$.

**Table 6.14.**[16]   Flow Mechanisms for Two-Phase Flow

| Flow mechanism | | Symbol | $N_{rel}$ | $N_{reg}$ |
|---|---|---|---|---|
| Liquid | Gas | | | |
| Turbulent | Turbulent | t-t | $> 2000$ | $> 2000$ |
| Viscous | Turbulent | v-t | $< 1000$ | $> 2000$ |
| Turbulent | Viscous | t-v | $> 2000$ | $< 1000$ |
| Viscous | Viscous | v-v | $< 1000$ | $< 1000$ |

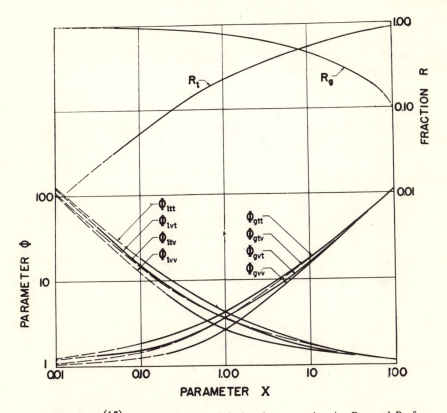

**Figure 6.14.**[17]   Curves showing relation between $\phi_1$, $\phi_g$, $R_1$, and $R_g$ for all flow mechanisms.

**Table 6.15.**[17] Values of Martinelli Functions with Independent Variable $X$ [a]

| X | All mechanisms | | Turbulent-turbulent | | Viscous-turbulent | | Turbulent-viscous | | Viscous-viscous | |
|---|---|---|---|---|---|---|---|---|---|---|
| | $R_l$ | $R_g$ | $\phi_{l,tt}$ | $\phi_{g,tt}$ | $\phi_{l,vt}$ | $\phi_{g,vt}$ | $\phi_{l,tv}$ | $\phi_{g,tv}$ | $\phi_{l,vv}$ | $\phi_{g,vv}$ |
| 0.01 | | | (128) | (1.28) | (120.0) | (1.20) | (112.0) | (1.12) | (105.0) | (1.05) |
| 0.02 | | | (68.4) | (1.37) | (64.0) | (1.28) | (58.0) | (1.16) | (53.5) | (1.07) |
| 0.04 | | | 38.5 | 1.54 | (34.0) | (1.36) | (31.0) | (1.24) | (28.0) | (1.12) |
| 0.07 | (0.04) | (0.96) | 24.4 | 1.71 | 20.7 | 1.45 | (19.3) | (1.35) | (17.0) | (1.19) |
| 0.10 | 0.05 | 0.95 | 18.5 | 1.85 | 15.2 | 1.52 | (14.5) | (1.45) | (12.4) | (1.24) |
| 0.2 | 0.09 | 0.91 | 11.2 | 2.23 | 8.90 | 1.78 | (8.70) | (1.74) | (7.00) | (1.40) |
| 0.4 | 0.14 | 0.86 | 7.05 | 2.83 | 5.62 | 2.25 | (5.50) | (2.20) | 4.25 | 1.70 |
| 0.7 | 0.19 | 0.81 | 5.04 | 3.53 | 4.07 | 2.85 | (4.07) | (2.85) | 3.08 | 2.16 |
| 1.0 | 0.23 | 0.77 | 4.20 | 4.20 | 3.48 | 3.48 | (3.48) | (3.48) | 2.61 | 2.61 |
| 2.0 | 0.31 | 0.69 | 3.10 | 6.20 | 2.62 | 5.25 | (2.62) | (5.24) | 2.06 | 4.12 |
| 4.0 | 0.40 | 0.60 | 2.38 | 9.50 | 2.05 | 8.20 | (2.15) | (8.60) | 1.76 | 7.00 |
| 7.0 | 0.48 | 0.52 | 1.96 | 13.7 | 1.73 | 12.1 | (1.83) | (12.8) | 1.60 | 11.2 |
| 10.0 | 0.53 | 0.47 | 1.75 | 17.5 | 1.59 | 15.9 | (1.66) | (16.6) | 1.50 | 15.0 |
| 20.0 | 0.66 | 0.34 | 1.48 | 29.5 | (1.40) | (28.0) | (1.44) | (28.8) | 1.36 | 27.3 |
| 40.0 | 0.76 | 0.24 | 1.29 | 51.5 | (1.25) | (50.0) | (1.25) | (50.0) | 1.25 | 50.0 |
| 70.0 | 0.84 | 0.16 | 1.17 | 82.0 | (1.17) | (82.0) | (1.17) | (82.0) | (1.17) | (82.0) |
| 100.0 | (0.90) | (0.10) | 1.11 | 111.0 | (1.11) | (111.0) | (1.11) | (111.0) | (1.11) | (111.0) |

[a] Parentheses indicate regions where little or no data were used.

**Momentum Pressure Loss ($\Delta P_m$)**

$$\Delta P_m = (G^2/g_0)\,(V_2 - V_1) \tag{6.5}$$

$V_2$ and $V_1$ are calculated on the assumption that the velocity of each phase is the same using Equation (6.6).

$$V = 1/\rho = (1 - z)V_1 + z V_g \tag{6.6}$$

The assumption that both phases are at the same velocity can be a source of error. If $\Delta P_m > 10\%$ of $\Delta P_{tp}$, the error can be significant.


## APPLICATION

The method outlined above is good for a section of pipe over which there has been very little change in the vapor/liquid ratio and $\Delta P_m$ is not a large fraction of $\Delta P_{tp}$. However, this ratio can be changing very rapidly due to line pressure drop and subsequent liquid flashing. Since the line pressure drop is what we are trying to calculate, it is impossible, at the outset, to predict the extent of flashing. A trial and error calculation results.

The calculations are much easier if the increments are taken as $\Delta P$ increments rather than length ($\Delta L$) increments. With this approach, the pressure at the start and finish of each increment is known, while the length is unknown.

Flashing is calculated by means of an energy balance over the increment. If $\Delta P_m < 10\%$ of $\Delta P_{tp}$, the energy balance can be based on constant enthalpy. It is assumed that the vapor/liquid ratio in the feed to the system, along with the feed temperature, pressure, and composition, is known. From these conditions, the enthalpy of both phases can be calculated for the inlet of the first increment. Since the $\Delta P$ across the increment is known, the vapor/liquid ratio can also be calculated for the outlet of the increment. This is done by assuming constant enthalpy across the increment, and doing an adiabatic flash calculation. This calculation is done at the downstream pressure on a system whose composition is the summation of the liquid and vapor phases, thereby assuring that enthalpy changes in both phases are taken into account.

It is suggested that the line in question be treated as a single increment for the first trial. On the second trial, treat the line as two increments. If the calculated $\Delta L$ on the second trial is significantly different from that on the first trial, repeat the calculations using four increments, etc. The liquid and vapor physical properties are averaged over each increment from their initial and final values. As the number of increments increases, consideration should be given to a computer program for the calculations.

## CHOKING

This procedure does not apply for choke flow. In choke flow, the momentum term becomes the largest factor in $\Delta P_{tp}$ since $\Delta L \to 0$. Choke flow can occur, for example, when the frangible on a vessel has ruptured.

A convenient way to check for choke flow is to look for a negative length $(-\Delta L)$ in the calculation results. Negative $\Delta L$ occurs when $\Delta P_m > \Delta P_{tp}$ for increment following the point at which choke flow occurs. Or,

$$\Delta L = \frac{\Delta P_{tp} - \Delta P_m}{(\Delta P / \Delta L)_{tpf} + (\Delta P / \Delta L)_{grav}} \leqslant 0 \qquad (6.7)$$

## 6.3 Flow through Orifices[20]

1. Liquids

$$W = 157.6 d_0^2 C (h_L \rho^2)^{1/2} \qquad (6.8)$$

where $C$ is given in Figure 6.15.

2. Gases and vapors

$$W = 1891 Y d_0^2 C (\Delta P / \overline{V}_1)^{1/2} \qquad (6.9)$$

where $C$ is given in Figure 6.15 and $Y$ is given in Figure 6.16.

3. Air flow through orifice: use Table 6.16.
4. For location of orifices relative to fittings, see Table 6.17.

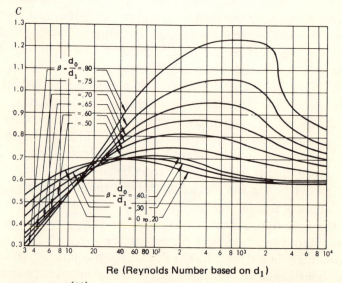

Re (Reynolds Number based on $d_1$)

**Figure 6.15.**[21] Flow coefficient $C$ for square-edged orifices.

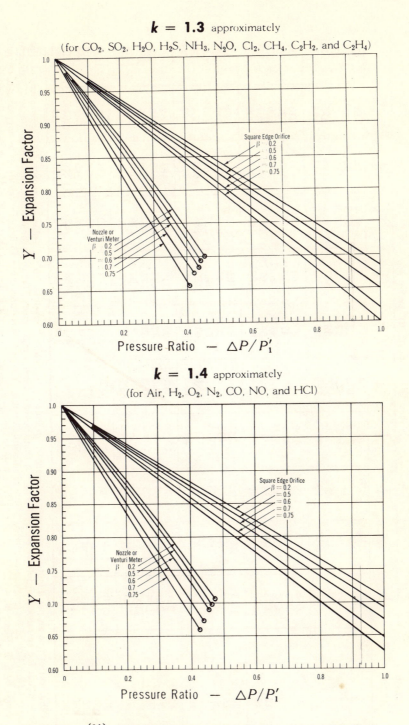

**Figure 6.16.**[21] Net expansion factor $Y$ for compressible flow through nozzles and orifices.

**Table 6.16.**[20]  Discharge of Air through an Orifice

| Gauge pressure (lb/in.²) | Diameter of orifice (in.) | | | | | | | | | | |
|---|---|---|---|---|---|---|---|---|---|---|---|
| | $\frac{1}{64}$ | $\frac{1}{32}$ | $\frac{1}{16}$ | $\frac{1}{8}$ | $\frac{1}{4}$ | $\frac{3}{8}$ | $\frac{1}{2}$ | $\frac{5}{8}$ | $\frac{3}{4}$ | $\frac{7}{8}$ | 1 |
| 1 | .028 | .112 | .450 | 1.80 | 7.18 | 16.2 | 28.7 | 45.0 | 64.7 | 88.1 | 115 |
| 2 | .040 | .158 | .633 | 2.53 | 10.1 | 22.8 | 40.5 | 63.3 | 91.2 | 124 | 162 |
| 3 | .048 | .194 | .775 | 3.10 | 12.4 | 27.8 | 49.5 | 77.5 | 111 | 152 | 198 |
| 4 | .056 | .223 | .892 | 3.56 | 14.3 | 32.1 | 57.0 | 89.2 | 128 | 175 | 228 |
| 5 | .062 | .248 | .993 | 3.97 | 15.9 | 35.7 | 63.5 | 99.3 | 143 | 195 | 254 |
| 6 | .068 | .272 | 1.09 | 4.34 | 17.4 | 39.1 | 69.5 | 109 | 156 | 213 | 278 |
| 7 | .073 | .293 | 1.17 | 4.68 | 18.7 | 42.2 | 75.0 | 117 | 168 | 230 | 300 |
| 9 | .083 | .331 | 1.32 | 5.30 | 21.2 | 47.7 | 84.7 | 132 | 191 | 260 | 339 |
| 12 | .095 | .379 | 1.52 | 6.07 | 24.3 | 54.6 | 97.0 | 152 | 218 | 297 | 388 |
| 15 | .105 | .420 | 1.68 | 6.72 | 26.9 | 60.5 | 108 | 168 | 242 | 329 | 430 |
| 20 | .123 | .491 | 1.96 | 7.86 | 31.4 | 70.7 | 126 | 196 | 283 | 385 | 503 |
| 25 | .140 | .562 | 2.25 | 8.98 | 35.9 | 80.9 | 144 | 225 | 323 | 440 | 575 |
| 30 | .158 | .633 | 2.53 | 10.1 | 40.5 | 91.1 | 162 | 253 | 365 | 496 | 648 |
| 35 | .176 | .703 | 2.81 | 11.3 | 45.0 | 101 | 180 | 281 | 405 | 551 | 720 |
| 40 | .194 | .774 | 3.10 | 12.4 | 49.6 | 112 | 198 | 310 | 446 | 607 | 793 |
| 45 | .211 | .845 | 3.38 | 13.5 | 54.1 | 122 | 216 | 338 | 487 | 662 | 865 |
| 50 | .229 | .916 | 3.66 | 14.7 | 58.6 | 132 | 235 | 366 | 528 | 718 | 938 |
| 60 | .264 | 1.06 | 4.23 | 16.9 | 67.6 | 152 | 271 | 423 | 609 | 828 | 1082 |
| 70 | .300 | 1.20 | 4.79 | 19.2 | 76.7 | 173 | 307 | 479 | 690 | 939 | 1227 |
| 80 | .335 | 1.34 | 5.36 | 21.4 | 85.7 | 193 | 343 | 536 | 771 | 1050 | 1371 |
| 90 | .370 | 1.48 | 5.92 | 23.7 | 94.8 | 213 | 379 | 592 | 853 | 1161 | 1516 |
| 100 | .406 | 1.62 | 6.49 | 26.0 | 104 | 234 | 415 | 649 | 934 | 1272 | 1661 |
| 110 | .441 | 1.76 | 7.05 | 28.2 | 113 | 254 | 452 | 705 | 1016 | 1383 | 1806 |
| 120 | .476 | 1.91 | 7.62 | 30.5 | 122 | 274 | 488 | 762 | 1097 | 1494 | 1951 |
| 125 | .494 | 1.98 | 7.90 | 31.6 | 126 | 284 | 506 | 790 | 1138 | 1549 | 2023 |

[a]In cubic feet of free air per minute at standard atmospheric pressure of 14.7 lb/in.² absolute, 70°F. Table is based on 100% coefficient of flow. For well-rounded entrance multiply values by 0.97. For sharp edged orifices a multiplier of 0.65 may be used for approximate results. Values for pressures from 1 to 15 lbs gauge calculated by standard adiabatic formula. Values for pressures above 15 lb gauge calculated by approximate formula proposed by S. A. Moss: $W_s = .5303 (aCP_1)/(T_1)^{1/2}$ where $W_s$ is the discharge (lb/sec); $a$ is the area of orifice (in.²); $C$ is the coefficient of flow; $P_1$ is the upstream total pressure (lb/in.², abs); $T_1$ is the upstream temperature (°F, abs). Values used in calculating the table were; $C = 1.0$, $P_1 =$ gauge pressure + 14.7 lb/in.², $T_1 = 530°F$ (abs). Weights ($W$) were converted to volumes using density factor of 0.07494 lb/ft³. This is correct for dry air at 14.7 lb/in.² absolute pressure and 70°F. The formula cannot be used where $P_1$ is less than two times the barometric pressure.

**Table 6.17.** Locations of Orifices and Nozzles Relative to Pipe Fittings—
Distances are in Pipe Diameters ($D$)

| | $D_2/D^a$ | Distances, upstream fitting to orifice | | Distance, vanes to orifice | Distances, nearest downstream fitting from orifice |
|---|---|---|---|---|---|
| | | Without straightening vanes | With straightening vanes | | |
| Single 90° ell, tee or | 0.2 | 6 | | | 1 |
| cross used as ell | 0.4 | 6 | | | |
| | 0.6 | 8 | | | |
| | 0.8 | 20 | 10 | 8 | 2 |
| 2 short radius 90° | 0.2 | $9^b$ | | | 1 |
| ells in form of S | 0.4 | $9^b$ | 8 | | |
| | 0.6 | $14^b$ | $10^b$ | 6 | |
| | 0.8 | $25^b$ | $16^b$ | 10 | 2 |
| 2 long or short radius 90° | 0.2 | $15^b$ | 5.5 | 5 | 1 |
| ells in perpendicular | 0.4 | $18^b$ | 6 | | |
| planes | 0.6 | $25^b$ | 8 | 6 | |
| | 0.8 | $40^b$ | 12 | 6.5 | 2 |
| Contraction or enlargement | 0.2 | 4 | Vanes | | 1 |
| | 0.4 | 6 | have no | | |
| | 0.6 | 9 | advantage | | |
| | 0.8 | 15 | | | 2 |
| Globe valve | 0.2 | 18 | 8 | 5 | 1 |
| | 0.4 | 22 | 8 | 5 | |
| | 0.6 | 30 | 9 | 6 | |
| | 0.8 | 50 | 15 | 9 | 2 |
| Gate valve, 1/3 open | 0.2 | 10 | Same as globe valve | | 1 |
| | 0.4 | 12 | | | |
| | 0.6 | 48 | | | |
| | 0.8 | >60 | | | 2 |

$^a D_2/D$ = orifice diameter/i.d. of pipe.
$^b$A.G.A. Gas Measurement Committee Report No. 2.

## NOMENCLATURE

### Single Phase (Section 6.1)

$A$     Internal cross-sectional area of pipe (ft$^2$)

$C_v$    Valve coefficient

$D$     Internal pipe diameter (ft)

$d$     Internal pipe diameter (in.)

$E$     Friction loss or head loss (ft-lb/lb$_f$)

$f$     Fanning friction factor (dimensionless)

$G$     Mass velocity (lb/sec-ft$^2$)

$g$     Acceleration of gravity (ft/sec$^2$)

$g_c$    Conversion factor (32.174 ft-lb$_m$/lb$_f$-sec$^2$)

$K$     Number of velocity heads lost (dimensionless)

$L$     Length of pipe (ft)

$M$     Molecular weight

$N$     Number of velocity heads lost in compressible flow (dimensionless)

$P$     Pressure (lb force/ft$^2$) (usually absolute pressure)

$P'$    Pressure (psia, or psi in $\Delta P'$)

$Q$     Actual volumetric flowrate (ft$^3$/sec or ft$^3$/hr)

$Q_a$    Actual volumetric flowrate (gpm)

$q_a$    Actual volumetric flowrate (ft$^2$/min)

$q_h$    Actual volumetric flowrate (gph)

$R$     Universal gas constant (1,545 ft-lb$_f$/lb-mole °R)

$N_{Re}$  Reynolds number (dimensionless, see Figure 6.7)

$T$     Absolute temperature (°R)

$V$     Fluid velocity (ft/sec)

$W$     Mass flowrate (lb/hr)

$w$     Mass flowrate (lb/sec)

$\gamma$    Heat capacity ratio

$\epsilon$    Pipe roughness (ft)

$\mu$    Viscosity [lb/(sec) (ft)]

$\mu'$   Viscosity (cP)

$\rho$    Density (lb/ft$^3$)

$\rho_v$    Vapor density

$\psi$    Pressure-drop correction in Figure 6.1

*Subscripts*

0     Refers to condition in upstream reservoir

1     Refers to upstream condition in pipe

2     Refers to downstream condition in pipe

100    Refers pressure drop to 100 total equivalent ft of pipe

cni    Maximum for isothermal compressible flow

G     Gas or vapor

L     Liquid

## Two Phase (Section 6.2)

$A$     Inside cross sectional area of the pipe (ft$^2$)

$D$     Inside pipe diameter (ft)

$g$     Acceleration due to gravity (ft/sec$^2$)

$g_0$     Gravitational constant (32.2 ft-lb$_m$/lb$_f$-sec$^2$)

$G$     Mass velocity (lb/ft$^2$-sec), calculated from the rate divided by the cross sectional area or $G = W/A$

$\Delta L$     Length of piping (ft)

$N_{Re}$     Reynolds number

$(\Delta P/\Delta L)_{grav}$     Pressure gradient due to gravity (psf/ft)

$(\Delta P/\Delta L)_v$     Pressure gradient for vapor alone flowing in pipe (psf/ft)

$(\Delta P/\Delta L)_l$     Pressure gradient for liquid alone flowing in pipe (psf/ft)

$P_m$     Pressure loss due to momentum (psf)

$(\Delta P/\Delta L)_{tpf}$     Two-phase pressure gradient due to friction (psf/ft)

$R_l, R_g$     Fraction of pipe filled with liquid or gas

$V_g$     Specific volume of the gas phase (ft$^3$/lb)

$V_l$     Specific volume of the liquid phase (ft$^3$/lb)

$V_1, V_2$     Specific volume at points 1 and 2 (ft$^3$/lb)

$w$     Total mass flowrate (lb/sec)

$X$     Lockhart–Martinelli parameter, see Equation 6.1

$z$     Quality equal to the weight fraction of gas flowing, dimensionless

$\emptyset_l, \emptyset_g$     Multiplying factor, see Figure 6.14

$\rho_g, \rho_l$     Fluid density (lb/ft$^3$)

$\theta$     Angle of flow (degrees)

## Flow through Orifices (Section 6.3)

$C$     Orifice flow coefficient

$d_1$     Pipe inside diameter (in.)

$d_0$     Orifice diameter (in.)

$h_L$     Pressure drop across orifice (ft of liquid)

$k$     Ratio of specific heats of gas ($c_p/c_v$) (Table 6.11)

$\Delta P$     Pressure drop across orifice (psi)

$P_1'$     Upstream pressure (psia)

$\overline{V}$     Specific volume of fluid upstream of orifice (ft$^3$/lb)

$W$     Flowrate (lb/hr)

$Y$     Expansion factor through orifices, compressible flow, Figure 6.16.

$\rho$     Density of fluid (lb/ft$^3$)

$\beta$     Ratio of $d_0/d_1$

# REFERENCES

1. E. E. Ludwig, *Applied Design for Chemical and Petrochemical Plants*, Vol. 1, Gulf Publishing Co., Houston, Texas, 1964, p. 52.

2. R. Kern, Size pump piping and components, *Hydrocarbon Processing*, 52, 83–85 (1973).

3. ITT, Engineering Manual EM-611, Marlow Pumps, Fluid Handling Division, International Telephone and Telegraph Corp., Midland Park, New Jersey, p. 26.

4. Tufline Division of Xomox Corporation, Tufline Catalog No. 5, Birmingham, Michigan, 1974, pp. L7 and 14 of Process Control Valves section.

5. Crane Company, Engineering Division, Flow of Fluids, Technical Paper No. 410, Chicago, Illinois, 1965, pp. B-14 and B-15.

6. G. A. Shaw, and A. W. Loomis, Cameron Hydraulic Data, 14 ed., Ingersoll-Rand Company, Woodcliff Lake, New Jersey, 1970, pp. 69–70.

7. L. L. Simpson, Process piping: Functional design, *Chem. Eng. (Deskbook Issue)* April 14, 169–172 (1969).

8. Dow Chemical USA, Dow Plastic Lined Pipe and Fittings Catalog, Midland, Michigan, 1972, pp. 8–11.

9. R. W. Gallant, Sizing pipe for liquids and vapors, *Chem. Eng.* February 24, 96–104 (1969).

10. Walworth Company, Valve Catalog No. 57, Section on Flow of Fluids in Pipes, Walworth Co., 1957, pp. E-53.

11. L. L. Simpson, Sizing process piping, *Chem. Eng.* June 17, 204 (1968).

12. R. G. Perry, C. H. Chilton, and S. D. Kirkpatrick, *Chemical Engineers' Handbook*, 4th ed., McGraw-Hill New York, 1963, pp. 3–131.

13. C. E. Lapple, Compressibility in gas flow problems, *Chem. Eng.* May, 125 (1949).

14. L. K. Spink, Principles and Practice of Flow Meter Engineering, The Foxboro Co., Foxboro, Massachusetts, 1973, pp. 288–297.

15. J. C. Stevens, Hydrographic Data Book, 7th ed., Leupold & Stevens Instruments, Inc., Portland, Oregon, 1968, pp. 18–19.

16. P. Griffith and W. S. Tong, Two Phase Flow Heat Transfer, American Institute of Chemical Engineers, Today Series, 1973, pp. 30–34.

17. R. W. Lockhart, and R. C. Martinelli, Proposed correlation of data for isothermal two-phase, two-component flow in pipes, *Chem. Eng. Prog. 45*, 39–45 (1949).

18. A. E. DeGrance, and R. W. Atherton, Chemical engineering aspects of two-phase flow, *Chem. Eng.*, July 13, 98 (1970).

19. P. Griffith, Two-phase flow, in Roshenow and Hartnett (eds.), *Handbook of Heat Transfer*, McGraw-Hill, New York, 1973, Section 14.

20. E. E. Ludwig, *Applied Process Design for Chemical and Petro-Chemical Plants*, Vol. 1, Gulf Publishing Co., Houston, Texas, 1964, pp. 72–74.

21. Crane Company, Flow of Fluids, Technical Paper No. 410, 15th ed., Chicago, Illinois, 1976, pp. A-20 and A-21.

## SELECTED READING

### Single Phase

Crane Company, Engineering Division, Flow of Fluids, 15th Printing, Technical Paper No. 410, Chicago, Ill., 1976.

Dow Plastic Lined Pipe and Fittings Catalog, Dow Chemical U.S.A., Midland, Mich., 1972.

R. W. Gallant, Sizing pipe for liquids and vapors, *Chem. Eng.* February 24 (1969).

R. Kern, Part 1, Useful properties of fluids for piping design, December 23 (1974); Part 2, How to compute pipe size, January 6 (1975); Part 3, Measuring flow in pipes with orifices and nozzles, February 3 (1975); Part 4, How to size flowmeters, March 3 (1975); Part 5, Control valves in process plants, April 14 (1975); Part 6, How to design piping for pump-suction conditions, April 28 (1975); Part 7, How to size piping for pump-discharge conditions, May 26 (1975); Part 8, piping design for two-phase flow, June 23 (1975); Part 9, How to design piping for reboiler systems, August 4, (1975); Part 10, How to design overheat condensing systems, September 15 (1975); Part 11, How to size piping and components as gas expands at flow conditions, October 13 (1975); Part 12, Piping systems for process plants, November 10 (1975); *Chem. Eng.*

R. Kern, Size pump piping and components, *Hydrocarbon Processing* Vol. 52, No. 3, March (1973).

C. E. Lapple, Compressibility in gas flow problems, *Chem. Eng.* May (1949).

E. E. Ludwig, *Applied Process Design for Chemical and Petrochemical Plants*, Vol. 1, Gulf Publishing Co., Houston, Texas, (1964).

R. G. Perry and C. H. Chilton, *Chemical Engineers' Handbook*, 5th ed., McGraw-Hill, New York, 1973.

G. A. Shaw and A. W. Loomis, *Cameron Hydraulic Data*, 15th ed., Ingersoll-Rand Company, Woodcliff Lake, N.J., 1977.

L. L. Simpson, Process piping: Functional design, Deskbook Issue, April 14 (1969); Sizing piping for process plants, June 17, 204 (1968), *Chem. Eng.*

L. K. Spink, *Principles and Practice of Flow Meter Engineering*, The Foxboro Co., Foxboro, Mass., 1973.

J. C. Stevens, *Hydrographic Data Book*, 7th ed., Leupold & Stevens Instruments, Inc., Portland, Oregon, May, 1968.

### Two Phase

A. E. DeGrance and R. W. Atherton, Part 1, Chemical engineering aspects of two-phase flow, March 23 (1970); Part 2, Phase equilibria, flow regimes, energy loss, April 20 (1970); Part 3, transferring heat in two-phase systems, May 4 (1970); Part 4, Horizontal flow correlations, July 13 (1970); Part 5, Mechanical-energy balance, August 10, (1970); Part 6, Vertical and inclined-flow correlations, October 5 (1970); Part 7, Pressure-drop sample calculations, November 2 (1970); Part 8, The coupled energy balances, February 22 (1971); *Chem. Eng.*

P. Griffith, "Two-Phase Flow," Section 14 of *Handbook of Heat Transfer* (Roshenow and Hartnett, eds.), McGraw-Hill Book, Co., New York, 1973.

P. Griffith and W. S. Tong, Two phase flow heat transfer, *A.I.Ch.E. Today Series*, American Institute of Chemical Engineers, 1973.

R. W. Lockhart and R. C. Martinelli, Proposed correlation of data for isothermal two-phase, two-component flow in pipes, *Chem. Eng. Prog.* Vol. 45, No. 1, January (1949).

G. W. Scovier and K. Aziz, *The Flow of Complex Mixtures in Pipes*, Van Nostrand Reinhold Co., New York, 1972.

G. B. Wallis, *One-Dimensional Two-Phase Flow*, McGraw-Hill, New York, 1969.

## Flow through Orifices

Crane Company, *Flow of Fluids*, Technical Paper No. 410, 15th ed. Chicago, Ill., 1976.

E. E. Ludwig, *Applied Process Design for Chemical and Petro-Chemical Plants*, Vol. 1, Gulf Publishing Co., Houston, Texas, 1964.

L. K. Spink, *Principles and Practice of Flow Meter Engineering*, The Foxboro Co., Foxboro, Mass., 1973.

# 7  Gas–Solid Separations

## SETTLING LAWS[1]–TERMINAL VELOCITY

When a particle falls under the influence of gravity, it will accelerate until the frictional drag in the fluid balances the gravitational forces. At this point it will continue to fall at constant velocity. This is the terminal velocity or free-settling velocity. The general formula for spherically shaped particles is

$$u_t = \left[ \frac{4g_L D_p(\rho_s - \rho)}{3\rho_v C} \right]^{1/2} \tag{7.1}$$

Depending on the size of the particle, various forms of Equation (7.1) are used to calculate the actual terminal velocity. Figure 7.1 summarizes the equations and their application.

## SEPARATION EQUIPMENT

Figure 7.2[2] classifies commonly known solids as to typical size. It also presents the type of equipment generally used to separate such solids from gases.

Figure 7.3[3] gives the typical collection efficiency that can be expected from various types of gas–solid separators.

Figure 7.4 can be used to evaluate the performance of a cyclone for removing solids from gases.

## NOMENCLATURE

$D_p$  Diameter of particle (ft)
$C$  Overall drag coefficient (dimensionless)
$g_L$  Acceleration due to gravity (32.2 ft/sec$^2$)
$\rho_s$  True particle density (lb/ft$^3$)
$\rho_v$  Vapor density (lb/ft$^3$)

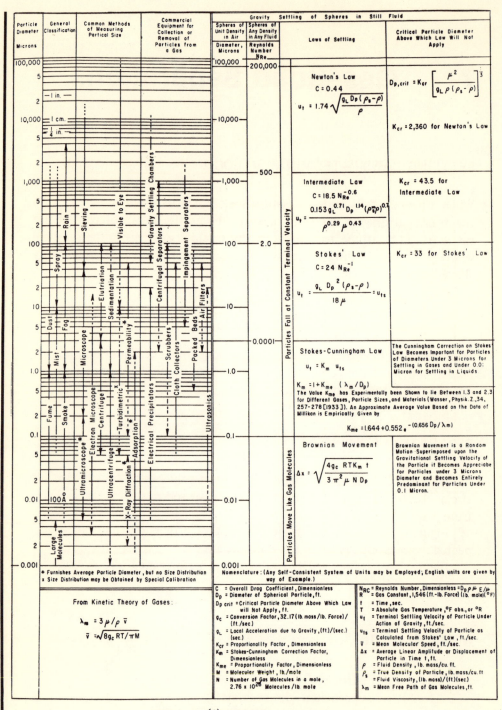

**Figure 7.1.**[4] Particle size classification.

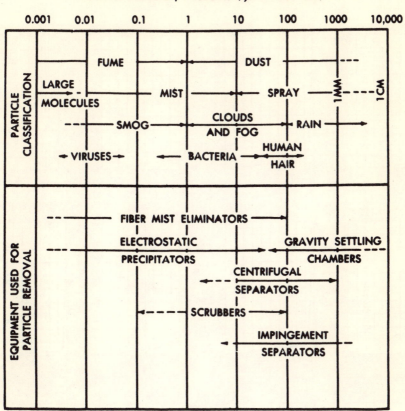

**Figure 7.2.**[2]  Particle classification and useful collection equipment vs. particle size.

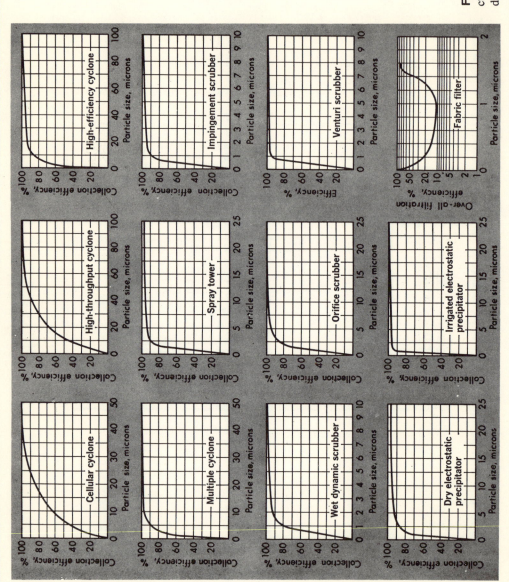

**Figure 7.3.**[3] Efficiency curves for various types of dust-collecting equipment.

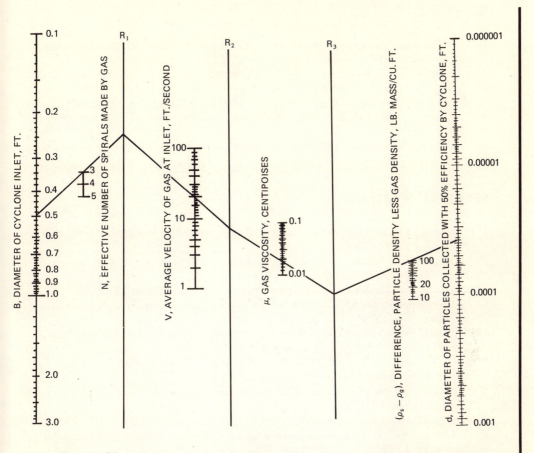

**Figure 7.4.**[5] Cut size diameter of particles collected by a cyclone separator. For example, calculate the cut size particle diameter for an entrained solid in air having a density of 100 lb/ft³. The solid is to be separated in a cyclone separator. The following data are available: $B = 0.5$ ft, $V = 20$ ft/sec, $N = 3$. The air flows at a temperature of 80°F. At 80°F, $\mu = 0.018$ cP, $\rho_g = 0.0765$ lb/ft³, $\rho_s - \rho_g = 100$ lb/ft³.

| Connect | With | Mark |
|---------|------|------|
| $B = 0.5$ ft | $N = 3$ | $R_1$ |
| $R_1$ | $V = 20$ | $R_2$ |
| $R_2$ | $\mu = 0.018$ | $R_3$ |
| $R_3$ | $\rho_s - \rho_g = 100$ | read, $d = 0.000038$ ft |

## REFERENCES

1. E. E. Ludwig, *Applied Process Design for Chemical and Petrochemical Plants*, Vol. 1, Gulf Publishing Co., Houston, Texas, 1964, p. 133.
2. J. A. Brink, W. F. Burggrabe, and L. E. Greenwell, Mist removal from compressed gases, *Chem. Eng. Prog. 62*, 61 (1966).
3. G. D. Sargent, Dust collection equipment, *Chem. Eng.* January 27, 141 (1969).
4. J. H. Perry, *Chemical Engineers' Handbook*, 3rd Ed., McGraw-Hill, New York, 1950, p. 1019.
5. J. F. Kuong, Nomograph finds cyclone particle size, *Hydrocarbon Processing* March, 205 (1967).

## SELECTED READING

J. H. Abbott, and D. C. Drehmel, Control of fine particulate emissions, *Chem. Eng. Prog.*, December (1976).

J. A. Brink, W. F. Burggrabe, and L. E. Greenwell, Mist removal from compressed gases, *Chem. Eng. Prog.*, Vol. 62, No. 4, April (1966).

S. Calvert, How to choose a particulate scrubber, *Chem. Eng.*, August 29 (1977).

G. J. Celenza, Designing air pollution control systems, Chem. Eng. Prog., Vol. 66, No. 11, November (1970).

J. F. Kuong, Nomograph finds cyclone particle size, *Hydrocarbon Processing*, March (1967).

E. E. Ludwig, *Applied Process Design for Chemical and Petro-Chemical Plants*, Vol. 1, Gulf Publishing Co., Houston, Texas, 1964.

W. L. O'Connell, How to attack air pollution control problems, *Chem. Eng.*, Deskbook Issue, October 18 (1976).

J. H. Perry, *Chemical Engineers' Handbook*, 3rd ed., McGraw-Hill, New York, 1950.

R. H. Perry, and C. H. Chilton, *Chemical Engineers' Handbook*, 5th ed., McGraw-Hill, New York, 1973.

J. N. Peters, Predicting efficiency of fine-particle collectors, *Chem. Eng.*, April 16 (1973).

G. D. Sargent, Dust collection equipment, *Chem. Eng.*, January 27 (1969).

# 8  Heat Transfer

## 8.1  Heat Transfer Coefficients

### INTRODUCTION

The approximate overall heat transfer coefficient $U$ can be determined from the following equation:

$$\frac{1}{U} = \frac{1}{h_i} + F_i + \frac{l_w}{k_w} + F_o + \frac{1}{h_o} \qquad (8.1)$$

where $U$ is the overall heat transfer coefficient [Btu/(ft$^2$-hr-°F)]; $h_i$ is the inside film coefficient [Btu/(ft$^2$-hr-°F)]; $F_i$ is the inside fouling factor [(ft$^2$-hr-°F)/Btu]. $l_w$ is the wall thickness (ft); $k_w$ is the thermal conductivity of the wall [(Btu-ft)/(ft$^2$-hr-°F)]; $F_o$ is the outside fouling factor [(ft$^2$-hr-°F)/Btu]; $h_o$ is the outside film coefficient [Btu/(ft$^2$-hr-°F)]. A graphical representation of the various terms is shown in Figure 8.1. In Figure 8.1, the fouling factors $F_i$ and $F_o$ are expressed as $L_i/k_i$ and $L_o/k_o$.

The charts and tables that follow present methods that can be used to obtain a preliminary value for $U$. However, final designs should be based on methods which more fully take into account heat exchanger geometry, fluid velocities, and fluid properties.

### Example (Figure 8.2):  Effect of Velocity on Heat Transfer Rates

1. If shell side coefficient of a unit is 25 Btu/(ft$^2$-hr-°F) and velocity in shell is doubled read new shell side coefficient, $h_o$, as 36 (line a).
2. If tube side coefficient is 25 and velocity is doubled, read new tube coefficient, $h_o$, as 43.1 (line a).
3. In either case, pressure drop would increase by factor of 4.
4. This procedure may be used in reverse for reduced flow.

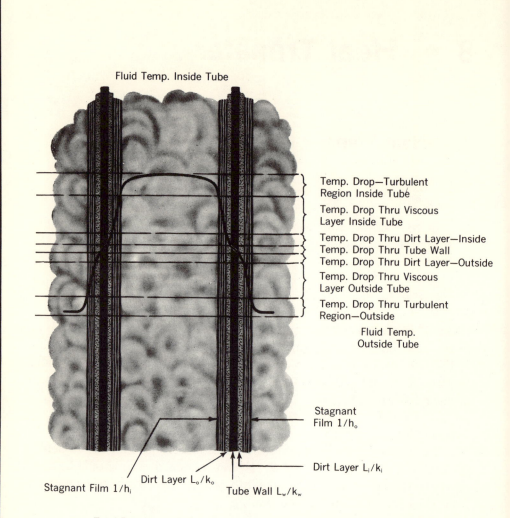

Fluid Temp. Inside Tube

Temp. Drop—Turbulent Region Inside Tube

Temp. Drop Thru Viscous Layer Inside Tube

Temp. Drop Thru Dirt Layer—Inside
Temp. Drop Thru Tube Wall
Temp. Drop Thru Dirt Layer—Outside

Temp. Drop Thru Viscous Layer Outside Tube

Temp. Drop Thru Turbulent Region—Outside

Fluid Temp. Outside Tube

Stagnant Film $1/h_o$

Dirt Layer $L_i/k_i$

Stagnant Film $1/h_i$
Dirt Layer $L_o/k_o$
Tube Wall $L_w/k_w$

Total Resistance To The Flow Of Heat, $1/U$, Is Equal To The Sum Of The

Individual Resistances Or $1/U = \dfrac{1}{h_i} + \dfrac{L_i}{k_i} + \dfrac{L_w}{k_w} + \dfrac{1}{h_o} + \dfrac{L_o}{k_o}$

**Figure 8.1.**[1]  Flow of heat through tube walls.  Total resistance to the flow of heat $1/U$ is equal to the sum of the individual resistances or $1/U = (1/h_i) + (L_i/k_i) + (L_w/k_w) + (1/h_o) + (L_o/k_o)$.

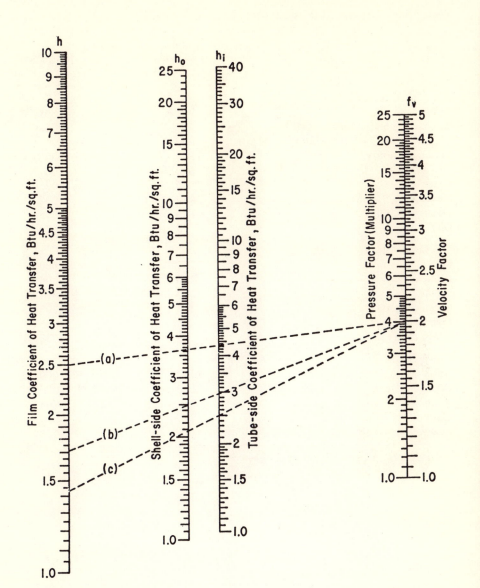

**Figure 8.2.**[6]   Effect of velocity on heat transfer rates.

**Table 8.1.**[1] Inside and Outside Film Coefficient, $h_i$ and $h_o$

|  | $h_i$ or $h_o$ Conductance [Btu/(hr-ft$^2$-°F)] | Resistance [(hr-ft$^2$-°F)/Btu] |
|---|---|---|
| No change of state |  |  |
| Water | 900 | 0.0011 |
| Organic liquids | 200 | 0.0050 |
| Organic liquid in water solution | 400 | 0.0025 |
| Liquid ammonia | 1500 | 0.0007 |
| 25% brine | 600 | 0.0017 |
| Gases | 40 | 0.0250 |
| except hydrogen | 450 | 0.0022 |
| except helium | 180 | 0.0056 |
| Oils | 35 | 0.0286 |
| Condensing |  |  |
| Steam | 1500[a] | 0.0007 |
| Diphenyl | 220 | 0.0045 |
| Dowtherm-A | 250 | 0.0040 |
| Organic vapors | 275 | 0.0036 |
| Light oils | 290 | 0.0034 |
| Heavy oils (vacuum) | 30 | 0.0333 |
| Evaporating |  |  |
| Water | 1300 | 0.0008 |
| Organic solvents | 170 | 0.0059 |
| Ammonia | 290 | 0.0034 |
| Light oils | 210 | 0.0048 |
| Heavy oils | 20 | 0.0500 |

[a]Film type; dropwise may be 4–5 times higher.

**Table 8.2.**[2]  Fouling Resistance, $F_o$ and $F_i$ [(hr-ft$^2$-$^\circ$F)/Btu]

| | Up to 240°F/ 125°F or less | | 240°F–400°F/ over 125°F[a] | |
|---|---|---|---|---|
| | Velocity 3 ft/sec and less | Velocity over 3 ft/sec | Velocity 3 ft/sec and less | Velocity over 3 ft/sec |
| *Fouling Resistances for Water* | | | | |
| Types of water | | | | |
| Sea water | 0.0005 | 0.0005 | 0.001 | 0.001 |
| Brackish water | 0.002 | 0.001 | 0.003 | 0.002 |
| Cooling tower and artificial spray pond | | | | |
|   Treated makeup | 0.001 | 0.001 | 0.002 | 0.002 |
|   Untreated | 0.003 | 0.003 | 0.005 | 0.004 |
| City or well water | | | | |
|   (such as the Great Lakes) | 0.001 | 0.001 | 0.002 | 0.002 |
| Great lakes | 0.001 | 0.001 | 0.002 | 0.002 |
| River water | | | | |
|   Minimum | 0.002 | 0.001 | 0.003 | 0.002 |
|   Mississippi | 0.003 | 0.002 | 0.004 | 0.003 |
|   Delaware, Schuylkill | 0.003 | 0.002 | 0.004 | 0.003 |
|   East River and New York Bay | 0.003 | 0.002 | 0.004 | 0.003 |
|   Chicago Sanitary Canal | 0.008 | 0.006 | 0.010 | 0.008 |
| Muddy or silty | 0.003 | 0.002 | 0.004 | 0.003 |
| Hard (over 15 grains/gal) | 0.003 | 0.003 | 0.005 | 0.005 |
| Engine jacket | 0.001 | 0.001 | 0.001 | 0.001 |
| Distilled | 0.0005 | 0.0005 | 0.0005 | 0.0005 |
| Treated boiler feedwater | 0.001 | 0.0005 | 0.001 | 0.001 |
| Boiler blowdown | 0.002 | 0.002 | 0.002 | 0.002 |
| *Fouling Resistances for Industrial Fluids* | | | | |
| Oils | | | | |
| Fuel oil | | | | 0.005 |
| Transformer oil | | | | 0.001 |
| Engine lube oil | | | | 0.001 |
| Quench oil | | | | 0.004 |
| Gases and vapors | | | | |
| Manufactured gas | | | | 0.01 |
| Engine exhaust gas | | | | 0.01 |
| Steam (non-oil-bearing) | | | | 0.0005 |
| Exhaust steam (oil bearing) | | | | 0.001 |
| Refrigerant vapors (oil bearing) | | | | 0.002 |
| Compressed air | | | | 0.002 |
| Industrial organic heat transfer media | | | | 0.001 |
| Liquids | | | | |
| Refrigerant liquids | | | | 0.001 |
| Hydraulic fluid | | | | 0.001 |
| Industrial organic heat transfer media | | | | 0.001 |
| Molten heat transfer salts | | | | 0.0005 |

**Table 8.2.**[2]  *Cont'd.*

*Fouling Resistances for Chemical Processing Streams*
Gases and vapors

| | |
|---|---|
| Acid gas | 0.001 |
| Solvent vapors | 0.001 |
| Stable overhead products | 0.001 |

Liquids

| | |
|---|---|
| MEA and DEA solutions | 0.002 |
| DEG and TEG solutions | 0.002 |
| Stable side draw and bottom product | 0.001 |
| Caustic solutions | 0.002 |
| Vegetable oils | 0.003 |

*Fouling Resistances for Natural Gas–Gasoline Processing Streams*
Gases and vapors

| | |
|---|---|
| Natural gas | 0.001 |
| Overhead products | 0.001 |

Liquids

| | |
|---|---|
| Lean oil | 0.002 |
| Rich oil | 0.001 |
| Natural gasoline and liquefied petroleum gases | 0.001 |

*Fouling Resistances for Oil Refinery Streams*
Crude and vacuum unit gases and vapors

| | |
|---|---|
| Atmospheric tower overhead vapors | 0.001 |
| Light naphthas | 0.001 |
| Vacuum overhead vapors | 0.002 |

*Crude and Vacuum Liquids*
Crude oil

| | 0–199°F | | | 200–299°F | | |
|---|---|---|---|---|---|---|
| | Velocity under 2 ft/sec | Velocity 2–4 ft/sec | Velocity 4 ft/sec and over | Velocity under 2 ft/sec | Velocity 2–4 ft/sec | Velocity 4 ft/sec and over |
| Dry salt[b] | 0.003 | 0.002 | 0.002 | 0.003 | 0.002 | 0.002 |
| | 0.003 | 0.002 | 0.002 | 0.005 | 0.004 | 0.004 |

| | 300–499°F | | | 500°F and over | | |
|---|---|---|---|---|---|---|
| | Velocity under 2 ft/sec | Velocity 2–4 ft/sec | Velocity 4 ft/sec and over | Velocity under 2 ft/sec | Velocity 2–4 ft/sec | Velocity 4 ft/sec and over |
| Dry salt[b] | 0.004 | 0.003 | 0.002 | 0.005 | 0.004 | 0.003 |
| | 0.006 | 0.005 | 0.004 | 0.007 | 0.006 | 0.005 |

**Table 8.2.**[2] *Cont'd.*

| | |
|---|---|
| Gasoline | 0.001 |
| Naphtha and light distillates | 0.001 |
| Kerosene | 0.001 |
| Light gas oil | 0.002 |
| Heavy gas oil | 0.003 |
| Heavy fuel oils | 0.005 |
| Asphalt and residuum | 0.010 |

*Cracking and Coking Unit Streams*

| | |
|---|---|
| Overhead vapors | 0.002 |
| Light cycle oil | 0.002 |
| Heavy cycle oil | 0.003 |
| Light coker gas oil | 0.003 |
| Heavy coker gas oil | 0.004 |
| Bottoms slurry oil (4½ ft/sec minimum) | 0.003 |
| Light liquid products | 0.002 |

*Catalytic Reforming, Hydrocracking, and Hydrodesulfurization Streams*

| | |
|---|---|
| Reformer charge | 0.002 |
| Reformer effluent | 0.001 |
| Hydrocracker charge and effluent[c] | 0.002 |
| Recycle gas | 0.001 |
| Hydrodesulfurization charge and effluent[c] | 0.002 |
| Overhead vapors | 0.001 |
| Liquid product over 50° A.P.I. | 0.001 |
| Liquid product 30–50° A.P.I. | 0.002 |

*Light Ends Processing Streams*

| | |
|---|---|
| Overhead vapors and gases | 0.001 |
| Liquid products | 0.001 |
| Absorption oils | 0.002 |
| Alkylation trace acid streams | 0.002 |
| Reboiler streams | 0.003 |

*Lube Oil Process Streams*

| | |
|---|---|
| Feed stock | 0.002 |
| Solvent feed mix | 0.002 |
| Solvent | 0.001 |
| Extract[d] | 0.003 |
| Raffinate | 0.001 |
| Asphalt | 0.005 |
| Wax slurries[d] | 0.003 |
| Refined lube oil | 0.001 |

[a] Ratings in columns 3 and 4 are based on a temperature of the heating medium of 240–400° F. If the heating medium temperature is over 400° F and the cooling medium is known to scale, these ratings should be modified accordingly.

[b] Normally desalted below this temperature range. (Footnote to apply to 200–299° F, 300–499° F, 500° F and over.)

[c] Depending on charge characteristics and storage history, charge resistance may be many times this value.

[d] Precautions must be taken to prevent wax deposition on cold tube walls.

**Table 8.3[4]** Overall Coefficients in Typical Petrochemical Applications: $U$, Btu/(hr-ft² - °F)

| In Tubes | Outside Tubes | Type Equipment | Velocities, Ft./Sec. | | Overall Coefficient | Temp. Range, °F | Estimated Fouling | | |
| --- | --- | --- | --- | --- | --- | --- | --- | --- | --- |
| | | | Tube | Shell | | | Tube | Shell | Overall |
| **A. Heating-Cooling** | | | | | | | | | |
| Butadiene mix. (Super-heating) | Steam | H | 25-35 | 1.0-1.8 | 12 | 400-100 | ... | ... | 0.04 |
| Solvent | Solvent | H | | | 35-40 | 110-30 | ... | ... | 0.0065 |
| Solvent | Propylene (vaporization) | K | 1-2 | | 30-40 | 40-0 | ... | ... | 0.006 |
| C4 Unsaturates | Propylene (Vaporization) | K | | | 13-18 | 100-35 | ... | ... | 0.005 |
| Solvent | Chilled Water | H | 20-40 | | 35-75 | 115-40 | ... | ... | ... |
| Oil | Oil | H | | | 60-85 | 150-100 | 0.003 | 0.001 | ... |
| Ethylene-vapor | Condensate & Vapor | K | | | 90-125 | 600-200 | 0.0015 | 0.0015 | ... |
| Ethylene vapor | Chilled water | H-U | | | 50-80 | 270-100 | 0.002 | 0.001 | ... |
| Condensate | Propylene (refrigerant) | H | | | 60-135 | 60-30 | 0.001 | 0.001 | ... |
| Chilled water | Transformer oil | H | | | 40-75 | 75-50 | 0.001 | 0.001 | ... |
| Calcium Brine-25% | Chlorinated C1 | K-U | | | 40-60 | -20-+10 | 0.002 | 0.005 | 0.002 |
| Ethylene liquid | Ethylene vapor | H | 1-2 | 0.5-1.0 | 10-20 | -170-(-100) | | | 0.002 |
| Propane vapor | Propane liquid | K-U | | | 6-15 | -25-100 | | | |
| Lights & chlor. HC | Steam | U | | | 12-30 | 30-260 | 0.001 | 0.001 | 0.3 |
| Unsat. light HC, CO, CO2, H2 | Steam | H | | | 15-25 | 400-100 | | | |
| Ethonolamine | Steam | H | | | 10-20 | -30-220 | 0.001 | 0.001 | 0.004 |
| Steam | Air mixture | U (in tank) | | | 50-60 | 190-230 | 0.0005 | 0.0015 | 0.005 |
| Steam | Styrene & Tars | H | 4-7 | | 100-130 | 90-25 | 0.001 | 0.002 | |
| Chilled Water | Lean Copper Solvent | H | 4-5 | 1-2 | 100-125 | 180-90 | | 0.001 | |
| Water* | Treated water | H | 3-5 | | 6-10 | 90-110 | | | |
| Water | C2-chlor. HC, lights | H | 2-3 | | 7-15 | 360-100 | 0.002 | 0.001 | |
| Water | Hydrogen chloride | H | | | 45-30 | 230-90 | 0.002 | 0.001 | |
| Water | Heavy C2-Chlor. | H | | | 55-35 | 300-90 | 0.001 | 0.001 | |
| Water | Perchlorethylene | H | | | 20-35 | 150-90 | 0.001 | 0.001 | |
| Water | Air & Water Vapor | H | | | 230-160 | 370-90 | 0.0015 | 0.0015 | |
| Water | Engine Jacket Water | H | | | 80-115 | 175-90 | 0.0015 | 0.001 | |
| Water | Absorption Oil | H | 4-7 | | 8-18 | 130-90 | 0.0015 | 0.001 | |
| Water | Air-Chlorine | U | 5-7 | | 170-225 | 250-90 | | | 0.005 |
| Water | Treated Water | H | | | | 200-90 | 0.001 | 0.001 | |
| **B. Condensing** | | | | | | | | | |
| C4 Unsat. | Propylene refrig. | K | ᐯ | | 58-68 | 60-35 | | | 0.005 |
| HC Unsat. lights | Propylene refrig. | K | ᐯ | | 50-60 | 45-3 | | | 0.0055 |
| Butadiene | Propylene refrig. | K | ᐯ | | 65-80 | 20-35 | | | 0.004 |

| Hot Fluid | Cold Fluid | Type | v | | Range | Range | | | |
|---|---|---|---|---|---|---|---|---|---|
| Lights & Chloro-ethanes | Propylene refrig. | KU | | | 15-25 | 130-(20) | 0.002 | 0.001 | |
| Ethylene | Propylene refrig. | KU | 7-8 | | 60-90 | 120-(10) | 0.001 | 0.001 | |
| Unsat. Chloro HC. | Water | H | 3-8 | | 90-120 | 145-90 | 0.001 | 0.001 | |
| Unsat. Chloro HC. | Water | H | 6 | | 180-140 | 110-90 | 0.002 | 0.001 | |
| Unsat. Chloro HC. | Water | H | | | 15-25 | 130-(20) | 0.001 | 0.001 | |
| Chloro-HC. | Water | KU | | | 20-30 | 110-(10) | 0.002 | 0.001 | |
| Solvent & Non Cond. | Water | H | | | 25-15 | 260-90 | 0.0015 | 0.004 | 0.003 |
| Water | Propylene Vapor | H | 2-3 | | 130-150 | 200-90 | 0.0015 | 0.001 | |
| Water | Propylene | H | | | 60-100 | 130-90 | 0.002 | 0.0001 | |
| Water | Steam | H | | | 225-110 | 300-90 | 0.0015 | 0.001 | |
| Water | Steam | H | | | 190-235 | 230-130 | 0.0001 | 0.0001 | |
| Treated Water | Steam (Exhaust) | H | | | 20-30 | 220-130 | 0.0001 | 0.0001 | |
| Oil | Steam | H | | | 70-110 | 375-130 | 0.003 | 0.001 | |
| Water | Propylene Cooling & Cond. | H | | | 25-50 | 30-45 (C) | 0.0015 | 0.001 | |
| Chilled Water | Air-Chlorine (Part. Cond.) | U | | | 110-150 | 15-20 (Co) | } 0.0015 | 0.005 | |
| | | | | | 8-15 | 8-15 (Co) | | 0.003 | |
| Water | Light HC, Cool & Cond. | H | | | 20-30 | 10-15 (Co) | 0.0015 | 0.001 | |
| Water | Ammonia | H | | | 35-90 | 270-90 | 0.001 | 0.001 | |
| Water | Ammonia | U | | | 140-165 | 120-90 | 0.001 | 0.001 | |
| | | | | | 280-300 | 110-90 | | | |
| Air-Water Vapor | Freon | KU | | | 10-20 | { 60-10 | 0.001 | 0.001 | 0.01 |
| **C. Reboiling** | | | | | | | | | |
| Solvent, Copper-NH3. | Steam | H | 7-8 | | 130-150 | 180-160 | 0.001 | 0.001 | 0.005 |
| C4 Unsat. | Steam | H | | | 95-115 | 95-150 | 0.001 | 0.001 | 0.0065 |
| Chloro. HC. | Steam | VT | | | 35-25 | 300-350 | 0.001 | 0.001 | |
| Chloro. Unsat. HC. | Steam | VT | | | 100-140 | 230-130 | 0.002 | 0.001 | |
| Chloro. ethane. | Steam | VT | | | 90-135 | 300-350 | 0.002 | 0.0005 | |
| Chloro. ethane. | Steam | U | | | 50-70 | 30-190 | 0.04 | 0.001 | |
| Solvent (heavy). | Steam | VT | | | 70-115 | 375-300 | 0.002 | 0.0005 | |
| Mono-di-ethanolamines | Steam | VT | | | 210-155 | 450-350 | 0.003 | 0.0005 | |
| Organics, acid, water. | Steam | VT | | | 60-100 | 450-300 | 0.002 | 0.0015 | |
| Amines and water. | Steam | VT | | | 120-140 | 360-250 | 0.002 | 0.0005 | |
| Steam | Naphtha frac. | Annulus, Long. F.N. | | | 15-20 | 270-220 | 0.0035 | 0.001 | |
| Propylene. | C2, C3= | KU | | 25-35 | 120-140 | 150-40 | 0.001 | 0.001 | |
| Propylene-Butadiene. | Butadiene, Unsat. | H | | | 15-18 | 400-100 | | | 0.02 |

\* Unless specified, all water is untreated, brackish, bay or sea.
Notes: H = Horizontal, Fixed or Floating Tube Sheet.
     U = U—Tube Horizontal Bundle.
     K = Kettle Type.
     V = Vertical.
     R = Reboiler.

T = Thermosiphon.
v = Variable.
HC = Hydrocarbon.
(C) = Cooling range Δt.
(Co) = Condensing range Δt.

## Table 8.4.[3]  Thermal Resistance of Pipes and Tubing[a]

### TABLE A
Tubing

| Tube Size (O.D.) In. | BWG | Factor |
|---|---|---|
| ¼ | 18 | 0.005185 |
|   | 20 | 0.003423 |
|   | 22 | 0.002645 |
|   | 24 | 0.002017 |
| ⅜ | 16 | 0.006651 |
|   | 18 | 0.004733 |
|   | 20 | 0.003228 |
|   | 22 | 0.002513 |
|   | 24 | 0.001950 |
| ½ | 12 | 0.011931 |
|   | 14 | 0.008405 |
|   | 16 | 0.006274 |
|   | 18 | 0.004545 |
|   | 20 | 0.003146 |
| ⅝ | 10 | 0.014503 |
|   | 12 | 0.011108 |
|   | 14 | 0.007995 |
|   | 16 | 0.006038 |
|   | 18 | 0.004417 |
| ¾ | 10 | 0.013816 |
|   | 12 | 0.010733 |
|   | 14 | 0.007817 |
|   | 16 | 0.005951 |
|   | 18 | 0.004376 |
| ⅞ | 9 | 0.015054 |
|   | 10 | 0.013332 |
|   | 12 | 0.010447 |
|   | 14 | 0.007670 |
|   | 16 | 0.005864 |
|   | 18 | 0.004330 |
| 1 | 8 | 0.016686 |
|   | 10 | 0.012998 |
|   | 12 | 0.010247 |
|   | 14 | 0.007562 |
|   | 16 | 0.005802 |
|   | 18 | 0.004296 |
| 1¼ | 8 | 0.015965 |
|   | 10 | 0.012568 |
|   | 12 | 0.009979 |
|   | 14 | 0.007420 |
|   | 16 | 0.005721 |
|   | 18 | 0.004254 |
| 1½ | 8 | 0.015529 |
|   | 10 | 0.012300 |
|   | 12 | 0.009813 |
|   | 14 | 0.007328 |
|   | 16 | 0.005649 |
|   | 18 | 0.004223 |

### TABLE B
Pipe

| Nominal Size | Sched. | Factor |
|---|---|---|
| ⅛ | 40 | 0.006905 |
|   | 80 | 0.010686 |
| ¼ | 40 | 0.008874 |
|   | 80 | 0.013075 |
| ⅜ | 40 | 0.008864 |
|   | 80 | 0.013144 |
| ½ | 40 | 0.010516 |
|   | 80 | 0.015078 |
|   | 160 | 0.020622 |
| ¾ | 5 | 0.005782 |
|   | 10 | 0.007529 |
|   | 40 | 0.010604 |
|   | 80 | 0.015190 |
|   | 160 | 0.023474 |
| 1 | 5 | 0.005705 |
|   | 10 | 0.009931 |
|   | 40 | 0.012383 |
|   | 80 | 0.017813 |
|   | 160 | 0.026212 |
| 1¼ | 5 | 0.005641 |
|   | 10 | 0.009738 |
|   | 40 | 0.012778 |
|   | 80 | 0.018090 |
|   | 160 | 0.024970 |
| 1½ | 5 | 0.005611 |
|   | 10 | 0.009647 |
|   | 40 | 0.013111 |
|   | 80 | 0.018715 |
|   | 160 | 0.027762 |
| 2 | 5 | 0.005569 |
|   | 10 | 0.009527 |
|   | 40 | 0.013745 |
|   | 80 | 0.020070 |
|   | 160 | 0.033730 |

### Thermal Conductivity
Effect of Temperature upon Thermal Conductivity of Metals and Alloys*
Main body of table is $k$ in Btu/(hr.) (sq. ft.) (°F./ft.)

| t, °F. | 32 | 212 | 392 | 572 | 752 | 932 | 1112 | Melting point, °C. |
| t, °C. | 0 | 100 | 200 | 300 | 400 | 500 | 600 | |
|---|---|---|---|---|---|---|---|---|
| Aluminum | 117 | 119 | 124 | 133 | 144 | 155 | ... | 660 |
| Brass (70-30) | 56 | 60 | 63 | 66 | 67 | ... | ... | 940 |
| Cast iron | 32 | 30 | 28 | 26 | 25 | ... | ... | 1,275 |
| Cast high silicon iron | 30 | ... | ... | ... | ... | ... | ... | 1,200 |
| Copper (pure) | 224 | 218 | 215 | 212 | 210 | 207 | 204 | 1,083 |
| Lead | 20 | 19 | 18 | 18 | ... | ... | ... | 327.5 |
| Nickel | 36 | 34 | 33 | 32 | ... | ... | ... | 1,452 |
| Silver | 242 | 238 | ... | ... | ... | ... | ... | 960.5 |
| Sodium | 81 | ... | ... | ... | ... | ... | ... | 97.5 |
| Steel (mild) | ... | 26 | 26 | 25 | 23 | 22 | 21 | 1,375 |
| Tantalum (at 18°C.) | 32 | ... | ... | ... | ... | ... | ... | 2,850 |
| Tin | 36 | 34 | 33 | ... | ... | ... | ... | 231.85 |
| Wrought iron (Swedish) | ... | 32 | 30 | 28 | 26 | 23 | ... | 1,505 |
| Zinc | 65 | 64 | 62 | 59 | 54 | ... | ... | 419.4 |

Thermal Conductivity of Chromium Alloys*
$k$ = Btu/(hr.) (sq. ft.) (°F./ft.)

| American Iron and Steel Institute Type No. | k at 212°F. | k at 932°F. |
|---|---|---|
| 301, 302, 302B, 303, 304, 316 | 9.4 | 12.4 |
| 308 | 8.8 | 12.5 |
| 309, 310 | 8.0 | 10.8 |
| 321, 347 | 9.3 | 12.6 |
| 403, 406, 410, 414, 416 | 14.4 | 16.6 |
| 430, 430F | 15.1 | 15.2 |
| 442 | 12.5 | 14.2 |
| 501, 502 | 21.2 | 19.5 |

From, "Chemical Engineers' Handbook" Third Edition, McGraw-Hill, 1950

[a]The resistance of the tube or pipe wall, referred to its outside surface, may be calculated from the following equation: $r_w$ = (factor)/$k$, where $r_w$ is the wall resistance (hr-ft$^2$-°F)/Btu, factor is from column (A) or (B), and $k$ is the thermal conductivity, Btu/hr-ft$^2$-(°F/ft).

## Table 8.5.[1]  Thermal Resistance of Glass-Lined Pipe[a]

| | $r_w$ |
|---|---|
| 0.05 in. glass | 0.008333 |
| 0.05 in. Nucerite | 0.006250 |

[a]The values for $r_w$ are for the glass layer *only*. The resistance of the metal wall must be added.

## 8.2   Heat Losses from Tanks

**PROCEDURE**

1. Calculate area exposed (ft$^2$).
2. Choose appropriate value of $U$ from Table 8.7.
3. Calculate heat loss from

$$q = UA(t_p - t_a) \qquad (8.2)$$

where $U$ is the overall coefficient from Table 8.7 [Btu/(ft$^2$-hr-°F)]; $A$ is the exposed area of the tank (ft$^2$); $t_p$ is the tank temperature (°F); $t_a$ is the ambient temperature (°F); $q$ is the heat loss (Btu/hr).

For other thermal conductivities, multiply values of $U$ in Table 8.7 by $(K'/0.23)$, where $K'$ is the new thermal conductivity [Btu/hr-ft$^2$-(°F/in.)]. See Table 8.8 for thermal conductivity data of other insulating materials.

## 8.3   Heating of Process Piping and Vessels: Heat Losses from Insulated Pipelines[8]

**PROCEDURE**

1. Calculate the thermal conductance:

$$C_i = \left[ \frac{1}{(12/2\pi K) \ln (d_o/d_i)} \right] \qquad (8.3)$$

where $C_i$ is the thermal conductance of the insulation [Btu/(hr-ft-°F)]; $K$ is the thermal conductivity of the insulation [Btu/hr-ft$^2$-(°F/in.)]; $d_o$ is the outside diameter of the insulation (in.); $d_i$ is the inside diameter of insulation (in.). Use a value of $K$ from Table 8.8.
2. Calculate heat loss per foot of pipe:

$$q = 1.15 C_i(t_p - t_a) \qquad (8.4)$$

where $q$ is the heat lost by the pipe per foot [Btu/(hr-ft)]; $t_p$ is the pipe temperature (°F); $t_a$ is the ambient temperature (°F).

## 8.4   Heating of Process Piping and Vessels: Steam Tracing

**VESSELS**

**Procedure**

1. Calculate heat losses from the vessel using the procedure in Section 8.2.
2. Use Figure 8.3 to obtain the heat output per foot for a 150 psig steam tracer.

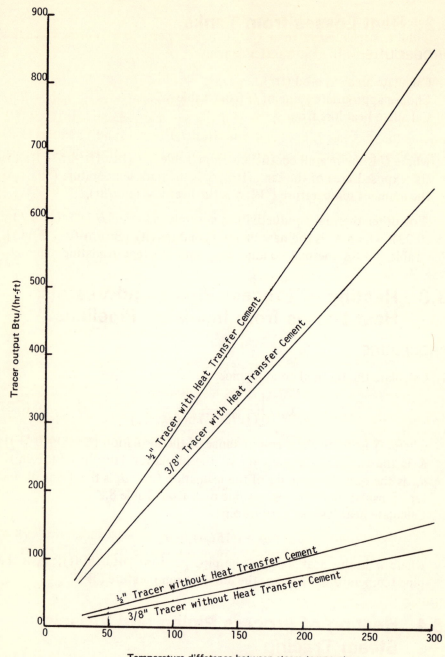

Tracer output Btu/(hr·ft)

½" Tracer with Heat Transfer Cement

3/8" Tracer with Heat Transfer Cement

½" Tracer without Heat Transfer Cement

3/8" Tracer without Heat Transfer Cement

Temperature difference between steam temperature
(366°F, 150 psig) and the process temperature, °F

**Figure 8.3.** Heat supplied by 150 psig tracer.

3. Calculate the length of the tracer needed.
4. If the tracer is longer than 150 ft, repeat steps 2 and 3 using value for steam tracer with heat transfer cement.

### Example

How much steam tracing will be needed to keep a storage tank 8 ft in diameter × 20 ft in length at 70°F if the ambient temperature is −20°F? Assume the tank has 2 in. of calcium silicate insulation and 150 psig steam is available for tracing. The tracer will be installed without heat transfer cement.

1. Heat losses from tank.

    a. Exposed area.

    $$\text{Heads} = (2)\,[\pi(8)^2/4] = 100 \text{ ft}^2$$
    $$\text{Side} = \pi(8)\,(20) = 503 \text{ ft}^2$$
    $$\text{Total area} = 603 \text{ ft}^2$$

    b. $K$ for calcium silicate = 0.372 (Btu-in.)/(ft$^2$-hr-°F)

    $$U = (0.11)\left(\frac{0.372}{0.23}\right) \quad \text{(From Table 8.7)}$$

    $$= 0.178 \text{ Btu/(hr-ft}^2\text{-°F)}$$

    c. $q = (0.178)\,(603)\,[70 - (-20)] = 9660$ Btu/hr

2. Heat output from steam tracer, assuming a $\frac{3}{8}$-in. tracer, $\Delta T = 296$°F.

    $$q_t = 115 \text{ Btu/(hr-ft)} \quad \text{(Figure 8.3, no heat transfer cement)}$$

3. Length of tracer needed = $q/q_t$

    $$\text{length} = \frac{9660}{115} = 84 \text{ ft}$$

This is equivalent to about 4 wraps around the circumference of the tank. Since this is reasonable, the tank can be traced with 84 ft of $\frac{3}{8}$-in. tubing without heat transfer cement.

### PIPING

### Procedure

1. Calculate heat losses $q$ from a pipeline using the procedure in Section 8.3.
2. Use Figure 8.3 to obtain the heat output per foot for a 150 psig steam tracer.

3. Calculate the number of parallel tracers needed, $q/q_t$. If less than one, use one tracer.
4. If more than one tracer is called for, repeat steps 2 and 3 above for a tracer with heat transfer cement. It is usually better to use a smaller number of tracers with heat transfer cement.

### Example

How many parallel tracers will be needed to keep a 4-in. pipeline at 200°F? The line is covered with $1\frac{1}{2}$ in. of insulation with a thermal conductivity of 0.35 Btu/(hr-ft$^2$-°F/in.). Steam at 150 psig will be used for tracing. Ambient temperature is $-20$°F.

1. Heat losses.

a. $C_i = \left[ \dfrac{1}{12/[2\pi(.35)] \ \ln (7/4)} \right]$    (see Equation 8.3)

$= 0.327$ Btu/(hr-ft-°F)

b. $q = (1.15) (0.327) [200 - (-20)]$
$= 82.7$ Btu/(hr-ft)

2. Heat output from steam tracer, assuming a $\frac{3}{8}$-in. tracer, $\Delta T = 166$°F.
$q_t = 64.7$ Btu/(hr-ft)    (Figure 8.3, no heat transfer cement)
3. The number of tracers needed is (82.7/64.7) = 1.278.
Therefore, 2 parallel tracers are called for.
4. Try a $\frac{3}{8}$-in. tracer with heat transfer cement
$q_t = 360$ Btu/(hr-ft)
The number of tracers needed equals (82.7/360), which is less than 1.
Therefore, one $\frac{3}{8}$-in. tracer with heat transfer cement is called for.
In conclusion, use one $\frac{3}{8}$-in. tracer with heat transfer cement.

# 8.5 Heating of Process Piping and Vessels: Dowtherm SR-1 Tracing*

The methods used for sizing Dowtherm SR-1 systems are identical to those outlined for steam tracing in Section 8.4. The only difference is that Figure 8.4 should be used to determine the heat supplied by the SR-1 tracers, rather than Figure 8.3.

---

*Dowtherm SR-1 heat transfer medium is a specially inhibited ethylene glycol based product designed for use in heat transfer systems.

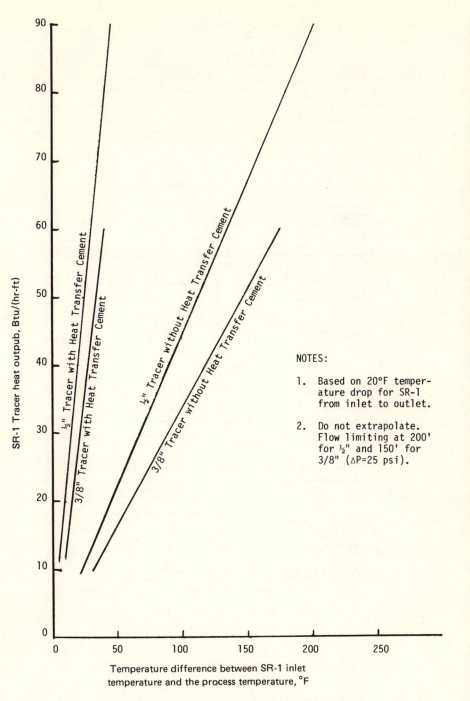

**Figure 8.4.** Heat supplied by Dowtherm SR-1 tracer.

**Table 8.6.**[5]   Overall Coefficients (*U*) for Platecoils

| Clamp on Platecoil | Water and solvents | | Viscous products | | Air and gases | |
|---|---|---|---|---|---|---|
| | Heating | Cooling | Heating | Cooling | Heating | Cooling |
| With heat transfer mastic | 30–40 | 20–30 | 12–20 | 5–12 | 1–3 | 1–3 |
| Without heat transfer mastic | 15–25 | 10–20 | 6–12 | 3–8 | 1–3 | 1–3 |

**Table 8.7.**[7]   Heat Loss from Storage Tanks and Product Correction Factors

Heat loss expressed as U (BTU/hr. sq. ft. F)
$\Delta T$ = Product temperature minus air temperature.

| Surface Condition | | Still Air | 10 mph | 15 mph | 20 mph | 25 mph | 30 mph |
|---|---|---|---|---|---|---|---|
| | | General Range of $\Delta T$ = 60 F | | | | | |
| | Uninsulated | 1.8 | 4.1 | 4.7 | 5.2 | 5.7 | 6.1 |
| 1″ | Insulation | 0.18 | 0.20 | 0.20 | 0.21 | 0.21 | 0.21 |
| 1½″ | Insulation | 0.13 | 0.14 | 0.14 | 0.14 | 0.14 | 0.14 |
| 2″ | Insulation | 0.10 | 0.11 | 0.11 | 0.11 | 0.11 | 0.11 |
| | | General Range of $\Delta T$ = 100 F | | | | | |
| | Uninsulated | 2.1 | 4.4 | 5.1 | 5.7 | 6.1 | 6.5 |
| 1″ | Insulation | 0.18 | 0.20 | 0.20 | 0.21 | 0.21 | 0.21 |
| 1½″ | Insulation | 0.13 | 0.14 | 0.14 | 0.14 | 0.14 | 0.14 |
| 2″ | Insulation | 0.10 | 0.11 | 0.11 | 0.11 | 0.11 | 0.11 |
| | | General Range of $\Delta T$ = 200 F | | | | | |
| | Uninsulated | 2.7 | 5.1 | 5.7 | 6.4 | 6.8 | 7.4 |
| 1″ | Insulation | 0.19 | 0.21 | 0.21 | 0.22 | 0.22 | 0.22 |
| 1½″ | Insulation | 0.13 | 0.15 | 0.15 | 0.15 | 0.15 | 0.15 |
| 2″ | Insulation | 0.11 | 0.11 | 0.11 | 0.11 | 0.11 | 0.11 |

A k value of 0.23 was used in calculating U for insulated tanks.

Calculated from data in Oil and Gas Journal's "The Refiner's Notebook," No. 125, by Prof. W. L. Nelson.

Product correction factors.  Apply to uninsulated U values only.

| Product | Approximate Product Temp. | | |
|---|---|---|---|
| | 75 F | 150 F | 250 F |
| Watery solutions ......................... | 1.00 | 1.00 | 1.00 |
| Gasoline, Kerosene, etc. ................ | 0.90 | 0.90 | 0.90 |
| Light oils ................................... | 0.80 | 0.85 | 0.90 |
| Medium oils ............................... | 0.70 | 0.75 | 0.80 |
| Heavy oils ................................. | 0.60 | 0.65 | 0.70 |
| Asphalts, Tars, etc. ...................... | 0.50 | 0.55 | 0.60 |
| Gases or Vapor spaces .................. | 0.50 | 0.50 | 0.50 |

U values as listed for insulated tanks, apply to all products without correction.

**Table 8.8.**[9]   Thermal Conductivities of Some Insulating Materials

| Material | Temperature range (°F) | | Temperature (°F) | Thermal conductivity (Btu-in.)/(ft$^2$-hr-°F) |
|---|---|---|---|---|
| | Minimum | Maximum | | |
| Calcium silicate | 100 | 1200 | 70 | 0.37 |
| | | | 150 | 0.41 |
| | | | 300 | 0.47 |
| Glass foam | −400 | 800 | 70 | 0.39 |
| | | | 150 | 0.43 |
| | | | 300 | 0.506 |
| Glass fiber | −300 | 600 | 70 | 0.25 |
| | | | 150 | 0.30 |
| | | | 300 | 0.39 |
| Magnesia | −20 | 600 | 70 | 0.41 |
| | | | 150 | 0.43 |
| | | | 300 | 0.49 |
| Perlite | −200 | 1500 | 300 | 0.42 |
| | | | 500 | 0.51 |
| Polystyrene foam | −400 | 175 | 70 | 0.26 |
| | | | 150 | 0.90 |
| Polyurethane foam | −50 | 230 | 70 | 0.17 |
| | | | 150 | 0.27 |

# 8.6   Double Pipe Exchangers[10]

To calculate the heat transfer coefficient of a fluid flowing between two pipes, or an annulus, it is necessary to use an equivalent diameter $d_e$. The equivalent diameter is defined as 4 times the hydraulic radius. The hydraulic radius is the radius of a pipe equivalent to the annulus cross section. The hydraulic radius is found as a ratio of the flow area to the wetted perimeter.

$$\text{flow area} = \frac{\pi}{4}(D_o^2 - D_i^2)$$

However, the wetted perimeter is different for heat transfer or pressure drop. For heat transfer the "wetted" perimeter is the outer circumference of the inner pipe, while for pressure drop, the wetted perimeter is the sum of the inner circumference of the outer pipe and the outer circumference of the inner pipe.

Heat transfer:

$$D_e = \frac{4\pi(D_o^2 - D_i^2)}{4\pi D_i} = \frac{D_o^2 - D_i^2}{D_i} \tag{8.5}$$

Pressure drop:

$$D_e' = \frac{4\pi(D_o^2 - D_i^2)}{4\pi(D_o + D_i)} = D_o - D_i \tag{8.6}$$

By using $D_e$, instead of $D$ one can calculate the outside film coefficient. $D_e'$ is used in pressure drop calculations.

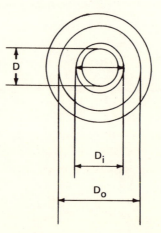

## PRESSURE DROP IN PIPES AND ANNULI

The pressure drop in tubes can be calculated by

$$\Delta F = \frac{2fG^2L}{g_c\rho^2 D} \quad \text{(ft of liquid)} \tag{8.6}$$

$$f = \frac{16}{DG/\mu} \text{ for } (DG/\mu) \leqslant 2,300 \tag{8.7}$$

$$f = 0.0035 + \frac{0.264}{(DG/\mu)^{0.42}} \text{ for } (DG/\mu) > 2,300 \tag{8.8}$$

$$\Delta P = (\Delta F \times \rho/144) \tag{8.9}$$

For an annulus, use $D'_e$ in place of $D$.

The only other pressure drop of consequence is that due to the reversal of flow in the annulus flow. For each hairpin reversal

$$\Delta F_r = \frac{V^2}{2g'} \tag{8.10}$$

where $V$ is the velocity (ft/sec) and $g'$ is the acceleration due to gravity (32.2 ft/sec$^2$). Again

$$\Delta P = (\Delta F_r \times \rho/144) \tag{8.11}$$

The total annulus pressure drop is calculated from

$$\Delta P_t = \left[\frac{(\Delta F + \Delta F_r)\rho}{144}\right]$$

## SUGGESTED ORDER OF CALCULATION FOR A DOUBLE PIPE EXCHANGER

Process conditions required:

Hot fluid: $T_1, T_2, W, C, \mu, k, \Delta P, R_D, S,$ or $\rho$

Cold fluid: $t_1, t_2, w, c, \mu, k, \Delta P, R_D, s,$ or $\rho$

The diameter of the pipes must be given or assumed. (See Table 12.1.) Try to keep velocities in the range of 3–10 ft/sec. The following assumes the cold fluid is in the inner pipe, as is most often the case.

1. Calculate duty

$$Q = WC(T_1 - T_2) = wc(t_2 - t_1)$$

2. Calculate log mean temperature difference (LMTD).

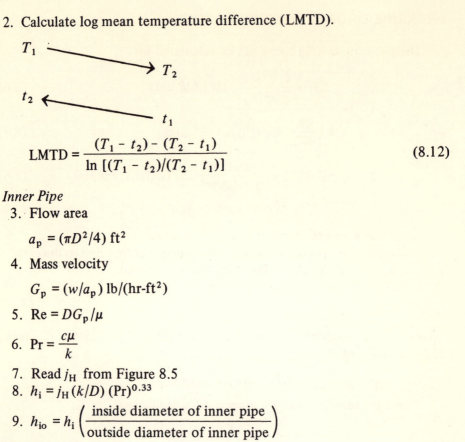

$$\text{LMTD} = \frac{(T_1 - t_2) - (T_2 - t_1)}{\ln\left[(T_1 - t_2)/(T_2 - t_1)\right]} \tag{8.12}$$

*Inner Pipe*

3. Flow area

$$a_p = (\pi D^2/4)\ \text{ft}^2$$

4. Mass velocity

$$G_p = (w/a_p)\ \text{lb}/(\text{hr-ft}^2)$$

5. $\text{Re} = DG_p/\mu$

6. $\text{Pr} = \dfrac{c\mu}{k}$

7. Read $j_H$ from Figure 8.5
8. $h_i = j_H\,(k/D)\,(\text{Pr})^{0.33}$

9. $h_{io} = h_i\left(\dfrac{\text{inside diameter of inner pipe}}{\text{outside diameter of inner pipe}}\right)$

*Annulus*

10. $a_a = \pi(D_o^2 - D_i^2)/4$
11. $D_e = (D_o^2 - D_i^2)/D_i$
12. $G_a = W/a_a$
13. $\text{Re} = D_e G_a/\mu$

14. $\text{Pr} = \dfrac{C\mu}{k}$

15. $h_o = j_H\,(k/D_e)\,(\text{Pr})^{1/3}$
    where $j_H$ is found in Figure 8.5.

*Overall Coefficients*

16. $U_C = \dfrac{1}{(1/h_{io}) + (1/h_o) + R_w}$
    for $R_w$ see Table 8.4.
17. $(1/U_D) = (1/U_C) + R_D$

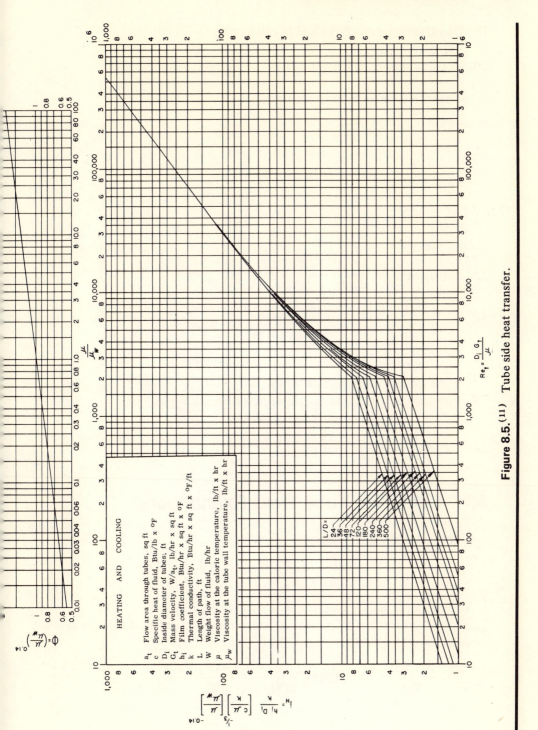

**Figure 8.5.**[11]  Tube side heat transfer.

18. Calculate area $A = Q/[U_D(\text{LMTD})]$
19. Calculate length $L = A/\pi D_i$
20. Decide number of hairpins

**Calculate the Change in Pressure Drop**

*Inner Pipe*

21. $\Delta F = \dfrac{2fG^2 L}{g_c \rho^2 D}$

    using Equation (8.7) or (8.8) for $f$.

    $\Delta P = \dfrac{\Delta F \rho}{144}$ psi

*Outer Pipe*

22. Calculate $D'_e = \dfrac{4\pi(D_o^2 - D_i^2)}{4\pi(D_o + D_i)} = D_o - D_i$

23. $\text{Re} = \dfrac{D'_e G_a}{\mu}$

24. $\Delta F = \dfrac{2fG_a^2 L}{g_c \rho^2 D'_e}$

    using Equation (8.7) or (8.8) for $f$.

25. Entrance and exit losses, one velocity head/hairpin.

    $\Delta F_r = (V^2/2g')$ (number of hairpins)

26. $\Delta P_t = \left[\dfrac{(\Delta F + \Delta F_r)\,\rho}{144}\right]$ psi

# 8.7   Shell and Tube Heat Exchangers*

Process conditions required:

$$\text{Hot fluid:}\quad T_1, T_2, W, C, s, \mu, k, R_d, \Delta P$$
$$\text{Cold fluid:}\quad t_1, t_2, w, c, s, \mu, k, R_d, \Delta P$$

1. Heat balance is $Q = WC(T_1 - T_2) = wc(t_2 - t_1)$.
2. True temperature difference, $\Delta t$ (assuming a number of tube passes) is calculated as follows:

$$\text{LMTD} = \frac{(T_1 - t_2) - (T_2 - t_1)}{\ln\,[(T_1 - t_2)/(T_2 - t_1)]}$$

$$R = \frac{T_1 - T_2}{t_2 - t_1} \qquad S = \frac{t_2 - t_1}{T_1 - t_1}$$

*After Ref. 11, with permission.

*F* from Figure 8.11 and 8.12

$$\Delta t = \text{LMTD} \times F$$

3. The mean temperatures of the hot fluid ($T_c$) and the cold fluid ($t_c$) must be calculated.

For the exchanger:

a. Assume a tentative value of $U_D$ with the aid of Table 8.5, and compute the surface from $A = Q/(U_D \Delta t)$. It is always better to assume $U_D$ too high than too low, as this practice ensures arriving at the minimum surface.

b. Assume a plausible number of tube passes for the pressure drop allowed, and select an exchanger for the nearest number of tubes from the tube counts of Table 8.9. For tube dimensions, see Table 17.1.

c. Correct the tentative $U_D$ to the surface corresponding to the actual number of tubes which can be contained in the shell (see Table 8.9).

The performance calculation for the film coefficients should start with the tube side. If the tube-side film coefficient is relatively greater than $U_D$ and the pressure-drop allowance is reasonably fulfilled and not exceeded, the calculation can proceed to the shell side. Whenever the number of tube passes is altered, the surface in the shell is also altered, changing the value of $A$ and $U_D$. For the remainder of the calculation shown here, it is assumed that the cold fluid flows in the tubes as it does in most cases.

**COLD FLUID: TUBE SIDE**

4. Flow area $a_t$. Flow area per tube $a_t'$ from Table 17.1.

$$a_t = N_t a_t'/144n \quad (\text{ft}^2)$$

5. Mass velocity, $G_t = w/a_t \quad [\text{lb}/(\text{hr-ft}^2)]$

6. $\text{Re}_t = DG_t/\mu$

   Obtain $D$ from Table 17.1 and obtain $\mu$ at $t_c$ (mean temperature).

7. Obtain $j_H$ from Figure 8.5.

8. At $t_c$ obtain $c$ and $k$.

9. $h_i = j_H (k/D)(c\mu/k)^{1/3} (\mu/\mu_w)^{0.14}$

   For water use Figure 8.7.

10. $h_{io} = h_i \left(\dfrac{\text{i.d.}}{\text{o.d.}}\right)$

**Pressure Drop**

a. For $\text{Re}_t$ in step 6, obtain $f$ from Figure 8.8.

b. $\Delta P_t = \dfrac{f G_t^2 Ln}{5.22 \times 10^{10} Ds\phi_t}$ psi

**Table 8.9.**[12]  Heat Exchanger Tube Sheet Layout Count.

| i.d. of shell (in.) | One-Pass Fixed Tubes | | | | | Two-Pass Fixed Tubes | | | | | Two-Pass U Tubes² | | | | | Four-Pass Fixed Tubes | | | | | Four-Pass U Tubes | | | | |
|---|---|---|---|---|---|---|---|---|---|---|---|---|---|---|---|---|---|---|---|---|---|---|---|---|---|
| arrangement → | ¾"on 15⁄16"△ | ¾"on 1"△ | ¾"on 1"□ | 1"on 1¼"△ | 1"on 1¼"□ | ¾"on 15⁄16"△ | ¾"on 1"△ | ¾"on 1"□ | 1"on 1¼"△ | 1"on 1¼"□ | ¾"on 15⁄16"△ | ¾"on 1"△ | ¾"on 1"□ | 1"on 1¼"△ | 1"on 1¼"□ | ¾"on 15⁄16"△ | ¾"on 1"△ | ¾"on 1"□ | 1"on 1¼"△ | 1"on 1¼"□ | ¾"on 15⁄16"△ | ¾"on 1"△ | ¾"on 1"□ | 1"on 1¼"△ | 1"on 1¼"□ |
| 8 | 33 | 33 | 33 | 15 | 17 | 32 | 28 | 26 | 16 | 12 | 8 | 8 | 12 | XX | XX | XX | XX | XX | XX | XX | XX | XX | XX | XX | XX |
| 10 | 69 | 57 | 53 | 33 | 33 | 58 | 56 | 48 | 32 | 26 | 34 | 26 | 30 | 8 | 12 | 48 | 44 | 48 | 24 | 24 | 28 | 20 | 24 | XX | XX |
| 12 | 105 | 91 | 85 | 57 | 45 | 94 | 90 | 78 | 52 | 40 | 64 | 60 | 52 | 26 | 22 | 84 | 72 | 72 | 44 | 40 | 56 | 52 | 44 | 20 | XX |
| 13¼ | 135 | 117 | 101 | 73 | 65 | 124 | 110 | 94 | 62 | 56 | 94 | 72 | 72 | 42 | 38 | 108 | 96 | 88 | 60 | 48 | 84 | 64 | 64 | 36 | XX |
| 15¼ | 193 | 157 | 139 | 103 | 83 | 166 | 154 | 126 | 92 | 76 | 134 | 108 | 100 | 58 | 58 | 154 | 134 | 126 | 78 | 74 | 122 | 98 | 90 | 50 | XX |
| 17¼ | 247 | 217 | 183 | 133 | 111 | 228 | 208 | 172 | 126 | 106 | 180 | 158 | 142 | 84 | 76 | 196 | 180 | 142 | 104 | 84 | 166 | 146 | 130 | 74 | XX |
| 19¼ | 307 | 277 | 235 | 163 | 139 | 300 | 264 | 222 | 162 | 136 | 234 | 212 | 188 | 120 | 100 | 266 | 232 | 192 | 138 | 110 | 218 | 198 | 174 | 110 | XX |
| 21¼ | 391 | 343 | 287 | 205 | 179 | 370 | 326 | 280 | 204 | 172 | 304 | 270 | 242 | 154 | 134 | 332 | 292 | 242 | 176 | 142 | 286 | 254 | 226 | 142 | XX |
| 23¼ | 481 | 423 | 355 | 247 | 215 | 452 | 398 | 346 | 244 | 218 | 398 | 336 | 304 | 192 | 180 | 412 | 360 | 308 | 212 | 188 | 378 | 318 | 286 | 178 | XX |
| 25 | 553 | 493 | 419 | 307 | 255 | 528 | 468 | 408 | 292 | 248 | 460 | 406 | 362 | 234 | 214 | 484 | 424 | 366 | 258 | 214 | 438 | 386 | 342 | 218 | XX |
| 27 | 663 | 577 | 495 | 361 | 303 | 626 | 556 | 486 | 346 | 298 | 558 | 484 | 436 | 284 | 256 | 576 | 508 | 440 | 308 | 260 | 534 | 462 | 414 | 266 | XX |
| 29 | 763 | 667 | 587 | 427 | 359 | 734 | 646 | 560 | 410 | 348 | 648 | 566 | 506 | 340 | 304 | 680 | 596 | 510 | 368 | 310 | 622 | 542 | 482 | 322 | XX |
| 31 | 881 | 765 | 665 | 481 | 413 | 846 | 746 | 644 | 462 | 402 | 768 | 674 | 586 | 396 | 356 | 788 | 692 | 590 | 422 | 360 | 740 | 648 | 560 | 376 | XX |
| 33 | 1019 | 889 | 765 | 551 | 477 | 964 | 858 | 746 | 530 | 460 | 882 | 772 | 688 | 466 | 406 | 904 | 802 | 688 | 486 | 414 | 852 | 744 | 660 | 444 | XX |
| 35 | 1143 | 1007 | 865 | 633 | 545 | 1088 | 972 | 840 | 608 | 522 | 1008 | 882 | 778 | 532 | 464 | 1024 | 912 | 778 | 560 | 476 | 976 | 852 | 748 | 508 | XX |
| 37 | 1269 | 1127 | 965 | 699 | 595 | 1242 | 1088 | 946 | 688 | 584 | 1126 | 1000 | 884 | 610 | 526 | 1172 | 1024 | 880 | 638 | 534 | 1092 | 968 | 852 | 584 | XX |

| Pass | Tubes | Arrangement | 8 | 10 | 12 | 13¼ | 15¼ | 17¼ | 19¼ | 21¼ | 23¼ | 25 | 27 | 29 | 31 | 33 | 35 | 37 |
|------|-------|-------------|---|----|----|-----|-----|-----|-----|-----|-----|----|----|----|----|----|----|----|
| Six-Pass | Fixed Tubes | 3/4" on 15/16" △ | XX | XX | XX | 80 | 116 | 174 | 230 | 294 | 372 | 440 | 532 | 632 | 732 | 844 | 964 | 1106 |
| | | 3/4" on 1" △□ | XX | XX | XX | 66 | 104 | 156 | 202 | 258 | 322 | 388 | 464 | 548 | 640 | 744 | 852 | 964 |
| | | 3/4" on 1" □ | XX | XX | XX | 54 | 78 | 116 | 158 | 212 | 266 | 324 | 394 | 460 | 536 | 634 | 724 | 818 |
| | | 1" on 1¼" △□ | XX | XX | XX | 34 | 56 | 82 | 112 | 150 | 182 | 226 | 274 | 338 | 382 | 442 | 514 | 586 |
| | | 1" on 1¼" □ | XX | XX | XX | XX | 44 | 66 | 88 | 116 | 154 | 184 | 226 | 268 | 318 | 368 | 430 | 484 |
| Six-Pass | U Tubes[2] | 3/4" on 15/16" △ | XX | XX | XX | 74 | 110 | 156 | 206 | 272 | 358 | 416 | 510 | 596 | 716 | 826 | 944 | 1058 |
| | | 3/4" on 1" △□ | XX | XX | XX | 56 | 88 | 134 | 184 | 238 | 300 | 366 | 440 | 518 | 626 | 720 | 826 | 940 |
| | | 3/4" on 1" □ | XX | XX | XX | 56 | 80 | 118 | 160 | 210 | 268 | 322 | 392 | 458 | 534 | 632 | 718 | 820 |
| | | 1" on 1¼" △□ | XX | XX | XX | 30 | 42 | 68 | 100 | 130 | 168 | 206 | 252 | 304 | 356 | 426 | 488 | 562 |
| | | 1" on 1¼" □ | XX | XX | XX | XX | 42 | 60 | 80 | 110 | 152 | 182 | 224 | 268 | 316 | 362 | 420 | 478 |
| Eight-Pass | Fixed Tubes | 3/4" on 15/16" △ | XX | XX | XX | XX | 94 | 140 | 198 | 258 | 332 | 398 | 484 | 576 | 682 | 790 | 902 | 1040 |
| | | 3/4" on 1" △□ | XX | XX | XX | XX | 82 | 124 | 170 | 224 | 286 | 344 | 422 | 496 | 588 | 694 | 798 | 902 |
| | | 3/4" on 1" □ | XX | XX | XX | XX | XX | 94 | 132 | 174 | 228 | 286 | 352 | 414 | 490 | 576 | 662 | 760 |
| | | 1" on 1¼" △□ | XX | XX | XX | XX | XX | 66 | 90 | 120 | 154 | 190 | 240 | 298 | 342 | 400 | 466 | 542 |
| | | 1" on 1¼" □ | XX | XX | XX | XX | XX | XX | 74 | 94 | 128 | 150 | 192 | 230 | 280 | 334 | 388 | 438 |
| Eight-Pass | U Tubes[2] | 3/4" on 15/16" △ | XX | XX | XX | 68 | 102 | 142 | 190 | 254 | 342 | 398 | 490 | 578 | 688 | 796 | 916 | 1032 |
| | | 3/4" on 1" △□ | XX | XX | XX | 52 | 82 | 122 | 170 | 226 | 286 | 350 | 422 | 498 | 600 | 692 | 796 | 908 |
| | | 3/4" on 1" □ | XX | XX | XX | 48 | 70 | 106 | 146 | 194 | 254 | 306 | 374 | 438 | 512 | 608 | 692 | 792 |
| | | 1" on 1¼" △□ | XX | XX | XX | 24 | 38 | 58 | 90 | 118 | 154 | 190 | 238 | 290 | 340 | 404 | 464 | 540 |
| | | 1" on 1¼" □ | XX | XX | XX | XX | 34 | 50 | 70 | 98 | 142 | 170 | 206 | 254 | 300 | 344 | 396 | 456 |

[a]Allowance made for tie rods.
[b]R. O. B. = 2½ × tube diameter actual number of "U" tubes is one-half the above figures.

c. $\Delta P_r = (4n/s)(V^2/2g')$
or use Figure 8.9.

d. $\Delta P_T = \Delta P_t + \Delta P_r$ psi

If the pressure drop is unacceptable, assume a new pass arrangement.

## HOT FLUID: SHELL SIDE

4'. Flow area (assume a plausible baffle spacing for the pressure drop allowed)

$$a_s = (\text{i.d.} \times C'B/144\, P_t)$$

5'. Mass velocity

$$G_s = W/a_s$$

6'. $\text{Re}_s = D_e G_s/\mu$. Obtain $D_e$ from Figure 8.6, using inserted table. Be sure to convert from $d_e$ (in.) to $D_e$ (ft). Obtain $\mu$ at $T_c$.

7'. Obtain $j_H$ from Figure 8.6.

8'. At $T_c$ obtain $c$ and $k$.

9'. $h_o = j_H \left(\dfrac{k}{D_e}\right)\left(\dfrac{c\mu}{k}\right)^{1/3}\left(\dfrac{\mu}{\mu_w}\right)^{0.14}$

Check pressure drop. If unsatisfactory, assume a new baffle spacing.

### Pressure Drop

a'. For $\text{Re}_s$ in step 6' obtain $f$ from Figure 8.10. (LMTD correction factors are presented in Figures 8.11 and 8.12.)

b'. Number of crosses is $N + 1 = 12L/B$.

$$\Delta P_s = \frac{fG_s^2 D_s(N+1)}{5.22 \times 10^{10} D_e s\phi_s}$$

If the pressure drop is unacceptable, assume a new baffle spacing. If both sides are satisfactory for film coefficients and pressure drop, the trial may be concluded.

11. Clean overall coefficient $U_c$:

$$U_c = \frac{h_{io}h_o}{h_{io} + h_o}$$

12. Dirt factor $R_d$: Use $U_D$ obtained at the start of the calculations.

$$R_d = \frac{U_c - U_D}{U_c U_D}$$

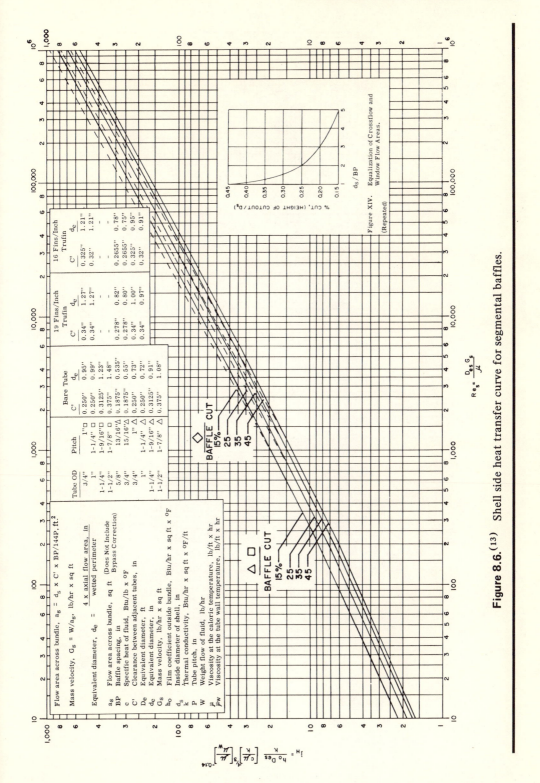

**Figure 8.6.** [13] Shell side heat transfer curve for segmental baffles.

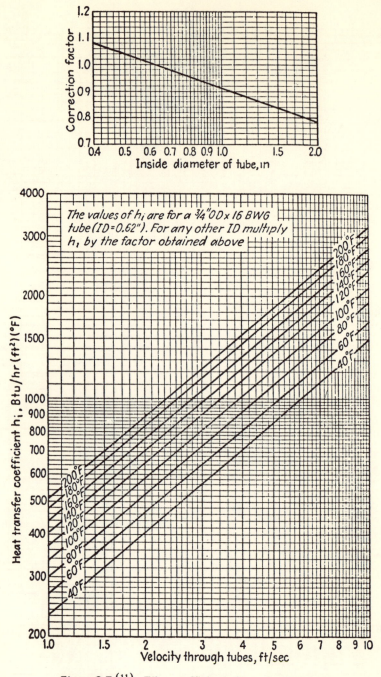

The values of $h_i$ are for a $\frac{3}{4}$" OD x 16 BWG tube (ID = 0.62"). For any other ID multiply $h_i$ by the factor obtained above

**Figure 8.7.**[11]   Film coefficients for water in tubes.

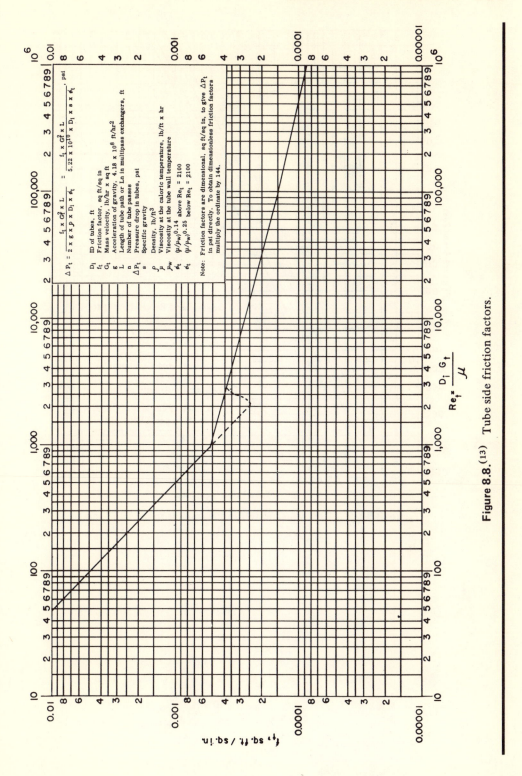

**Figure 8.8.** [13] Tube side friction factors.

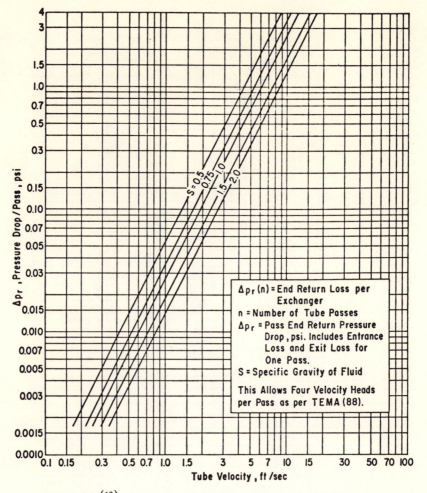

**Figure 8.9.**[12]   Tube side and return pressure drop per tube pass.

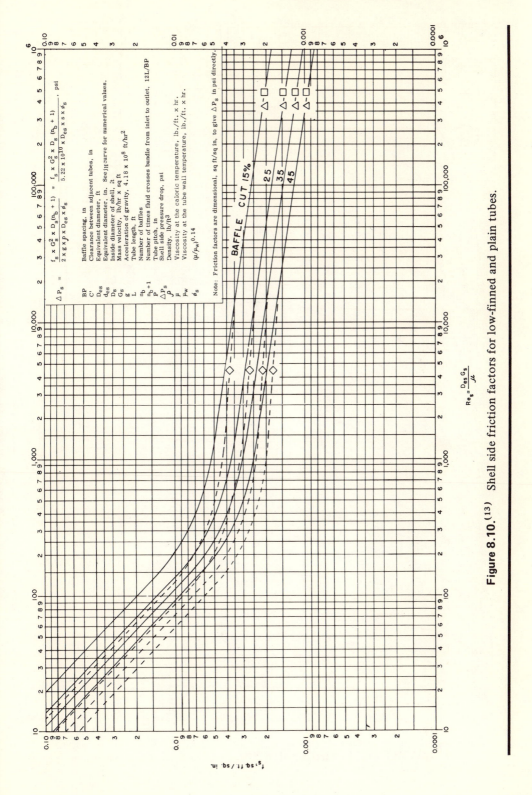

**Figure 8.10.**[13]  Shell side friction factors for low-finned and plain tubes.

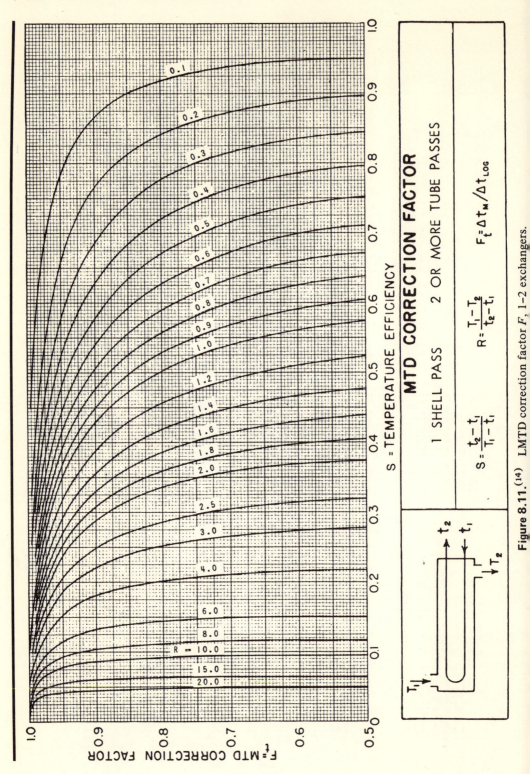

**Figure 8.11.**[(14)] LMTD correction factor $F$, 1–2 exchangers.

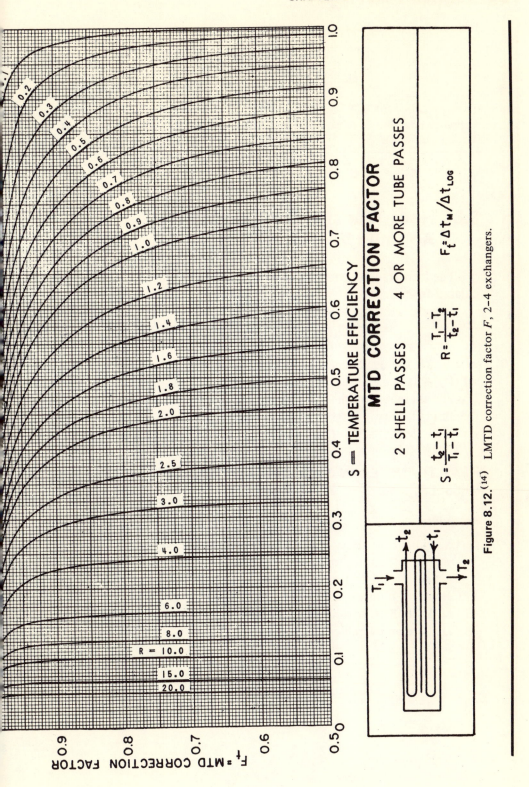

**Figure 8.12.**[(14)]   LMTD correction factor $F$, 2–4 exchangers.

**Table 8.10.**  Jacketed Glass-Lined Steel Vessel Heat Transfer

|  | Jacket coefficient ($h_0$) | Dirt factor ($R_d$) | Wall resistance ($R_w$) | Overall coefficient ($U$) |
|---|---|---|---|---|
| Heating/cooling with water using agitating nozzles | 100–300 | 0.001–0.003 | 0.01 | 33–60 |
| Heating with steam | 1,000 | .001 | 0.01 | 50–70 |

## 8.8   Heat Transfer Coefficient in Agitated Vessels[15]

The heat transfer coefficient inside jacketed agitated vessels can be calculated from:

$$\frac{h_i D_i}{k} = 0.36 \left(\frac{L^2 N \rho}{\mu}\right)^{2/3} \left(\frac{C\mu}{k}\right)^{1/3} \left(\frac{\mu}{\mu_w}\right)^{0.14} \tag{8.13}$$

where $h_i$ is the inside coefficient [Btu/(hr-ft²-°F)]; $D_i$ is the i.d. of the vessel (ft); $k$ is the thermal conductivity of the liquid [(Btu-ft)/(ft²-hr-°F)]; $L$ is the diameter of the agitator (ft); $N$ is the agitator speed (rev/hr); $\rho$ is the fluid density (lb/ft³); $\mu$, $\mu_w$ is the fluid viscosity, bulk and wall, respectively [(lb/(ft-hr)]; $C$ is the specific heat of fluid [Btu/(lb-°F)].

### HEAT TRANSFER IN JACKETED AND AGITATED GLASS-LINED VESSELS

**Procedure**

1. Use Table 8.11 to get dimensions of vessel and agitator.
2. Determine agitator speed by either measuring if system is existing, or use Section 1.1 to establish a speed.
3. Calculate an inside coefficient using Equation (8.13).
4. Use Table 8.10 to determine the other resistances and jacket coefficient.
5. Calculate an overall coefficient from

$$\frac{1}{U} = \frac{1}{h_i} + \frac{1}{h_o} + R_d + R_w$$

**Table 8.11.**[16]  Typical Glassed Steel Reactor Dimensions (Pfaulder RA Series)

| Capacity (gal) | Straight side (ft) | Shell i.d. (ft) | Agitator[a] diameter (ft) | Jacket area (ft²) |
|---|---|---|---|---|
| 300 | 2.75 | 4.0 | 3.0 | 53 |
| 500 | 4.83 | 4.0 | 3.0 | 80 |
| 750 | 4.33 | 5.0 | 3.7 | 96 |
| 1,000 | 6.0 | 5.0 | 3.7 | 122 |
| 1,500 | 5.4 | 6.5 | 3.7 | 158 |
| 2,000 | 7.0 | 6.5 | 3.7 | 191 |
| 3,000 | 6.4 | 8.0 | 4.5 | 236 |
| 4,000 | 9.1 | 8.0 | 4.5 | 300 |

[a]Assume three-blade retreating curve type.

# 8.9   Falling Film Coefficients[17]

The following film coefficients are for the inside of vertical tubes. The film coefficient for water is[17]

$$h_i = 120 \left( \frac{W}{\pi D} \right)^{1/3} \tag{8.14}$$

where $h_i$ is the inside film heat transfer coefficient [Btu/(ft²-hr-°F)]; $W$ is the weight flow of hot liquid (lb/hr); $D$ is the inside diameter of the tubes (ft). For other liquids in turbulent flow, i.e., a Reynolds number greater than 2,000,

$$\frac{h_i}{(k^3 \rho^2 g / \mu_f^2)^{1/3}} = 0.01 \left[ \left( \frac{c\mu}{k} \right) \left( \frac{4G'}{\mu_f} \right) \right]^{1/3} \tag{8.15}$$

where $G' = W/(\pi D)$; $k$ is the thermal conductivity [(Btu-ft)/(ft²-hr-°F)]; $g$ is the acceleration of gravity (ft/hr²); $c$ is the specific heat of the fluid

[Btu/(lb-°F)]; $\mu$ is the viscosity [cP $\times$ 2.42 = lb/(ft-hr)]; $\mu_f$ is the viscosity at average film temperature [lb/(ft-hr)]; $\rho$ is the fluid density (lb/ft³). For liquids in streamline flow, i.e., a Reynolds number less than 2,000,

$$h_i = 0.67 \left[ \left( \frac{k^3 \rho^2 g}{\mu_f^2} \right) \left( \frac{c \mu_f^{5/3}}{k L \rho^{2/3} g^{1/3}} \right) \right]^{1/3} \left( \frac{4G'}{\mu_f} \right)^{1/9} \tag{8.16}$$

where $L$ is the tube length (ft).

## 8.10  Reboilers and Vaporizers[18]

### BASIC DESIGN CONSIDERATIONS

#### Flux

For organics in forced circulation, the maximum allowable flux is 20,000 Btu/(hr-ft²). For organics in natural circulation, the maximum allowable flux is 12,000 Btu/(hr-ft²). For vaporization of water from aqueous solutions of low concentrations, the maximum allowable flux for natural and forced circulation is 30,000 Btu/(hr-ft²).

#### Film Coefficient

For organics in forced or natural circulation the maximum heat transfer film coefficient is 300 Btu/(hr-ft²-°F). For water, the maximum vaporizing coefficient in forced or natural circulation is 1,000 Btu/(hr-ft²-°F). See also Section 8.1.

Film coefficients and required areas for forced circulation reboilers or vaporizers are calculated exactly as for forced convection heat exchangers except that the flux and film coefficients must not exceed those listed in the above sections. See Section 8.7 for forced convection heat exchangers.

Pressure drop in forced circulation reboilers and vaporizers is calculated as for forced convection heat exchangers except that an average specific gravity must be used. The average specific gravity to be used in the pressure drop calculation should be based on inlet and outlet conditions, taking into account the vapor fraction at the outlet of the exchanger. It is recommended[19] that the log mean average specific gravity be used. Let

$V_i$ = specific volume of liquid at inlet of exchanger
$V_o$ = specific volume of fluid at outlet

1. Calculate the mean specific volume

$$V_{mean} = \frac{V_o - V_i}{\ln(V_o/V_i)} \tag{8.17}$$

2. Calculate the mean specific gravity

$$S_{\text{mean}} = \frac{1}{62.4 \times V_{\text{mean}}}$$ (8.18)

3. Use $S_{\text{mean}}$ for calculating the pressure drop for the circulating fluid.

## KETTLE REBOILERS[18]

For vaporization of a fluid from a pool, Figure 8.13 can be used for calculating the boiling film coefficient. The lower curve can be used for calculating the preheat zone if the feed is subcooled. For organics the flux is limited to 12,000 Btu/(hr-ft²). For the vaporization of water, the flux is limited to 30,000 Btu/(hr-ft²) and the film coefficient is limited to 1,000 Btu/(ft²-hr-°F). The pressure drop in kettle reboilers is usually negligible.

Normally, a triangular pitch is used for the reboiler tubes. However, if

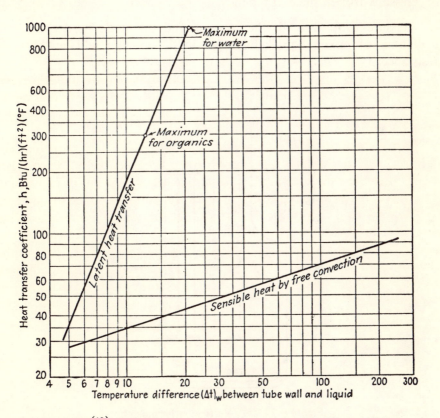

**Figure 8.13.**[18]  Natural-circulation boiling and sensible film coefficients.

a particularly dirty service is expected, a square pitch should be considered. This allows the solid material to settle out readily, resulting in easier cleaning.

## VERTICAL THERMOSYPHON REBOILERS[18]

### Recirculating Rate

The recirculating rate is attained when the total resistance in the vaporizing circuit is equal to the hydrostatic driving force on the liquid. Referring to Figure 8.14, the five resistances are:

1. Pressure drop in the inlet piping
2. Pressure drop through the reboiler
3. Expansion loss due to vaporization in the reboiler
4. Static head pressure of a column of mixed liquid and vapor ($Z_3$) in the reboiler
5. Pressure drop in the exit piping

Pressure drop through the reboiler and the static head pressure of a column of mixed liquid and vapor in the reboiler will be the major resistances.

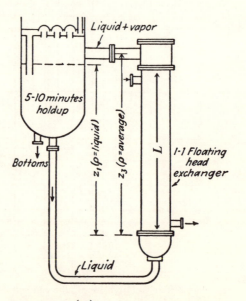

**Figure 8.14.**[18] Vertical thermosiphon reboiler connected to tower.

### Expansion Loss Due to Vaporization

This loss can be taken at two velocity heads based on the mean of the inlet and outlet densities.[18]

$$\Delta P_1 = \left( \frac{G^2}{144 g \rho_{avg}} \right) \tag{8.19}$$

### Weight of a Column of Mixed Liquid and Vapor

$$\frac{Z_3 \rho_{avg}}{144} = \frac{Z_3}{144 (V_o - V_i)} \ln \left( \frac{V_o}{V_i} \right), \text{ psi} \tag{8.20}$$

### Boiling Point Elevation

A check must be made for boiling point elevation due to excessive liquid head. If the liquid head is large, the liquid boiling point will be significantly elevated. This must be corrected for by: (1) doing the calculations in increments; or (2) reducing the liquid head to the point where it is negligible.

### Calculation Procedure

1. Make a heat balance.
2. Calculate the log mean temperature difference.
3. Start trial calculations.
4. Estimate area required using the maximum allowable flux; 12,000 Btu/(hr-ft$^2$) for organics; 30,000 Btu/(hr-ft$^2$) for water.
5. Assume an exchanger to fit estimated area (one tube pass, one shell pass).
6. Assume a circulation ratio of $4:1$ (or greater), that is the circulating liquid is four times the amount of vapor generated.
7. Calculate the specific volume of the vapor, $V_{vapor}$. Let the specific volume of the liquid be $V_{liquid}$.
8. Calculate the specific volume of the mixture leaving the reboiler.
9. Calculate the static pressure of the reboiler leg using Equation (8.20).
10. Calculate the flow area in the exchanger

$$a_t = N_t \frac{a_t'}{144 n} \tag{8.21}$$

11. Calculate the mass flowrate

$$G = W / a_t \tag{8.22}$$

12. Calculate the Reynolds number

$$N_{Re} = D G / \mu \tag{8.23}$$

13. Read the friction factor from Figure 8.8.
14. Calculate a mean specific gravity of the vapor–liquid mixture between inlet and outlet conditions per Equation (8.18).
15. Calculate the pressure drop in the exchanger.
16. Calculate the other resistances in the circuit. Add all resistances.
17. Calculate the driving force

$$\text{static pressure of leg} = Z_1 \rho / 144 \qquad (8.24)$$

18. If the sum of the resistances do not match the driving force, modify the exchanger and repeat steps 6–17.
19. After a match is made, calculate the boiling film coefficient using the method for forced convection heat exchangers (Section 8.7). If it is larger than 300 for organics or 1,000 for water, use 300 to 1,000.
20. Correct the film coefficient to the outside area.

$$h_{io} = h_i \, (\text{i.d.}/\text{o.d.}) \qquad (8.25)$$

21. Calculate the shell side coefficient.
22. Calculate the clean and dirty coefficients.
23. Calculate the dirt factor as

$$R_d = (U_c - U_d)/(U_c U_d) \qquad (8.26)$$

The exchanger calculations are now complete.


## EXAMPLE CALCULATION OF A VERTICAL THERMOSIPHON REBOILER

A reboiler is needed to generate 475 lb/hr of organic vapor at 800 mm Hg. The boiling point is 380°F. Heat will be supplied by hot oil at 500°F. Assume the hot oil leaves at 490°F. Also assume the heat transfer coefficient for the hot oil side (shell side) is 300 Btu/(hr-ft²-°F).

A circulation rate greater than 4:1 will be used. Also the exchanger is to be made of $\frac{3}{4}$ in. tubes, 16 BWG, with 1-in. triangular pitch. Referring to Figure 8.14, $Z_1 = 48$ in. and $Z_3 = 49.7$ in.

### Solution

1. Heat balance.
   Heat of vaporization at 800 mm Hg = 210.5 Btu/lb
   $Q = 210.5 \times 475 = 100,000$ Btu/hr
2. Log mean temperature difference (LMTD). Assume isothermal boiling.

$$\text{LMTD} = \frac{(500 - 380) - (490 - 380)}{\ln\left[(500 - 380)/(490 - 380)\right]} = 115°F$$

3. Trial 1.
4. Estimate trial area with flux of 12,000 Btu/(hr-ft$^2$)
5. Assume an exchanger with tubes 2.0 ft long.

Area per tube = (0.1963 ft$^2$/ft) (2.0) = 0.39 ft$^2$

Number of tubes needed = 8.33 ft$^2$/0.39 ft$^2$

= 21.4

However, from the tube sheet layout table (Table 8.9) for a minimum size shell (8 in.) we can get 33 tubes. Since the exchanger will cost the same whether we use 33 tubes or 21 tubes, continue the calculations with 33 tubes.

Area = 33 × 0.39 = 12.9 ft$^2$

Overall coefficient = $U_D$ = $Q/(A \times \text{LMTD})$

= 100,000/(12.9 × 115)

= 67.4 Btu/(hr-ft$^2$-°F)

6. Circulation ratio. Try 30:1.

Circulating liquid = 30 × 475 = 14250 lb/hr

7. Specific volume of vapor, $V_{\text{vapor}}$

$$V_{\text{vapor}} = \frac{359}{94.1} \times \frac{840}{492} \times \frac{760}{800}$$

$$= 6.19 \text{ ft}^3/\text{lb}$$

$$\rho_{\text{vapor}} = 1/V = 0.162 \text{ lb/ft}^3$$

$$\rho_{\text{liq}} = 58.5 \text{ lb/ft}^3$$

$$V_{\text{liq}} = \frac{1}{\rho} = 0.0171 \text{ ft}^3/\text{lb}$$

8. Specific volume of mixture, $V_o$

Total volume out of reboiler,

liquid, 14,250 × 0.0171 = 243 ft$^3$/hr

vapor, 475 × 6.19 = 2940 ft$^3$/hr

total, 243 + 2940 = 3183 ft$^3$/hr

Total weight of fluid out of reboiler

14,250 + 475 = 14725 lb/hr

$$V_o = \frac{2491}{14725} = 0.2162 \text{ ft}^3/\text{lb}$$

$$V_i = V_{\text{liq}} = 0.0171 \text{ ft}^3/\text{lb}$$

9. Static pressure in reboiler leg (Equation 8.20)

$$\frac{Z_3 \rho_{\text{avg}}}{144} = \frac{Z_3}{144(V_o - V_i)} \ln\left(\frac{V_o}{V_i}\right), \text{ psi}$$

$$= \frac{49.72/12}{144(0.2162 - 0.0171)} \ln\left(\frac{0.2162}{0.0171}\right)$$

$$= 0.367 \text{ psi}$$

10. Calculate the flow area in the exchanger

$$a_t = N_t \frac{a_t'}{144n}$$

$$= 33 \times 0.302/144 = 0.0692 \text{ ft}^2$$

11. Calculate the mass flow rate

$$G = W/a_t$$
$$= 14725/.0692$$
$$= 212789$$

12. Calculate the Reynolds number

$$N_{Re} = DG/\mu$$
$$\mu = 0.5 \text{ cP} \times 2.42 = 1.21 \text{ lb/ft-hr}$$
$$D = 0.62/12 = 0.0517 \text{ ft}$$
$$N_{Re} = (0.0517)(212789)/(1.21)$$
$$= 9086$$

13. Friction factor (Figure 6.7)

$$f = 0.008$$

14. Calculate the mean specific gravity

$$S_{mean} = \left(\frac{1}{V_{mean}}\right)\left(\frac{1}{62.4}\right)$$

$$= \frac{1}{62.4} \times \frac{\ln(V_o/V_i)}{V_o - V_i}$$

$$= \frac{1}{62.4} \times \left(\frac{\ln(0.2162/0.0171)}{0.2162 - 0.0171}\right)$$

$$= 0.2042$$

15. Calculate the pressure drop in the exchanger

$$\Delta P_t = \frac{fG_t^2 nL}{5.22 \times 10^{10} Ds\phi_t}$$

$$\Delta P_t = \frac{(0.008)(212789)^2(2.0)(1)}{(5.22 \times 10^{10})(0.0517)(0.2042)(1)}$$

$$= 1.31 \text{ psi}$$

16. Sum of the resistances, neglecting piping
    Losses = 1.31 + 0.367
    $$= 1.677 \text{ psi}$$

17. Calculate the driving force ($Z_1$ assumed at 48 in.)

$Z_1 \rho_i / 144 = 1.62$ psi; if $Z_1 = 49.72$ in., $Z_1 \rho_i / 144 = 1.68$ psi

18. The exchanger is marginal, as the driving force exactly equals the exchanger $\Delta P$. There is nothing left for piping losses. To be correct, the calculations should be restarted at step 6 with a new circulation rate.

19. Calculate the heat transfer coefficient

$$N_{Re} = 9086$$

$$j_H = 36$$

$$k = 0.08 \text{ Btu/(hr-ft-}^\circ\text{F)}$$

$$c = 0.61 \text{ Btu/(lb-}^\circ\text{F)}$$

$$h_i = j_H \left(\frac{k}{D}\right)\left(\frac{c\mu}{k}\right)^{1/3}\left(\frac{\mu}{\mu_w}\right)^{0.14}$$

$$= 36\left(\frac{.08}{.0517}\right)\left(\frac{0.61 \times 1.21}{0.08}\right)^{1/3}(1)^{0.14}$$

$$= 117 \text{ Btu/(hr-ft}^2)$$

20. Correct to the outside area

$$h_{io} = h_i \text{ (inside diameter/outside diameter)}$$
$$= 117 (0.62/0.75)$$
$$= 97.72$$

21. Shell side coefficient, 300 Btu/(hr-ft²-°F)
22. Clean coefficient, $U_c$,

$$U_c = \frac{h_{io} h_o}{h_{io} + h_o} = \left(\frac{(96.72 \times 300)}{396.72}\right)$$

$$= 73.14 \text{ Btu/(hr-ft}^2\text{-}^\circ\text{F)}$$

$$U_D = 67.4 \text{ Btu/(hr-ft}^2\text{-}^\circ\text{F)}$$

23. Dirt factor

$$R_d = (U_c - U_D)/(U_c U_D)$$

$$R_d = \frac{73.14 - 67.4}{(73.14)(67.4)}$$

$$= 0.001$$

which is reasonable.

The exchanger calculations are now complete. However, in practice the liquid level in the column can be raised or lowered until an optimum position is found which gives the best heat transfer.

## 8.11 Condensers

### GENERAL PERFORMANCE

Figure 8.15, which was published by the Andale Co.,[20] gives the general performance that can be expected from a condenser with cooling water at 30°C.

### Film Coefficients (Pure Vapors)

Figures 8.16 and 8.17 can be used for estimating condensing film coefficients for total condensation of a pure vapor.

### Film Coefficients (Steam and Organics)[21]

For condensing an organic–steam mixture, as in steam distillations, use (a) for vertical tubes,

$$h = 79 \left[ \frac{(\text{wt.\%})_A \lambda_A + (\text{wt.\%})_B \lambda_B}{(\text{wt.\%})_A L} \right]^{1/4} \tag{8.27}$$

(b) for horizontal tubes,

$$h = 61 \left[ \frac{(\text{wt.\%})_A \lambda_A + (\text{wt.\%})_B \lambda_B}{(\text{wt.\%})_A D_o} \right]^{1/4} \tag{8.28}$$

Where A and B refer to the organic and water in the condensate, respectively, $L$ and $D_o$ refer to the tube length (ft) and tube o.d. (ft). $\lambda$ is the latent heat of vaporization at the condenser design temperature (Btu/lb).

### Film Coefficients for Mixed Vapors or Partial Condensation

See Kern[21] or Ludwig[22] for a full treatment of the subject.

### CONDENSER CONFIGURATIONS FOR SPECIAL CONDITIONS

1. A special configuration (Figure 8.18) is used with very few or no light ends and has a condensation curve as shown in Figure 8.19.
2. A special configuration (Figure 8.20) must be used for a mixture with a broad range in the condensation curve (Figure 8.21).

In both of these cases the cooling water enters the condenser at the top. To prevent air from accumulating in the tubes, a hydraulic vented seal should be included.

NOTE:—So many factors enter into the design and construction of a vapor condenser that it is very difficult to draw a performance chart for vapor condensers with any degree of accuracy and not have the chart too complicated. This chart is for broad-estimate purposes only. It is designed to show at a glance the amount of outside tube surface required, under the conditions specified, for most of the commercial organic compounds. For other organic compounds the accuracy of the chart may vary within limits of plus or minus 25% of the indicated values.

CONDITIONS:—This chart is based on the use of 30-deg. C. cooling water, in the quantities specified below; (the resistance to flow should not exceed 4 to 6 pounds); the condensing of vapors on the inside of vertical tubes; the vapors having latent heats of 80 to 200 B.T.U.'s per pound; with the specific heat of the liquid varying from .3 to .6; and the liquid viscosities at the condensate outlet up to one centipoise.

WATER REQUIREMENTS,—per square foot of outside tube surface:—(1) One gallon per minute; (2), (3) and (4) one-half gallon per minute.

FACTORS:

Copper tubes—⅝" No. 14 ga........1.00
Steel tubes—⅝" No. 14 ga.........0.82
Use above tubes for clean vapors coming to the condenser at pressure of one atmosphere and above
Copper tubes—1" No. 14 ga........0.95
Steel tubes—1" No. 14 ga..........0.78
Use above tubes for vapors carrying solids in suspension and for pressures considerably below atmosphere.
Tube lengths in either case will vary from 8 to 10 feet.

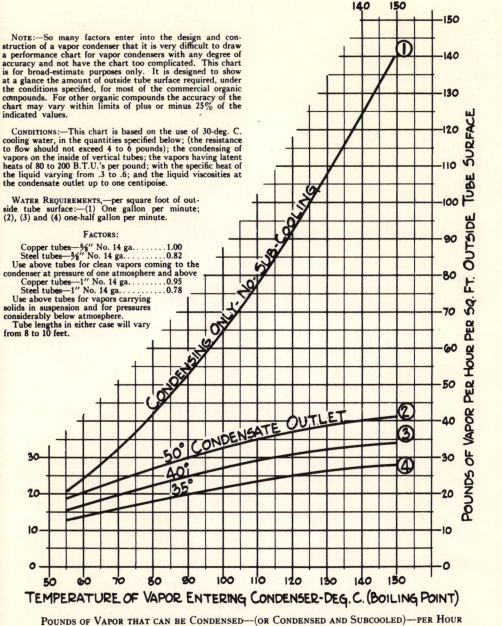

POUNDS OF VAPOR THAT CAN BE CONDENSED—(OR CONDENSED AND SUBCOOLED)—PER HOUR
PER SQUARE FOOT OF OUTSIDE TUBE SURFACE.

For example;—A typical organic compound vapor entering at 130° C.; the condensate and non-condensible vapors being subcooled to 40° C.; the tubes being ⅝" No. 14 gauge copper; and with other conditions as stated above;—32 pounds of the vapor will be condensed per hour per square foot of outside tube surface. If the tubes were 1" No. 14 gauge copper, the amount of vapor condensed per hour per square foot of outside tube surface, on the one inch tubes, would be 32 times 0.95.

**Figure 8.15.**[20]   Condenser performance chart. This figure is for the *preliminary* estimation of the performance of condensers where organic vapors are condensing inside vertical tubes.

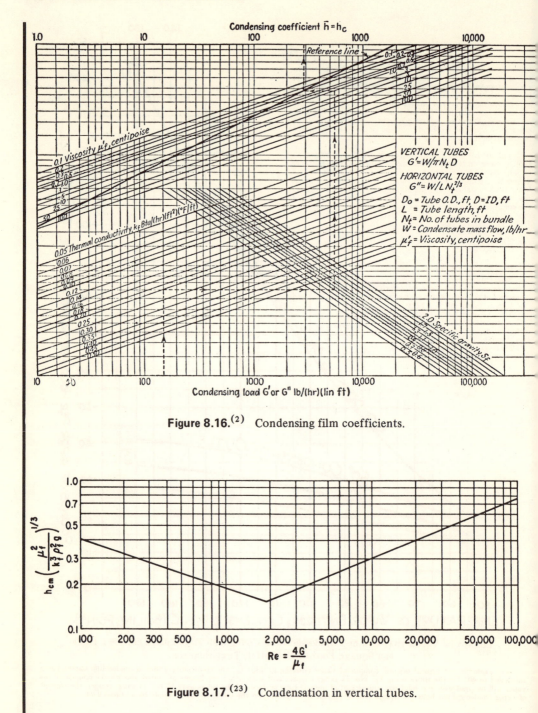

**Figure 8.16.**[2] Condensing film coefficients.

**Figure 8.17.**[23] Condensation in vertical tubes.

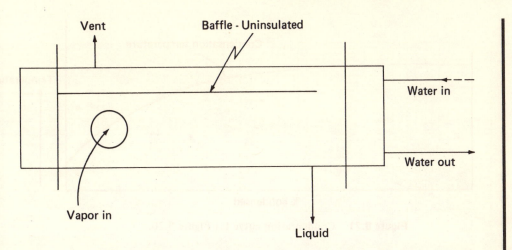

Figure 8.18. Condenser for material low in light ends.

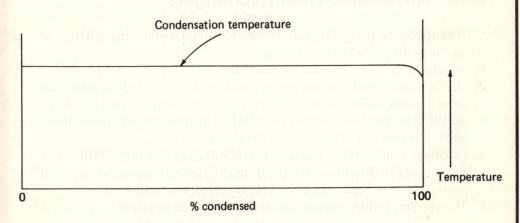

Figure 8.19. Condensation curve for Figure 8.18.

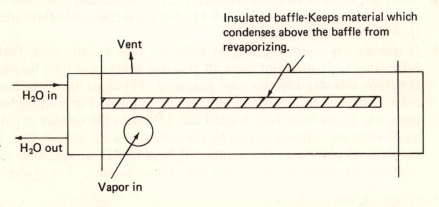

Figure 8.20. Condenser for material with broad condensing curve.

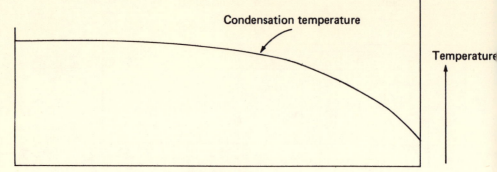

**Figure 8.21.** Condensation curve for Figure 8.20.

# 8.12 Air-Cooled Heat Exchangers*

The following procedure can be used to obtain a preliminary design or rating for an air cooled heat exchanger.

1. Select an overall coefficient $U$ from Table 8.12 or Figure 8.22.
2. (a) Calculate fluid cooling range (inlet–outlet, °F); (b) calculate approach (fluid outlet–air inlet, °F); (c) Select number of rows of tubes—usually 3–6; (d) Read surface per MMBtu/hr from proper curve. Interpolate for mid $U$'s. Use Figure 8.23(a)–(z).
3. Calculate total surface (surface per MMBtu/hr) × (duty MMBtu/hr). Apply LMTD (log mean temperature difference) correction factor, if applicable. See Figure 8.11 for LMTD correction factors.
4. The air temperature rise can be calculated by the equation

$$\Delta t_a = B/S$$

where $\Delta t_a$ is the air temperature rise (°F); $B$ is the constant depending on number of tube rows from Table 8.13. $S$ is the surface (ft²/MMBtu/hr) from Figure 8.23(a)–(z).
5. Determine bay width, number of bays and tube length, using Table 8.14. This table does not show all available widths and tube lengths; therefore, this step requires some judgment. For single bay units, look for a reasonable fit consistent with using long tubes. For multiple bay units, use 32- or 40-ft tube lengths and determine the number of bays by dividing the required surface by the surface of a 12- or 16-ft bay. If the decimal portion of the number of bays is 0.25 or less, drop the fraction; otherwise, round to the next whole number. For example,

*After Ref. 24, with permission.

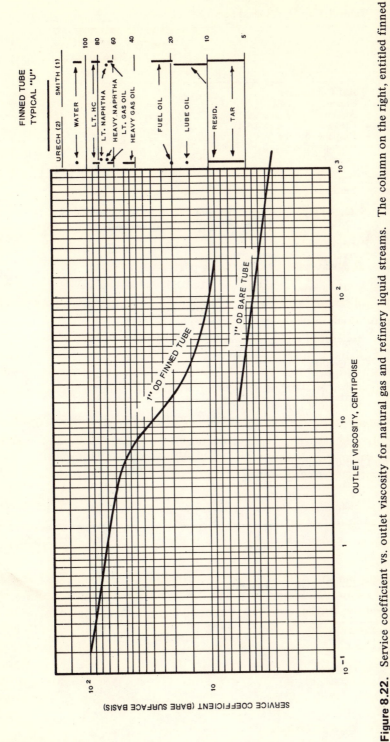
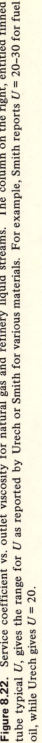

**Figure 8.22.** Service coefficient vs. outlet viscosity for natural gas and refinery liquid streams. The column on the right, entitled finned tube typical $U$, gives the range for $U$ as reported by Urech or Smith for various materials. For example, Smith reports $U = 20$–30 for fuel oil, while Urech gives $U = 20$.

**Table 8.12.**[24] Typical Service Coefficients

| Service | Typical $U$ bare surface basis |
|---|---|
| **Total Condenser** | |
| Steam........................................ | 130 |
| Ammonia..................................... | 110 |
| Freon 12.................................... | 75 |
| Propane, butane............................ | 90 |
| Pentanes through light naphtha, 70 psig..... | 90 |
| 20 psig..... | 85 |
| 10 psig..... | 75 |
| Heavy naphtha.............................. | 70 |
| Kerosene.................................... | 65 |

Note: For subcooled liquid, estimate condensing section with above coefficient and subcooling section as liquid cooler.

| Service | Typical $U$ bare surface basis |
|---|---|
| **Partial Condenser** | |
| Hydrogen-rich reactor effluent | |
| Hydrocracker, 1400–3000 psig psig....... | 80 |
| Cat. Reformer, 350–450 psig............. | 75 |
| Hydrotreaters, naphtha.................. | 70 |
| gas oil, 1150 psig......... | 70 |
| 960 psig......... | 60 |
| Crude column or stabilizer overhead........ | 65 |
| Cat. cracker fractionator overhead.......... | 75 |
| Amine still overhead, to 160°F............. | 100 |
| below 160°F.......... | 60 |
| **Gas Cooling** | |
| Ethylene.................................. | 80 |
| Hydrogen, 250 psig (3 psi $\Delta P$).............. | 60 |
| Natural gas, 0–50 psig (1 psi $\Delta P$)........... | 35 |
| 50–200 psig (3 psi $\Delta P$)......... | 50 |
| 200–1500 psig (1 psi $\Delta P$)....... | 60 |
| (3 psi $\Delta P$)....... | 70 |
| (5 psi $\Delta P$)....... | 85 |
| (10 psi $\Delta P$)...... | 95 |
| **Liquid Cooling** | |
| Lean carbonate solution................... | 80 |
| Lean amine, 15–20 wt. %................. | 100 |
| 20–25 wt. %................. | 95 |
| Sulfinol, 7 cp. outlet..................... | 70 |

**Table 8.13.**[24] The $B$ Constant

| Number of tube rows | 3 | 4 | 5 | 6 |
|---|---|---|---|---|
| $B$ | 5,160 | 7,590 | 10,230 | 13,060 |

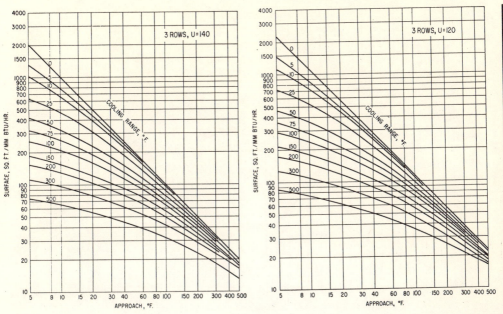

**Fig. a**—Required surface area for three rows, U = 140.

**Fig. b**—Required surface area for three rows, U = 120.

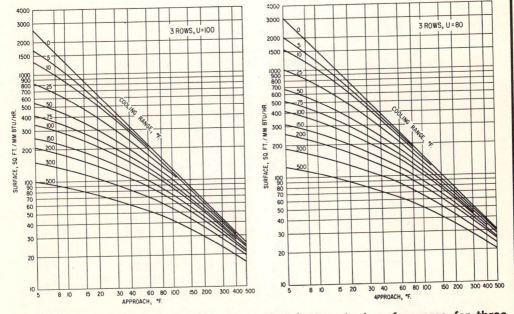

**Fig. c**—Required surface area for three rows, U = 100.

**Fig. d**—Required surface area for three rows, U = 80.

**Figure 8.23(a–z).** Required surface area for air-cooled heat exchangers as a function of the number of rows, overall $U$, approach, and cooling range.

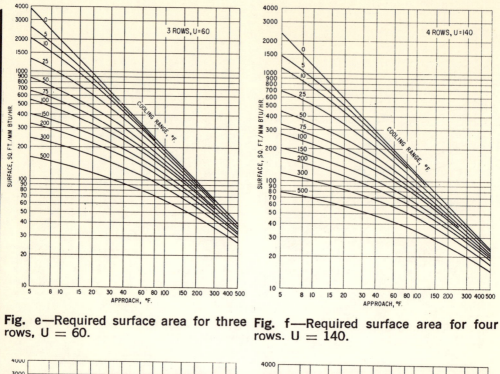

**Fig. e**—Required surface area for three rows, U = 60.

**Fig. f**—Required surface area for four rows. U = 140.

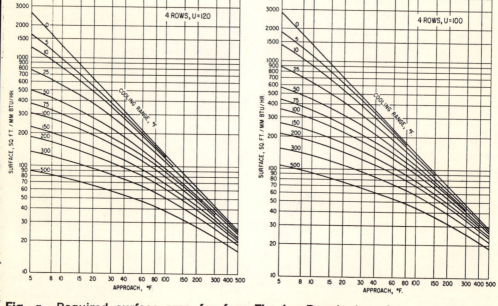

**Fig. g**—Required surface area for four rows, U =120.

**Fig. h**—Required surface area for four rows, U = 100.

**Figure 8.23(a–z).** *Cont'd.*

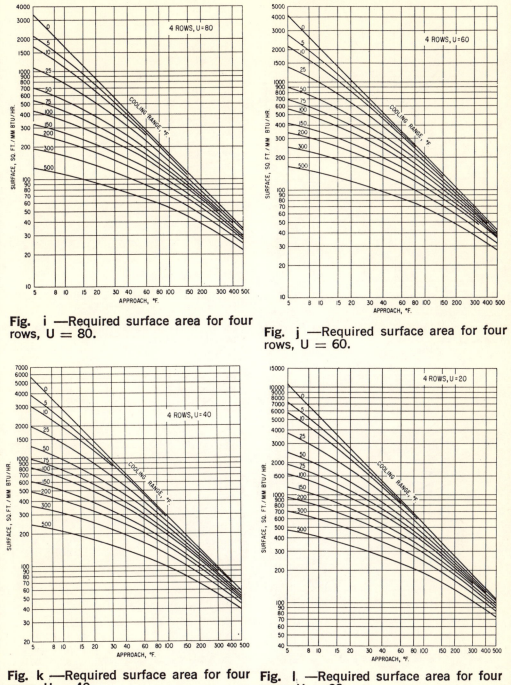

Fig. i —Required surface area for four rows, U = 80.

Fig. j —Required surface area for four rows, U = 60.

Fig. k —Required surface area for four rows, U = 40.

Fig. l —Required surface area for four rows, U = 20.

**Figure 8.23(a–z).** *Cont'd.*

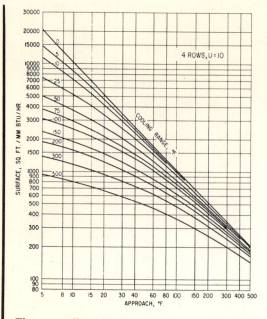

**Fig. m—Required surface area for four rows, U = 10.**

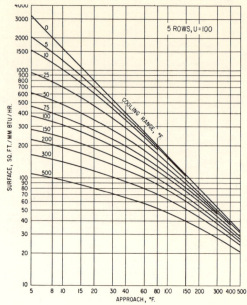

**Fig. n —Required surface area for five rows, U = 100.**

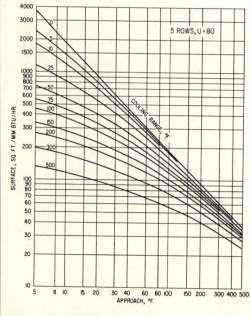

**Fig. o —Required surface area for five rows. U = 80.**

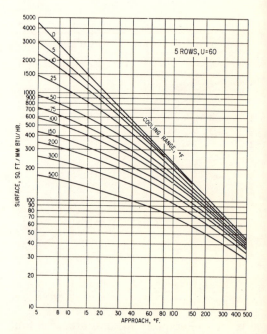

**Fig. p —Required surface area for five rows, U = 60.**

**Figure 8.23(a–z).** *Cont'd.*

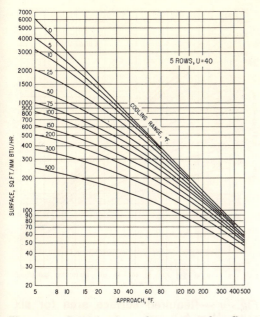

**Fig. q** —Required surface area for five rows, U = 40.

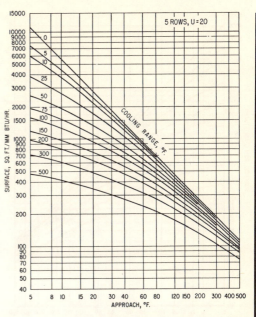

**Fig. r** —Required surface area for five rows, U = 20.

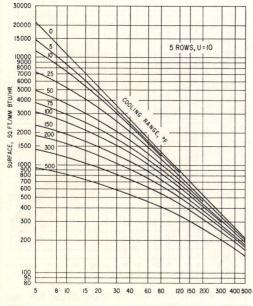

**Fig. s** —Required surface area for five rows, U = 10.

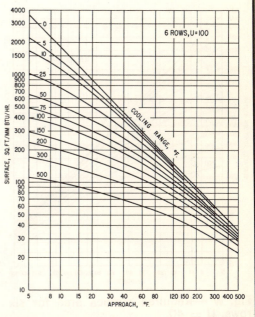

**Fig. t** —Required surface area for six rows, U = 100.

**Figure 8.23(a–z).** *Cont'd.*

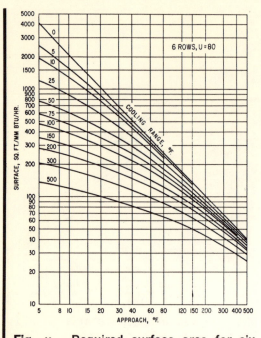

Fig. u —Required surface area for six rows, U = 80.

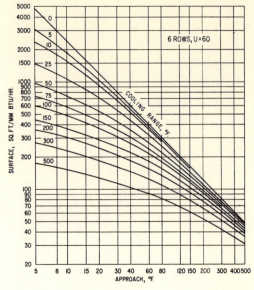

Fig. v —Required surface area for six rows, U = 60.

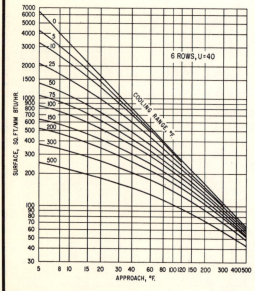

Fig. w —Required surface area for six rows, U = 40.

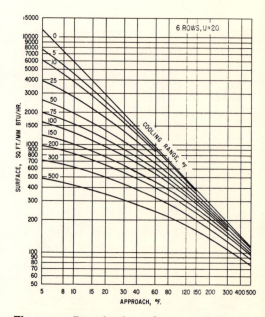

Fig. x —Required surface area for six rows, U = 20.

Figure 8.23(a–z). Cont'd.

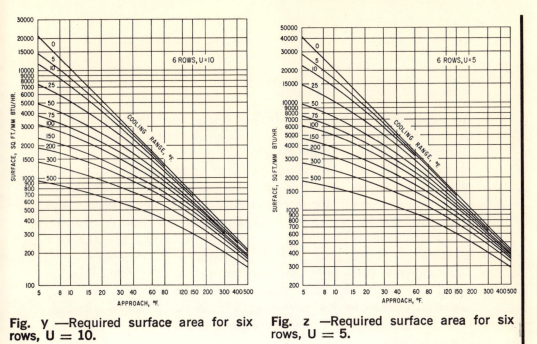

**Fig. y** —Required surface area for six rows, U = 10.

**Fig. z** —Required surface area for six rows, U = 5.

**Figure 8.23(a–z).** *Concluded*

**Table 8.14.**[(24)] Layout Information for Air-Cooled Exchangers ($\frac{5}{8}$-in. Fin on $2\frac{5}{16}$-in. Triangular Pitch 1-in. o.d. Tubes)[a]

| Section Width, Feet | Number of Rows | | | | Typical Tube Length, Feet |
|---|---|---|---|---|---|
| | **3** | **4** | **5** | **6** | |
| 6 | 24.09 | 31.94 | 40.06 | 47.91 | 6,10,15,20,24,30 |
| 8 | 32.20 | 42.94 | 53.67 | 64.40 | 10,15,20,24,30 |
| 12 | 48.18 | 63.88 | 80.12 | 95.82 | 12,16,24,32,40 |
| 16 | 64.40 | 85.88 | 107.34 | 128.80 | 16,24,32,40 |

[a]In ft$^2$/ft tube length, bare tube basis.

Table 8.15.[24]   Face Velocities

| Number of rows | 3 | 4 | 5 | 6 |
|---|---|---|---|---|
| Face velocity (ft/min) | 605 | 550 | 515 | 485 |

2.2 = 2 and 2.5 = 3.  For either of these cases, divide the calculated surface by the number of bays to get square feet per bay.  Calculate tube length to bay width ratio and establish the number of fans per bay.

6. The number of tube rows should be based on an economic evaluation of air cooler cost vs. operating cost.  There is no simple correlation between the number of tube rows and the heat transfer coefficient.  For quick estimates, select either 4 or 6 tube rows and go from there.  Otherwise, evaluate several different tube rows and use the best selection.

7. The number of fans can be estimated on the basis that fanned section length divided by bay width seldom exceeds 1.8.  Thus, a 16-ft wide bay with 24-ft tubes would have one fan (ratio = 1.5), whereas the same 16-ft wide bay with 32-ft tubes would have two fans (ratio ≐ 2.0).

8. Fan horsepower:
   a. Preliminary hp = (face area/bay) (number of bays) (7.5 hp/100 ft² of face area)
   b. Elevation correction
      multiply hp from (a) by $C_e$.
      $C_e = 1. +$ elevation above sea level (ft)/31,000.
   c. Inlet air temperature correction
      multiply hp from (b) by $C_t$.
      $C_t =$ (inlet air, °R)/530
   d. The above calculations are for design fan horsepower based on summer operation.  If two-speed or adjustable-pitch fans or louvers are used, no correction is necessary for winter operation and driver size is estimated by rounding the fan Bhp figure to a standard motor size.  However, if single-speed fans without the above features are used, the driver must be sized to handle the denser air.

$$\text{maximum driver Bhp} = \left(\begin{array}{c}\text{design fan}\\\text{horsepower}\end{array}\right) \times \left[\frac{\text{summer design (°R)}}{\text{winter design (°R)}}\right]$$

Driver size is estimated by rounding the maximum Bhp figure to a standard motor size.

## Example

Cool a heavy naphtha stream from 300 to 150°F with 90°F air.  Duty is 25.0 MMBtu/hr.  Outlet viscosity is 0.93 cP.  Design conditions are

150 psig and 350°F.  Elevation is 50 ft.  Use 14 BWG, welded steel tubes and painted hood and structure, suspended V-belt drivers with two-speed TEFC motors, tension-wrapped fins.

1. From Figure 8.22, $U = 73$.
2. Cooling range is $300 - 150 = 150°F$.  Approach is $150 - 90 = 60°F$. Use 4 rows.
3. Determine values for $U$ vs. ft²/MMBtu/hr from Figure 8.23(h)–(j). From a plot such as Figure 8.24, determine the value of ft²/MMBtu/hr (138) for $U = 73$.

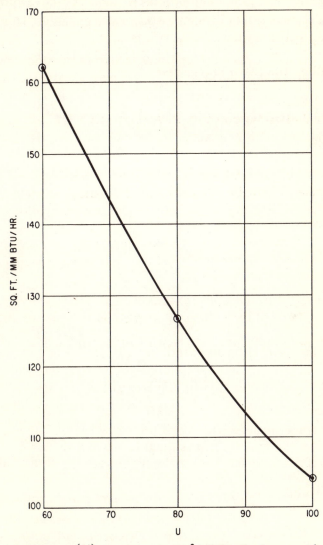

**Figure 8.24.**[24]  Curve to find ft²/MMBtu/hr for example problem.

| $U$ | ft$^2$/MMBtu/hr |
|-----|-----------------|
| 100 | 106 |
| 80  | 127 |
| 60  | 162 |
| 73  | 138 |

Total surface is (138) (25) = 3450 ft$^2$.

4. For a 16-ft bay the length of tubes equals 3450/85.88 = 40.2 ft.
   For a 12-ft bay, the length of tubes equals 3450/63.88 = 54.0 ft.
   A 16-ft bay should be used with 40-ft tubes [85.88 (40) = 3435 ft$^2$].
   Fanned section length divided by bay width equals 40/16 = 2.5. Since
   the ratio is greater than 1.8, use two fans.
5. Total hp = 16 × 40 × 1 × .075 × (1.+50./31000.) × (550./530.) =
   48.89 Bhp. Use two 30-hp drives.

# 8.13  Unsteady-State Heat Transfer[25]

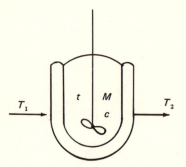

To *heat* the vessel from $t_1$ to $t_2$ with an isothermal heating medium at
$T_1$ (e.g., condensing steam):

$$\ln\left(\frac{T_1 - t_1}{T_1 - t_2}\right) = \frac{UA\theta}{Mc} \tag{8.29}$$

where $U$ is the overall coefficient [Btu/(hr-ft$^2$-°F)]; $A$ is the heat transfer
area (ft$^2$); $\theta$ is the time (hr); $M$ is the total mass, including the vessel (lb);
$c$ is the specific heat of the mass, including the vessel [Btu/(lb-°F)].

To *cool* a vessel from $t_1$ to $t_2$ with an isothermal cooling medium at
$T_1$:

$$\ln\left(\frac{t_1 - T_1}{t_2 - T_1}\right) = \frac{UA\theta}{Mc} \tag{8.30}$$

*To heat* a vessel from $t_1$ to $t_2$ with a nonisothermal heating medium whose inlet temperature is $T_1$ and outlet temperature is $T_2$:

$$\ln\left(\frac{T_1 - t_1}{T_1 - t_2}\right) = \frac{WC\theta}{Mc}\left(\frac{K_1 - 1}{K_1}\right) \tag{8.31}$$

where

$$K_1 = \exp\left(UA/WC\right) \tag{8.32}$$

and $W$ is the flow of heating or cooling medium (lb/hr); $C$ is the heat capacity of heating or cooling medium, $[\text{Btu}/(\text{lb-}^\circ\text{F})]$; $T$ is the inlet temperature of heating or cooling medium ($^\circ$F).

*To cool* a vessel from $t_1$ to $t_2$ with a nonisothermal cooling medium whose inlet temperature is $T_1$ and outlet temperature is $T_2$:

$$\ln\left(\frac{t_1 - T_1}{t_2 - T_1}\right) = \frac{WC\theta}{Mc}\left(\frac{K_1 - 1}{K_1}\right) \tag{8.33}$$

*To heat* a vessel with an external heat exchanger and an isothermal heating medium ($T_1 = T_2$) from $t_1$ to $t_2$:

$$\ln\left(\frac{T_1 - t_1}{T_1 - t_2}\right) = \frac{wc\theta}{Mc}\left(\frac{K_2 - 1}{K_2}\right) \tag{8.34}$$

$$K_2 = \exp\left(UA/wc\right)$$

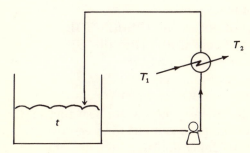

*To cool* a vessel with an external heat exchanger, and an isothermal cooling medium at $T_1$, from $t_1$ to $t_2$:

$$\ln\left(\frac{t_1 - T_1}{t_2 - T_1}\right) = \frac{wc\theta}{Mc}\left(\frac{K_2 - 1}{K_2}\right) \tag{8.35}$$

*To heat* a vessel from $t_1$ to $t_2$ with an external heat exchanger and a nonisothermal heating medium whose inlet is $T_1$ and outlet is $T_2$:

$$\ln\left(\frac{T_1 - t_1}{T_1 - t_2}\right) = \left(\frac{K_3 - 1}{M}\right)\left(\frac{wWC}{K_3 WC - wc}\right)\theta \qquad (8.36)$$

$$K_3 = \exp\left[UA\left(\frac{1}{wc} - \frac{1}{WC}\right)\right] \qquad (8.37)$$

*To cool* a vessel from $t_1$ to $t_2$ with an external heat exchanger and a nonisothermal cooling medium whose inlet is $T_1$ and oulet is $T_2$:

$$\ln\left(\frac{t_1 - T_1}{t_2 - T_1}\right) = \frac{(K_4 - 1)}{M}\left(\frac{wWC}{K_4 WC - wc}\right)\theta \qquad (8.38)$$

$$K_4 = \exp\left[UA\left(\frac{1}{wc} - \frac{1}{WC}\right)\right] \qquad (8.39)$$

where $W$ is the flow of cooling or heating medium (lb/hr); $C$ is the heat capacity of the cooling or heating medium [Btu/(lb-°F)]; $w$ is the flow through exchanger (lb/hr); $c$ is the heat capacity of mass to be heated or cooled [Btu/(lb-°F)]; $M$ is the total mass to be heated or cooled (lb).

## NOMENCLATURE

### Double Pipe Exchangers (Section 8.6)

$A$    Heat transfer area (ft$^2$)
$A_a$   Flow area of annulus (ft$^2$)
$a_p$   Flow area of inner pipe (ft$^2$)
$C$    Specific heat of hot stream [Btu/(lb-°F)]
$c$    Specific heat of cold stream [Btu/(lb-°F)]
$D$    Inner diameter of inner pipe (ft)
$D_i$   Outer diameter of inner pipe (ft)
$D_o$   Inner diameter of outer pipe (ft)
$D_e$   Equivalent diameter of annulus for heat transfer (in.)
$D_e'$   Equivalent diameter of annulus for pressure drop (in.)
$f$    Friction factor (dimensionless)
$\Delta F$   Pressure drop (ft of liquid)
$\Delta F_r$   Return pressure drop (ft of liquid)
$G$    Mass flow [lb/(hr-ft$^2$)]
$G_a$   Mass flowrate in annulus [lb/(hr-ft$^2$)]
$G_p$   Mass flowrate of inner pipe [lb/(hr-ft$^2$)]
$g_c$   Conversion factor ($4.18 \times 10^8$ ft/hr$^2$)

$g'$ Conversion factor (32.2 ft/sec²)

$h_i$ Inside heat transfer coefficient [Btu/(hr-ft²-°F)]

$h_{io}$ Inside coefficient, corrected to outside area [Btu/(hr-ft²-°F)]

$h_o$ Outside heat transfer coefficient [Btu/(hr-ft²-°F)]

$j_H$ Heat transfer factor (Figure 8.5)

$k$ Thermal conductivity [Btu/hr-ft²-(°F/ft)]

$L$ Pipe length (ft)

Pr Prandtl number (dimensionless)

$\Delta P$ Pressure drop (psi)

$\Delta P_t$ Total pressure drop in annulus (psi)

$Q$ Exchanger duty (Btu/hr)

$R_D$ Fouling resistance (Table 8.2)

Re Reynolds number (dimensionless)

$R_w$ Wall resistance (Table 8.3)

$S, s$ Specific gravity hot or cold fluid, respectively

$T$ Temperature of hot stream (°F) ($T_1$–inlet, $T_2$–outlet)

$t$ Temperature of cold stream (°F) ($t_1$–inlet, $t_2$–outlet)

$U_c$ Clean overall heat transfer coefficient [Btu/(hr-ft²-°F)]

$U_D$ Dirty overall heat transfer coefficient [Btu/(hr-ft²-°F)]

$V$ Fluid velocity (ft/sec)

$W$ Flowrate of hot stream (lb/hr)

$w$ Flowrate of cold stream (lb/hr)

$\rho$ Fluid density (lb/ft³)

$\mu$ Fluid viscosity (lb/ft-hr)

## Shell and Tube Heat Exchangers (Section 8.7)

$A$ Heat-transfer surface (ft²)

$a_t, a_t'$ Flow area and flow area per tube (ft²)

$a_s$ Flow area of the shell

$B$ Baffle spacing (in.)

$C$ Specific heat of hot fluid in derivations [Btu/(lb-°F)]

$C'$ Clearance between tubes (in.)

$c$ Specific heat of fluid [Btu/(lb-°F)]

$D$ Inside diameter of tubes (ft)

$D_s$ Inside diameter of shell (ft)

$D_e, D_e'$ Equivalent diameter for heat transfer and pressure drop (ft)

$F$ Temperature-difference factor, $\Delta t = F \times$ LMTD (dimensionless)

$f$ Friction factor (ft.²/in.²)

$G_t, G_s$ Mass velocity in tubes and shell [lb/(hr-ft²)]

| | |
|---|---|
| $g$ | Acceleration of gravity (ft/hr$^2$) |
| $g'$ | Acceleration of gravity (ft/sec$^2$) |
| $h_i, h_o$ | Heat transfer coefficient for inside fluid, and for outside fluid, respectively [Btu/(ft$^2$-hr-°F)] |
| $h_{io}$ | Value of $h_i$ when referred to the tube outside diameter, [Btu/(ft$^2$-hr-°F)] |
| $j_H$ | Factor for heat transfer (dimensionless) |
| $k$ | Thermal conductivity [(Btu-ft)/(ft$^2$-hr-°F)] |
| $L$ | Tube length (ft) |
| LMTD | Log mean temperature difference (°F) |
| $N$ | Number of shell-side baffles |
| $N_t$ | Number of tubes |
| $n$ | Number of tube passes |
| $P_r$ | Prandtl number (dimensionless) |
| $P_t$ | Tube pitch (in.) |
| $\Delta P_T, \Delta P_t, \Delta P_r$ | Total, tube, and return pressure drops, respectively (psi) |
| $\Delta P_s$ | Shell side pressure drop |
| $Q$ | Heat flow (Btu/hr) |
| $R$ | Temperature group $(T_1 - T_2)/(t_2 - t_1)$ (dimensionless) |
| $R_d$ | Dirt factor [(ft$^2$-hr-°F)/Btu] |
| $Re_t, Re_s$ | Reynolds number for heat transfer and pressure drop, respectively (dimensionless) |
| $S$ | Temperature group $(t_2 - t_1)/(T_1 - t_1)$ (dimensionless) |
| $s$ | Specific gravity (dimensionless) |
| $T_1, T_2$ | Inlet and outlet temperature of hot fluid, respectively (°F) |
| $T_c$ | Mean temperature of hot fluid (°F) |
| $t_1, t_2$ | Inlet and outlet temperature of cold fluid (°F) |
| $t_c$ | Mean temperature of cold fluid (°F) |
| $\Delta t$ | True temperature difference in $Q = U_D A \Delta t$ (°F) |
| $U_c, U_D$ | Clean and dirty overall coefficient of heat transfer, respectively [Btu/(ft$^2$-hr-°F)] |
| $V$ | Velocity (ft/sec) |
| $W$ | Weight flow of hot fluid (lb/hr) |
| $w$ | Weight flow of cold fluid (lb/hr) |
| $\mu$ | Viscosity [cP × 2.42 = lb/(ft-hr)] |
| $\mu_w$ | Viscosity at tube wall temperature, [cP × 2.42 = lb/(ft-hr)] |
| $(\mu/\mu_w)^{0.14}$ | Viscosity ratio |

*Subscripts (except as noted above)*

| | |
|---|---|
| s | Shell |
| t | Tube |

## Reboilers and Vaporizers (Section 8.10)

$A$    Heat transfer surface (ft$^2$)

$a'_t$    Cross sectional area per tube (in.$^2$)

$a_t$    Tube side flow area of exchanger (ft$^2$)

$D$    i.d. of tube (ft)

$G$    Mass flowrate [lb/(hr-ft$^2$)]

$g$    Acceleration due to gravity (ft/hr$^2$)

$h_i$    Inside coefficient [Btu/(ft$^2$-hr-°F)]

$h_{io}$    Inside coefficient corrected to outside area

$c$    Specific heat [Btu/(lb-°F)]

$k$    Thermal conductivity [Btu/(ft-hr-°F)]

$L$    Tube length (ft)

LMTD    Log mean temperature difference (°F)

$N_t$    Number of tubes

$n$    Number of tube passes

$Q$    Exchanger duty (Btu/hr)

$P$    Expansion loss due vaporization (psi)

$R_d$    Dirt factor [(ft$^2$-hr-°F)/Btu]

Re    Reynolds number

$S_i$    Specific gravity of liquid at inlet of exchanger

$S_{mean}$    Log mean specific gravity between inlet and outlet of exchanger

$S_{mix}$    Specific gravity of mixture at outlet of exchanger

$S_o$    Specific gravity of liquid at outlet of exchanger

$S_v$    Specific gravity of vapor at outlet of exchanger

$U_c$    Overall clean coefficient [Btu/(ft$^2$-hr-°F)]

$U_D$    Overall dirty coefficient [Btu/(ft$^2$-hr-°F)]

$V_i$    Specific volume of liquid at inlet to reboiler (ft$^3$/lb)

$V_{mean}$    Log mean specific volume between the exchanger inlet and outlet (ft$^3$/lb)

$V_o$    Specific volume at reboiler outlet (ft$^3$/lb)

$W$    Flowrate of circulating liquid to reboiler (lb/hr)

$Z_1$    Liquid height which acts as the driving force in thermosyphon reboiler (ft)

$Z_2$    Height of liquid–vapor mixture which acts as a static head opposing the driving force (ft)

$\mu$    Viscosity of liquid at inlet to exchanger [lb/(ft-hr)]

$\mu_M$    Viscosity at tube wall temperature [lb/(ft-hr)]

$\rho_e$    Liquid density of liquid in leg that provides the driving force (lb/ft$^2$)

$\rho_{avg}$    Average density of vapor–liquid mixture (lb/ft$^3$)

$\rho_L$    Density of liquid (lb/ft$^3$)

$\rho_V$    Density of vapor (lb/ft$^3$)

**Condensers (Section 8.11)**

$h, h_{cm}$ Film coefficient [Btu/(hr-ft²-°F)]

$\lambda$ Latent heat of vaporization (Btu/lb)

$G', G''$ Condensing load [lb/(hr-ft)] (see Figure 8.16)

$G'$ Mass flowrate per tube inside circumference, $W/\pi D$ [lb/(hr-ft)] (see Figure 8.17)

$L$ Tube length (ft)

$N$ Number of tubes

$W$ Condensing load (lb/hr)

$D$ Tube o.d. (ft)

$k_f$ Thermal conductivity at film temperature, [(Btu-ft)/(ft²-hr-°F)]

$\rho_f$ Density of condensate (lb/ft³)

$g$ Acceleration due to gravity ($4.17 \times 10^8$ ft/hr²)

$\mu_t$ Viscosity of condensate film [lb/(hr-ft)]

## REFERENCES

1. Pfaudler Heat Exchanger Data Book, Bulletin 1056, Pfaudler Division, Ritter Pfaudler Corporation, Rochester, New York, 1967, pp. 9–10.
2. Standards of Tubular Exchanger Manufacturers Association, 5th ed., New York, 1968, pp. 124–127.
3. Dowtherm Heat Transfer Fluids, Dow Chemical Co., Midland, Michigan, 1971, p. 104.
4. E. E. Ludwig, *Applied Process Design for Chemical and Petrochemical Plants*, Vol. 3, Gulf Publishing Co., Houston, Texas, 1965, p. 61.
5. Platecoil Catalog No. 5-63, Platecoil Division, Tranter Manufacturing, Inc., Lansing, Michigan, 1974, p. 68.
6. P. D. Shroff, *Chem. Processing* No. 4, 60–61 (1960).
7. Platecoil Catalog No. 5-63, Platecoil Division, Tranter Manufacturing, Inc., Lansing, Michigan, 1974, p. 70.
8. W. H. Holstein, What it costs to steam and electrically trace pipelines, *Chem. Eng. Prog. 62*, 107 (1966).
9. F. S. Chapman and F. A. Holland, Keeping piping hot, *Chem. Eng.*, December 20, 80–81 (1965).
10. D. Q. Kern, *Process Heat Transfer*, McGraw-Hill, New York, 1950, p. 102.
11. D. Q. Kern, *Process Heat Transfer*, McGraw-Hill, New York, 1950, pp. 221–245, 835.
12. E. E. Ludwig, *Applied Process Design for Chemical and Petrochemical Plants*, Vol. 3, Gulf Publishing Co., Houston, Texas, 1965, pp. 24, 135.
13. Engineering Data Book Section, Wolverine Division, Universal Oil Products, Inc., 1961, pp. 74–78.
14. Standards of Tubular Exchanger Manufacturers Association, 5th ed., New York, 1968, pp. 129–130.
15. D. Q. Kern, *Process Heat Transfer*, McGraw-Hill, New York, 1950, p. 718.

16. Pfaudler RA Series Glasteel Reactors, Bulletin 1086, Pfaudler Co., Division of Sybron Corp., Rochester, New York, 1976, pp. 6–7.
17. D. Q. Kern, *Process Heat Transfer*, McGraw-Hill, New York, 1950, p. 747.
18. D. Q. Kern, *Process Heat Transfer*, McGraw-Hill, New York, 1950, pp. 453–491.
19. E. E. Ludwig, *Applied Process Design for Chemical and Petrochemical Plants*, Vol. 3, Gulf Publishing Co., Houston, Texas, 1965, p. 120.
20. The Physics of Process Vapor Condenser Construction, The Andale Co., Bulletin 351, Philadelphia, Pennsylvania, 1935.
21. D. Q. Kern, *Process Heat Transfer*, McGraw-Hill, New York, 1950, pp. 267, 338.
22. E. E. Ludwig, *Applied Process Design for Chemical and Petrochemical Plants*, Vol. 3, Gulf Publishing Co., Houston, Texas, 1965, pp. 81, 89.
23. A. P. Colburn, *Trans. AICHE 30*, 187 (1934).
24. J. E. Lerner, Simplified air cooler estimating, *Hydrocarbon Processing*, February, 93–100 (1972).
25. D. Q. Kern, *Process Heat Transfer*, McGraw-Hill, New York, 1950, pp. 624–631.

## SELECTED READING

### Heat Transfer Coefficients and Fundamentals

N. H. Chen, Method to find tubeside heat transfer coefficient, *Chem. Eng.* June 30 (1958).

N. H. Chen, Save time in heat exchanger design, *Chem. Eng.* October 20 (1958).

N. H. Chen, Tubeside heat transfer coefficients for gases and vapors, *Chem. Eng.* January 12 (1959).

N. H. Chen, Condensing and boiling heat transfer coefficients, *Chem. Eng.* March 9 (1959).

Century Heat Exchanger Tube Manual, Century Brass Products, Waterbury, Conn. (1977).

O. Frank, Estimating overall heat transfer coefficients, *Chem. Eng.* May 13 (1974).

E. E. Ludwig, *Applied Process Design for Chemical and Petro-Chemical Plants*, Vol. 3, Gulf Publishing Co., Houston, Texas, 1965.

R. H. Perry and C. H. Chilton, *Chemical Engineers' Handbook*, 5th ed., McGraw-Hill, New York, 1973.

*Pfaudler Heat Exchanger Data Book*, Bulletin 1056, Pfaudler Division, Ritter Pfaudler Corporation, Rochester, New York, 1967.

P. D. Shroff, *Chemical Processing*, No. 4, 60–61 (1960).

Standards of Tubular Exchanger Manufacturers Association, 5th ed.   New York, 1968.

### Heat Losses from Insulated Pipelines

F. S. Chapman and F. A. Holland, Keeping piping hot, *Chem. Eng.* December 20 (1965).

W. H. Holstein, What it costs to steam and electrically trace pipelines, *Chem. Eng. Prog.*, Vol. 62, No. 3, March (1966).

### Double Pipe Exchangers

J. P. Holman, *Heat Transfer*, 3rd ed., McGraw-Hill, New York, 1972.

D. Q. Kern, *Process Heat Transfer*, McGraw-Hill, New York, 1950.

## Shell and Tube Heat Exchangers

Engineering Data Book Section, Wolverine Div., Universal Oil Products, Inc., 1961.
J. P. Holman, *Heat Transfer*, 3rd ed., McGraw-Hill, New York, 1972.
D. Q. Kern, *Process Heat Transfer*, McGraw-Hill, New York, 1950.
E. E. Ludwig, *Applied Process Design for Chemical and Petrochemical Plants*, Vol. 3, Gulf Publishing Co., Houston, Texas, 1967.
W. H. McAdams, *Heat Transmission*, 3rd ed., McGraw-Hill, New York, 1954.
*Standards of Tubular Exchanger Manufacturers Association*, 5th ed., New York, 1968.

## Heat Transfer Coefficient in Agitated Vessels

E. J. Ackley, Film coefficients of heat transfer for agitated process vessels, *Chem. Eng.*, August 22 (1960).
Bulletin 1086, Pfaudler RA Series Glasteel Reactors, Pfaudler Co., Division of Sybron Corp., Rochester, New York, 1976.
F. S. Chapman and F. A. Holland, Heat-transfer correlations for agitated liquids in process vessels, *Chem. Eng.* January 18, 1965.
F. S. Chapman and F. A. Holland, Heat transfer correlations in jacketed vessels, *Chem. Eng.* February 15 (1965).
T. H. Chilton, T. B. Drew, and R. H. Jebens, Heat transfer coefficients in agitated vessels, *Ind. Eng. Chem.* Vol. 36, No. 6, June (1944).
G. H. Cummings and A. S. West, Heat transfer data for kettles with jackets and coils, *Ind. Eng. Chem.* Vol. 42, No. 11, November (1950).
D. Q. Kern, *Process Heat Transfer*, McGraw-Hill, New York, 1950.
W. R. Penney and K. J. Bell, Close clearance agitators, *Ind. Eng. Chem.* Vol. 59, No. 4, April (1967).

## Reboilers and Vaporizers

G. K. Collins, Horizontal thermosiphon reboiler design, *Chem. Eng.* July 19 (1976).
J. R. Fair, What you need to design thermosiphon reboilers, *Petroleum Refiner*, Vol. 29, No. 2, February (1960).
O. Frank and R. D. Prickett, Designing vertical thermosiphon reboilers, *Chem. Eng.* September 3 (1973).
D. Q. Kern, *Process Heat Transfer*, McGraw-Hill, New York, 1950.
E. E. Ludwig, *Applied Process Design for Chemical and Petrochemical Plants*, Vol. 3, Gulf Publishing Co., Houston, Texas, 1965.
J. W. Palen and W. M. Small, A new way to design kettle and internal reboilers, *Hydrocarbon Processing*, Vol. 43, No. 11, November (1964).

## Condensers

The Andale Co., Bulletin 351, The Physics of Process Vapor Condenser Construction, Philadelphia, Pa., 1935.
K. J. Bell, Temperature profiles in condensers, *Chem. Eng. Prog.* Vol. 63, No. 7, July (1972).
A. P. Colburn, *Trans. AIChE*, Vol. 30, 187 (1934).
W. Gloyer, Thermal design of mixed vapor condensers, Part 1 and Part 2, *Hydrocarbon Processing*, June (1970).
D. Q. Kern, *Process Heat Transfer*, McGraw-Hill, New York, 1950.

E. E. Ludwig, *Applied Process Design for Chemical and Petrochemical Plants*, Vol. 3, Gulf Publishing Co., Houston, Texas, 1965.

D. E. Steinmeyer, Fog formation in partial condensers, *Chem. Eng. Prog.* Vol. 68, No. 7, July (1972).

D. E. Steinmeyer and A. C. Mueller, Why condensers don't operate as they are supposed to, *Chem. Eng. Prog.* Vol. 70, No. 7, July (1974).

## Air-Cooled Heat Exchangers

Air cooled heat exchange, *Chem. Eng. Prog.* Vol. 55, No. 4, April (1959).

G. M. Franklin and W. B. Munn, Problems with heat exchangers in low temperature environments, *Chem. Eng. Prog.* Vol. 70, No. 7, July (1974).

A. Y. Gunter and K. V. Shipes, Hot air recirculation by air coolers, *Chem. Eng. Prog.* Vol. 68, No. 2, February (1972).

J. E. Lerner, Simplified air cooler estimating, *Hydrocarbon Processing*, February (1972).

K. V. Shipes, Air-cooled exchangers in cold climates, *Chem. Eng. Prog.* Vol. 70, No. 7, July (1974).

## Unsteady-State Heat Transfer

T. R. Brown, Heating and cooling in batch processes, *Chem. Eng.* May 28 (1973).

S. H. Davis and W. W. Akers, Unsteady state heat transfer—I, *Chem. Eng.* April 18 (1960).

S. H. Davis and W. W. Akers, Unsteady state heat transfer—II, *Chem. Eng.* May 16 (1960).

D. Q. Kern, *Process Heat Transfer*, McGraw-Hill, New York, 1950.

# 9    Hydroclones

Typical design dimensions for commercial hydroclones are shown in Figure 9.1.

## PRESSURE DROP

Pressure drop across the hydroclone can be found using Equation (9.1):

$$\Delta P = \left(\frac{\rho V_i^2}{288 g_c}\right)\left(\frac{\alpha^2}{n}\right)\left[\left(\frac{D_c}{D_o}\right)^{2n} - 1\right] \tag{9.1}$$

## EFFICIENCY

The $d_{50}$ (the particle diameter which goes 50% to the underflow and 50% to the overflow) can be determined from Equation (9.2):

$$d_{50} = 2.7 \left[\frac{\tan(\theta/2)(1 - R_f)\mu}{D_c Q(\sigma - \rho)}\right]^{1/2} \left(\frac{2.3 D_o}{D_c}\right)^n \left(\frac{D_i^2}{\alpha}\right) \tag{9.2}$$

$$d_{95} = 1.7(d_{50}) \tag{9.3}$$

## NOMENCLATURE

$D_c, D_o, D_i$  Diameter (ft) (see Figure 9.1)
$\quad \Delta P$  Pressure drop (psi)
$\quad Q$  Feedrate (ft$^3$/sec)
$\quad R_f$  Bottoms flowrate/feed flowrate (often 0.10)
$\quad V_i$  Inlet velocity (ft/sec)
$\quad V_c$  Velocity at $D_c$ (ft/sec)
$\quad d_{50}$  That particle diameter which goes 50% to the overflow and 50% to the underflow
$\quad d_{95}$  That particle diameter which goes 95% to the underflow
$\quad g_c$  Gravitational constant (32.2 lb$_m$-ft/lb$_f$-sec$^2$)
$\quad n$  Dimensionless coefficient, usually $n = 0.8$ for $D_i = D_c/7$ and $D_o = D_c/5$
$\quad \rho$  Fluid density (lb/ft$^3$)

$\sigma$  Particle density (lb/ft$^3$)
$\mu$  Viscosity (lb/ft-sec)
$\theta$  Cyclone cone angle (degrees)
$\alpha$  Ratio $V_c/V_i$, usually 0.45–0.50

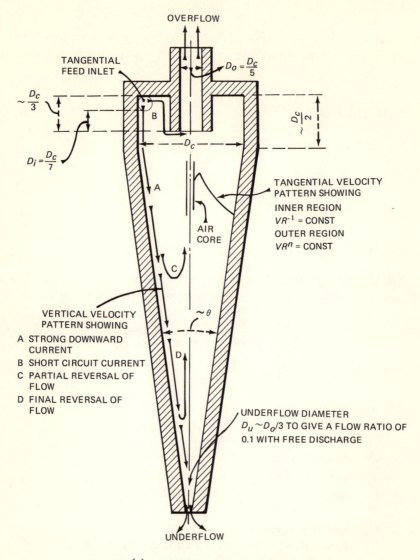

**Figure 9.1.**[1]  Cyclone design and flow patterns.

## REFERENCES

1. D. Bradley, A theoretical study of the hydraulic cyclone, *The Industrial Chemist*, September, 473–480 (1958).

## SELECTED READING

D. Bradley, A theoretical study of the hydraulic cyclone, *The Industrial Chemist*, September (1958).

D. Bradley, *The Hydroclone*, Pergamon Press, Ltd., London, England, 1965.

# 10    Materials

## METALS[1]

The chemical composition of metals commonly used in the process industries is listed in Table 10.1. Typical applications for these metals are given in Table 10.2.

**Table 10.1.**[1]   Chemical Composition of Selected Metals

| Material | C, % | Cr, % | Cu, % | Fe, % | Mn, % | Mo, % | Ni, % | Si, % | Other Elements, % |
|---|---|---|---|---|---|---|---|---|---|
| Admiralty metal | — | — | 70 | — | — | — | — | — | Zn, 29; Sn, 1 |
| Aluminum 2S | — | — | — | — | — | — | — | — | Al, 99.5+ |
| Aluminum 3S | — | — | 0.1 | 0.5 | 1.25 | — | — | 0.5 | Al, 97.7 |
| Brass, admiralty | — | — | 71 | — | — | — | — | — | Zn, 28; Sn, 1 |
| Brass, aluminum | — | — | 76 | — | — | — | — | — | Al, 2; Zn, 22 |
| Bronze, silicon | — | — | 96 | — | — | — | — | 3 | |
| Carpenter 20 alloy | — | 20 | 4 | 44 | — | 3 | 29 | — | |
| Copper, arsenical | — | — | 99.9+ | — | — | — | — | — | As, 0.04 |
| Croloy 2¼ alloy | 0.15 | 2.25 | — | 96.1 | — | 1.0 | — | 0.5 | |
| Cupro-nickel, 70–30 | — | — | 69.3 | 0.75 | — | — | 30 | — | |
| Cupro-nickel, 90–10 | — | — | 88.8 | 1.25 | — | — | 10 | — | |
| Durimet 20 alloy | 0.07 | 20 | 3.5 | 45.7 | — | 1.75 | 29 | — | |
| Duriron | 0.85 | — | — | 84.7 | — | — | — | 14.5 | |
| Hastelloy B alloy | 0.1 | — | — | 5 | — | 28 | 62 | — | |
| Hastelloy C alloy | 0.1 | 16 | — | 5 | — | 16 | 55 | — | W, 4 |
| Incoloy alloy 800 | 0.04 | 21 | 0.3 | 45.3 | 1 | — | 32 | 0.4 | |
| Inconel alloy 600 | 0.04 | 15.8 | 0.1 | 7 | 0.20 | — | 76 | 20 | |
| Iron, cast | 3.4 | — | — | 94.3 | 0.5 | — | — | 1.8 | |
| Lead | — | — | — | — | — | — | — | — | Pb, 99.95 |
| Monel alloy 400 | 0.12 | — | 31.5 | 1.4 | 1 | — | 66 | 0.15 | |
| Monel alloy K-500 | 0.15 | — | 29.5 | 1.0 | 0.60 | — | 65 | 0.15 | |
| Nickel 200 | 0.06 | — | 0.05 | 0.15 | 0.25 | — | 99.4 | — | |
| Nickel silver, 18% | — | — | 65 | — | — | — | 18 | — | Zn, 17 |
| NI-Hard alloy | 3.4 | 1.5 | — | 89.5 | 0.5 | — | 4.5 | 0.6 | |
| NI-Resist, Type I | 2.8 | 2.5 | 6.5 | 69.6 | 1.3 | — | 15.5 | 1.8 | |
| NI- Resist, Type II | 2.8 | 2.5 | — | 71.9 | 1.0 | — | 20.0 | 1.8 | |
| Steel, carbon-molybdenum | 0.15 | — | — | 99.4 | — | 0.5 | — | — | |
| Steel, low-carbon, electric-resitance-welded, ASTM A587–68 | 0.1 | — | — | 99.5 | 0.4 | — | ... | — | P, 0.01; S, 0.02; Al, 0.05 |
| Steel, mild (SAE 1020) | 0.2 | — | — | 99.1 | 0.45 | — | — | 0.25 | |
| Steel, Ni-Cr (SAE 3140) | 0.4 | 0.65 | — | 96.7 | 0.8 | — | 1.25 | 0.25 | |
| Steel, stainless<br>  AISI Type 30, or<br>  ASTM A312–64, Gr. TP-304 | 0.06 | 18 | — | 72 | — | — | 10 | — | |
|   AISI Type 316, or<br>  ASTM A312–64, Gr. TP-316 | 0.07 | 18 | — | 67.4 | — | 2.5 | 12 | — | |
|   AISI Type 347 | 0.07 | 18 | — | 70.2 | — | — | 11 | — | Cb, 0.7 |
| Tantalum | — | — | — | — | — | — | — | — | Ta, 99.9+ |
| Titanium (Ti-50A) | 0.08 max. | — | — | 0.20 max. | — | — | — | — | N, 0.05 max.; H, 0.015 max. |
| Worthite alloy | 0.07 | 20 | 1.75 | 46.7 | — | 3 | 25 | 3.5 | |

**Table 10.2.**[1] Applications of Various Metals in the Chemical Industry

*Admiralty Metal*—Used in exchanger tubes in contact with fresh and salt water, steam, oil and other liquids below 500 F.

*Aluminum 2S*—Low in strength, most readily welded aluminum type. Used in food plants. Resistant to formaldehyde, ammonia, dyes, phenol, hydrogen sulfide.

*Aluminum Alloy 3S*—Better weldability and mechanical properties because of manganese content.

*Brass, Admiralty*—Resistant to corrosive waters. Same applications as Admiralty metal.

*Brass, Aluminum*—Good resistance to corrosion and good retention of hardness at elevated temperatures.

*Bronze, Silicon*—Used for processing tanks and equipment because of high strength and toughness. Resists brines, sulfite solutions, sugar solutions, and organic acids.

*Copper*—Resistant to corrosive waters. Used for condenser- and heat-exchanger tubing.

*Cupro-Nickel, 70-30*—Similar to U.S. nickel coin. Resists corrosion of salt water passing through tubes in heat exchangers at high velocity.

*Duriron*—Available in cast form only. Resists acids such as nitric, sulfuric, and acetic, but not hydrofluric or fuming sulfuric.

*Durimet 20*—Extremely resistant to sulfuric acid at all concentrations except between 60 to 90% at boiling temperatures.

*Durimet T*—Austenitic Cr-Ni alloy with good corrosion-resistance properties.

*Hastelloy B*—Used for handling boiling hydrochloric acid and wet hydrochloric-acid gas.

*Hastelloy C*—Withstands strong oxidizing agents such as nitric acid and free chlorine. Also resistant to phosphoric, acetic, formic and sulfurous acids.

*Inconel*[a]—Ni-Cr alloy with small percentage Fe. Chromium content makes it resistant to oxidizing as well as reducing solutions at high temperatures. Prevents contamination and tarnishing by substances encountered in the soap and food industries.

*Iron, Cast*—Good resistance to internal and external corrosion. Extensively used for water and gas-distribution, and for sewage systems in cities.

*Lead*—High resistance to corrosion when insoluble coating forms such as lead sulfate, carbonate, or phosphate. Not good when soluble salts, such as lead nitrate, form through the action of nitric acid.

*Monel*[a] *400*—High-strength Ni-Cu alloy used primarily for handling alkaline solutions whenever copper contamination is not a problem. It has excellent resistance to corrosion by many airfree acids, caustic solutions, alkalies, salt solutions, food products, and other organic substances. Generally recommended—like commercial nickel—under reducing conditions, rather than those that are oxidizing.

*Monel*[a] *K-500*—Retains strength at elevated temperatures.

*Nickel 200*—Highly resistant to corrosion. Most frequently used where copper content of Monel is undesirable. Handles high concentrations of hot caustic and neutral salts.

*Nickel, Silver, 18%*—Called German silver. Highly resistant to corrosion. Used for foodstuffs with atmosphere contact.

*NI-Resist, Type 1*—For sulfuric acid service. Available in pipe and valve bodies.

*Steel, ASTM A587-68*—Highly ductile and aging-resistant. Can be bent considerably more than other steels.

*Steel, Stainless; AISI Type 304 or ASTM A312-64, Gr.TP-304*—The most common of the stainless steels. Resists corrosion by many materials; provides sanitary conditions for food and drug industries.

*Steel, Stainless; AISI Type 316 or ASTM A312-64, Gr.TP-316*—Most corrosion resistant of the stainless steels; also higher in price.

*Tantalum*—Resistant to nitric and other acids.

*Titanium, Ti-50A*—High corrosion resistance in oxidizing media. Resists hypochlorites, 30% sulfuric acid, perchlorates. Resists also abrasion and erosion by cavitation.

*Worthite*—Used for handling sulfuric acid, and concentrated solutions of acetic and phosphoric acids.

---

[a]Nickel, Monel and Inconel pipe can be readily welded by the arc or oxyacetylene methods. Since coiling and bending require soft tubing, Monel and nickel seamless tubing are supplied in four tempers: *as drawn; low-temperature, stress-relieved* (or equalized); *high-temperature, stress-relieved*; and *soft annealed.* Unless otherwise specified, the low-temperature, stress-relieved kind is furnished. Inconel is supplied in three temper grades: *as drawn; normalized;* and *soft.* Unless specified, the normalized type temper is furnished.

# REFERENCE

1. J. A. Masek, Metallic piping, *Chem. Eng.*, 216–217, June 17 (1968).

# SELECTED READING

Corrosion Data Survey of Tantalum, Fansteel Inc., N. Chicago, Ill.

Corrosion Engineering Bulletins, The International Nickel Co., Inc., New York Plaza, New York.

*Corrosion, The Journal of Science and Engineering*, a monthly publication of NACE, The Corrosion Society.

Corrosion Resistance of Hastelloy Alloys, Cabot Corp., Stellite Division, Kokomo, Ind.

Fontana and Greene, *Corrosion Engineering*, McGraw-Hill, New York, 1967.

N. E. Hamner, *Corrosion Data Survey, Metals Section*, 5th ed., National Association of Corrosion Engineers, Houston, Texas, 1974.

N. E. Hamner, *Corrosion Data Survey, Non Metals Section*, National Association of Corrosion Engineers, Houston, Texas, 1974.

*Materials Protection and Performance*, a monthly publication of NACE, The Corrosion Society.

R. H. Perry and C. H. Chilton, *Chemical Engineers' Handbook*, 5th ed., McGraw-Hill, New York.

M. Pourbaix, *Atlas of Electrochemical Equilibria in Aqueous Solutions*, National Association of Corrosion Engineers, Houston, Texas, 1977.

E. Rabald, *Corrosion Guide*, 2nd ed., Elsevier, New York, 1968.

Resistance to Corrosion, Huntington Alloy Products Division, The International Nickel Company, Inc., Huntington, W. Virginia.

L. L. Shrier, *Corrosion, Metal/Environment Reactions*, Vols. 1 and 2, Newnes-Butterworths, Boston, 1976.

R. W. Staehle, A. J. Forty, and D. VanRooyen, Proceedings of Conference, *Fundamental Aspects of Stress Corrosion Cracking*, The Ohio State University, Department of Metallurgical Engineering; National Association of Corrosion Engineers, Houston, Texas, 1969.

N. D. Tomashov, *Theory of Corrosion and Protection of Metals*, MacMillan, New York, 1966.

H. H. Uhlig, *Corrosion and Corrosion Control*, 2nd ed., John Wiley & Sons, 1971.

N. E. Woldman and R. D. Gibbons, *Engineering Alloys*, 5th ed., Van Nostrand Reinhold, New York, 1973.

# 11    Physical Properties*

**Table 11.1**[1]    Thermal Conductivities of Some Building and Insulating Materials [$k$ = Btu/(hr) (ft$^2$) ($^\circ$F/ft)]

| Material | Apparent density $\rho$, lb/ft$^3$ at room temperature | $^\circ$F | $k$ |
|---|---|---|---|
| Asbestos-cement boards..................... | 120 | 68 | 0.43 |
| Asbestos sheets........................... | 55.5 | 124 | 0.096 |
| Asbestos slate............................ | 112 | 32 | 0.087 |
| | 112 | 140 | 0.114 |
| Asbestos................................. | 29.3 | −328 | 0.043 |
| | 29.3 | 32 | 0.090 |
| | 36 | 32 | 0.087 |
| | 36 | 212 | 0.111 |
| | 36 | 392 | 0.120 |
| | 36 | 752 | 0.129 |
| | 43.5 | −328 | 0.090 |
| | 43.5 | 32 | 0.135 |
| Aluminum foil, 7 air spaces per 2.5 in........ | 0.2 | 100 | 0.025 |
| | | 351 | 0.038 |
| Ashes, wood.............................. | ...... | 32–212 | 0.041 |
| Asphalt.................................. | 132 | 68 | 0.43 |
| Bricks | | | |
|   Alumina (92–99% Al$_2$O$_3$ by weight) fused... | ...... | 801 | 1.8 |
|   Alumina (64–65% Al$_2$O$_3$ by weight)........ | ...... | 2399 | 2.7 |
|   (See also Bricks, fire clay)................ | 115 | 1472 | 0.62 |
| | 115 | 2012 | 0.63 |
|   Building brickwork...................... | ...... | 68 | 0.4 |
|   Chrome brick (32% Cr$_2$O$_3$ by weight)...... | 200 | 392 | 0.67 |
| | 200 | 1202 | 0.85 |
| | 200 | 2399 | 1.0 |
|   Diatomaceous earth, natural, across strata | 27.7 | 399 | 0.051 |
| | 27.7 | 1600 | 0.077 |
|   Diatomaceous, natural, parallel to strata | 27.7 | 399 | 0.081 |
| | 27.7 | 1600 | 0.106 |
|   Diatomaceous earth, molded and fired..... | 38 | 399 | 0.14 |
| | 38 | 1600 | 0.18 |
|   Diatomaceous earth and clay, molded and fired................................. | 42.3 | 399 | 0.14 |
| | 42.3 | 1600 | 0.19 |

*Tables 11.1–11.11 and 11.15, and Figures 11.1, 11.2, and 11.4–11.6 are reproduced with permission of McGraw-Hill Publishing Co. See Reference section for individual citations.

**Table 11.1.**[1]  *Cont'd.*

| Material | Apparent density $\rho$, lb/ft³ at room temperature | °F | k |
|---|---|---|---|
| Bricks: (*Continued*) | | | |
| Diatomaceous earth, high burn, large pores .............................. | 37 | 392 | 0.13 |
| | 37 | 1832 | 0.34 |
| Fire clay, Missouri....................... | ...... | 392 | 0.58 |
| | | 1112 | 0.85 |
| | | 1832 | 0.95 |
| | | 2552 | 1.02 |
| Kaolin insulating brick ................... | 27 | 932 | 0.15 |
| | 27 | 2102 | 0.26 |
| Kaolin insulating firebrick ................ | 19 | 392 | 0.050 |
| | 19 | 1400 | 0.113 |
| Magnesite (86.8% MgO, 6.3% $Fe_2O_3$, 3% CaO, 2.6% $SiO_2$ by weight)............. | 158 | 399 | 2.2 |
| | 158 | 1202 | 1.6 |
| | 158 | 2192 | 1.1 |
| Silicon carbide brick, recrystallized ........ | 129 | 1112 | 10.7 |
| | 129 | 1472 | 9.2 |
| | 129 | 1832 | 8.0 |
| | 129 | 2192 | 7.0 |
| | 129 | 2552 | 6.3 |
| Calcium carbonate, natural.................. | 162 | 86 | 1.3 |
| White marble........................... | ...... | ...... | 1.7 |
| Chalk.................................. | 96 | ...... | 0.4 |
| Calcium sulphate ($4H_2O$), artificial........... | 84.6. | 104 | 0.22 |
| Plaster, artificial......................... | 132 | 167 | 0.43 |
| Building.............................. | 77.9 | 77 | 0.25 |
| Cardboard, corrugated..................... | ...... | ...... | 0.037 |
| Celluloid................................. | 87.3 | 86 | 0.12 |
| Charcoal flakes............................ | 11.9 | 176 | 0.043 |
| | 15 | 176 | 0.051 |
| Clinker, granular.......................... | ...... | 32–1292 | 0.27 |
| Coke, petroleum........................... | ...... | 212 | 3.4 |
| | | 932 | 2.9 |
| Coke, powdered........................... | ...... | 32–212 | 0.11 |
| Concrete, cinder........................... | ...... | ...... | 0.20 |
| 1:4 dry................................. | ...... | ...... | 0.44 |
| Stone.................................. | ...... | ...... | 0.54 |

**Table 11.1.**[(1)]   *Cont'd.*

| Material | Apparent density $\rho$, lb/ft³ at room temperature | °F | $k$ |
|---|---|---|---|
| Cotton wool........................... | 5 | 86 | 0.024 |
| Cork board........................... | 10 | 86 | 0.025 |
| Cork, ground......................... | 9.4 | 86 | 0.025 |
|    Regranulated..................... | 8.1 | 86 | 0.026 |
| Diatomaceous earth powder, coarse ......... | 20.0 | 100 | 0.036 |
|  | 20.0 | 1600 | 0.082 |
| Molded pipe covering ................... | 26.0 | 399 | 0.051 |
|  | 26.0 | 1600 | 0.088 |
| 4 vol. calcined earth and 1 vol. cement, poured and fired ....·................. | 61.8 | 399 | 0.16 |
|  | 61.8 | 1600 | 0.23 |
| Dolomite.............................. | 167 | 122 | 1.0 |
| Ebonite.............................. | ...... | ...... | 0.10 |
| Enamel, silicate....................... | 38 | ...... | 0.5–0.75 |
| Felt, wool............................ | 20.6 | 86 | 0.03 |
| Fiber insulating board.................. | 14.8 | 70 | 0.028 |
| Fiber, red............................ | 80.5 | 68 | 0.27 |
|    With binder, baked................. | ...... | 68–207 | 0.097 |
| Glass................................ | ...... | ...... |  |
|    Boro-silicate type................. | 139 | 86–167 | 0.63 |
|    Soda glass........................ | ...... | ...... | 0.3–0.44 |
|    Window glass..................... | ...... | ...... | 0.3–0.61 |
| Granite.... ........................... | ...... | ...... | 1.0–2.3 |
| Graphite, dense, commercial............. | ...... | 32 | 86.7 |
|    Powdered, through 100 mesh........... | 30 | 104 | 0.104 |
| Gypsum, molded and dry................. | 78 | 68 | 0.25 |
| Hair, felt, perpendicular to fibers......... | 17 | 86 | 0.021 |
| Ice.................................. | 57.5 | 32 | 1.3 |
| Infusoria. |  |  |  |
| Kapok................................ |  |  |  |
| Lampblack............................ | 10 | 104 | 0.038 |
| Limestone (15.3 vol % $H_2O$)............... | 103 | 75 | 0.54 |
| Magnesia, powdered.................... | 49.7 | 117 | 0.35 |
| Magnesia, light carbonate............... | 19 | 70 | 0.04 |
| Magnesium oxide, compressed............ | 49.9 | 68 | 0.32 |
| Marble............................... | ...... | ...... | 1.2–1.7 |

**Table 11.1.**[1] *Cont'd.*

| Material | Apparent density $\rho$, lb/ft³ at room temperature | °F | k |
|---|---|---|---|
| Mica, perpendicular to planes............... | ...... | 122 | 0.25 |
| Mineral wool........................... | 9.4 | 86 | 0.0225 |
|  | 19.7 | 86 | 0.024 |
| Paper........................... | ...... | ...... | 0.075 |
| Paraffin wax........................... | ...... | 32 | 0.14 |
| Petroleum coke........................... | ...... | 212 | 3.4 |
|  | ...... | 932 | 2.9 |
| Porcelain........................... | ...... | 392 | 0.88 |
| Portland cement (see Concrete)............... | ...... | 194 | 0.17 |
| Pumice stone........................... | ...... | 70–151 | 0.14 |
| Rubber, hard........................... | 74.8 | 32 | 0.087 |
| Para........................... | ...... | 70 | 0.109 |
| Soft........................... | ...... | 70 | 0.075–0.092 |
| Sand, dry........................... | 94.6 | 68 | 0.19 |
| Sandstone........................... | 140 | 104 | 1.06 |
| Sawdust........................... | 12 | 70 | 0.03 |
| Silk........................... | 6.3 | ...... | 0.026 |
| Varnished........................... | ...... | 100 | 0.096 |
| Slag, blast furnace........................... | ...... | 75–261 | 0.064 |
| Slag wool........................... | 12 | 86 | 0.022 |
| Slate........................... | ...... | 201 | 0.86 |
| Snow........................... | 34.7 | 32 | 0.27 |
| Sulphur, monoclinic........................... | ...... | 212 | 0.09–0.097 |
| Rhombic........................... | ...... | 70 | 0.16 |
| Wallboard, insulating type................... | 14.8 | 70 | 0.028 |
| Wallboard, stiff pasteboard................... | 43 | 86 | 0.04 |
| Wood shavings........................... | 8.8 | 86 | 0.034 |
| Wood, across grain |  |  |  |
| Balsa........................... | 7–8 | 86 | 0.025–0.03 |
| Oak........................... | 51.5 | 59 | 0.12 |
| Maple........................... | 44.7 | 122 | 0.11 |
| Pine, white........................... | 34.0 | 59 | 0.087 |
| Teak........................... | 40.0 | 59 | 0.10 |
| White fir........................... | 28.1 | 140 | 0.062 |
| Wood, parallel to grain |  |  |  |
| Pine........................... | 34.4 | 70 | 0.20 |
| Wool, animal........................... | 6.9 | 86 | 0.021 |

**Table 11.2.**[2]  Thermal Conductivities, Specific Heats, and Specific Gravities of Metals and Alloys [$k$ = Btu/(hr) (ft$^2$) ($^\circ$F/ft)]

| Substance | Temp, °F | $k$ | Specific heat, Btu/(lb)(°F) | Specific gravity |
|---|---|---|---|---|
| Aluminum | 32 | 117 | 0.183 | 2.55–7.8 |
| Aluminum | 212 | 119 | 0.1824 | |
| Aluminum | 932 | 155 | 0.1872 | |
| Antimony | 32 | 10.6 | 0.0493 | |
| Antimony | 212 | 9.7 | 0.0508 | |
| Bismuth | 64 | 4.7 | 0.0294 | 9.8 |
| Bismuth | 212 | 3.9 | 0.0304 | |
| Brass (70-30) | 32 | 56 | 0.1315‡ | 8.4–8.7 |
| Brass | 212 | 60 | 0.1488‡ | |
| Brass | 752 | 67 | 0.2015‡ | |
| Copper | 32 | 224 | 0.1487 | 8.8–8.95 |
| Copper | 212 | 218 | 0.1712 | |
| Copper | 932 | 207 | 0.2634 | |
| Cadmium | 64 | 53.7 | 0.0550 | 8.65 |
| Cadmium | 212 | 52.2 | 0.0567 | |
| Gold | 64 | 169.0 | 0.030 | 19.25–19.35 |
| Gold | 212 | 170.8 | 0.031 | |
| Iron, cast | 32 | 32 | 0.1064 | 7.03–7.13 |
| Iron, cast | 212 | 30 | 0.1178 | |
| Iron, cast | 752 | 25 | 0.1519 | |
| Iron, wrought | 64 | 34.6 | See Iron | 7.6–7.9 |
| Iron, wrought | 212 | 27.6 | See Iron | |
| Lead | 32 | 20 | 0.0306 | 11.34 |
| Lead | 212 | 19 | 0.0315 | |
| Lead | 572 | 18 | 0.0335 | |
| Magnesium | 32–212 | 92 | 0.255 | 1.74 |
| Mercury | 32 | 4.8 | 0.0329 | 13.6 |
| Nickel | 32 | 36 | 0.1050 | 8.9 |
| Nickel | 212 | 34 | 0.1170 | |
| Nickel | 572 | 32 | 0.1408 | |
| Silver | 32 | 242 | 0.0557 | 10.4–10.6 |
| Silver | 212 | 238 | 0.0571 | |
| Steel | 32 | 26 | See Iron | 7.83 |
| Steel | 212 | 26 | See Iron | |
| Steel | 1112 | 21 | See Iron | |
| Tantalum | 64 | 32 | 0.0342 | 16.6 |
| Zinc | 32 | 65 | 0.0917 | 6.9–7.2 |
| Zinc | 212 | 64 | 0.0958 | |
| Zinc | 752 | 54 | 0.1082 | |

‡ Weighted value for copper and zinc.

**Table 11.3.**[3]   Thermal Conductivities of Liquids $[k = \text{Btu}/(\text{hr})(\text{ft}^2)(°F/\text{ft})]^a$

| Liquid | $t$, °F | $k$ | Liquid | $t$, °F | $k$ |
|---|---|---|---|---|---|
| Acetic acid 100% | 68 | 0.099 | Dichlorodifluoromethane | 20 | .057 |
| 50% | 68 | .20 | | 60 | .053 |
| Acetone | 86 | .102 | | 100 | .048 |
| | 167 | .095 | | 140 | .043 |
| Allyl alcohol | 77–86 | .104 | | 180 | .038 |
| Ammonia | 5–86 | .29 | Dichloroethane | 122 | .082 |
| Ammonia, aqueous 26% | 68 | .261 | Dichloromethane | 5 | .111 |
| | 140 | .29 | | 86 | .096 |
| Amyl acetate | 50 | .083 | | | |
| alcohol (*n*-) | 86 | .094 | Ethyl acetate | 68 | .101 |
| | 212 | .089 | alcohol 100% | 68 | .105 |
| (iso-) | 86 | .088 | 80% | 68 | .137 |
| | 167 | .087 | 60% | 68 | .176 |
| Aniline | 32–68 | .100 | 40% | 68 | .224 |
| | | | 20% | 68 | .281 |
| Benzene | 86 | .092 | 100% | 122 | .087 |
| | 140 | .087 | benzene | 86 | .086 |
| Bromobenzene | 86 | .074 | | 140 | .082 |
| | 212 | .070 | bromide | 68 | .070 |
| Butyl acetate (*n*-) | 77–86 | .085 | ether | 86 | .080 |
| alcohol (*n*-) | 86 | .097 | | 167 | .078 |
| | 167 | .095 | iodide | 104 | .064 |
| (iso-) | 50 | .091 | | 167 | .063 |
| Calcium chloride brine 30% | 86 | .32 | Ethylene glycol | 32 | .153 |
| 15% | 86 | .34 | Gasoline | 86 | .078 |
| Carbon disulfide | 86 | .093 | Glycerol 100% | 68 | .164 |
| | 167 | .088 | 80% | 68 | .189 |
| tetrachloride | 32 | .107 | 60% | 68 | .220 |
| | 154 | .094 | 40% | 68 | .259 |
| Chlorobenzene | 50 | .083 | 20% | 68 | .278 |
| Chloroform | 86 | .080 | 100% | 212 | .164 |
| Cymene (para-) | 86 | .078 | | | |
| | 140 | .079 | Heptane (*n*) | 86 | .081 |
| | | | | 140 | .079 |
| Decane (*n*-) | 86 | .085 | Hexane (*n*-) | 86 | 0.080 |
| | 140 | .083 | | 140 | .078 |

**Table 11.3.**[3]  *Cont'd.*

| Liquid | $t, °F$ | $k$ | Liquid | $t, °F$ | $k$ |
|---|---|---|---|---|---|
| Heptyl alcohol (*n*-) | 86 | .094 | Perchloroethylene | 122 | .092 |
| | 167 | .091 | Petroleum ether | 86 | .075 |
| Hexyl alcohol (*n*-) | 86 | .093 | | 167 | .073 |
| | 167 | .090 | Propyl alcohol (*n*-) | 86 | .099 |
| | | | | 167 | .095 |
| Kerosene | 68 | .086 | alcohol (iso-) | 86 | .091 |
| | 167 | .081 | | 140 | .090 |
| Lauric acid | 212 | .102 | | | |
| Mercury | 82 | 4.83 | Sodium | 212 | 49. |
| Methyl alcohol 100% | 68 | 0.124 | | 410 | 46. |
| 80% | 68 | .154 | Sodium chloride brine 25.0% | 86 | 0.33 |
| 60% | 68 | .190 | 12.5% | 86 | .34 |
| 40% | 68 | .234 | Stearic acid | 212 | .0786 |
| 20% | 68 | .284 | Sulfuric acid 90% | 86 | .21 |
| 100% | 122 | .114 | 60% | 86 | .25 |
| chloride | 5 | .111 | 30% | 86 | .30 |
| | 86 | .089 | Sulfur dioxide | 5 | .128 |
| | | | | 86 | .111 |
| Nitrobenzene | 86 | .095 | | | |
| | 212 | .088 | Toluene | 86 | .086 |
| Nitromethane | 86 | .125 | | 167 | .084 |
| | 140 | .120 | β-Trichloroethane | 122 | .077 |
| Nonane (*n*-) | 86 | .084 | Trichloroethylene | 122 | .080 |
| | 140 | .082 | Turpentine | 59 | .074 |
| Octane (*n*-) | 86 | .083 | | | |
| | 140 | .081 | Vaseline | 59 | .106 |
| Oils | | | Water | 32 | .343 |
| castor | 68 | .104 | | 100 | .363 |
| | 212 | .100 | | 200 | .393 |
| olive | 68 | .097 | | 300 | .395 |
| | 212 | .095 | | 420 | .376 |
| Oleic acid | 212 | .0925 | | 620 | .275 |
| Palmitic acid | 212 | .0835 | Xylene (ortho-) | 68 | .090 |
| Paraldehyde | 86 | .084 | (meta-) | 68 | .090 |
| | 212 | .078 | | | |
| Pentane (*n*-) | 86 | .078 | | | |
| | 167 | .074 | | | |

[a]A linear variation with temperature may be assumed. The extreme values given constitute the temperature over which the data are recommended.

**Table 11.4.**[4] Thermal Conductivities of Gases and Vapors
$[k = \text{Btu}/(\text{hr})\,(\text{ft}^2)\,(^\circ\text{F}/\text{ft})]^{a,\,b}$

| Substance | $t, ^\circ\text{F}$ | $k$ | Substance | $t, ^\circ\text{F}$ | $k$ |
|---|---|---|---|---|---|
| Acetone | 32 | 0.0057 | Chloroform | 32 | .0038 |
| | 115 | .0074 | | 115 | .0046 |
| | 212 | .0099 | | 212 | .0058 |
| | 363 | .0147 | | 363 | .0077 |
| Acetylene | −103 | .0068 | Cyclohexane | 216 | .0095 |
| | 32 | .0108 | | | |
| | 122 | .0140 | Dichlorodifluoromethane | 32 | .0048 |
| | 212 | .0172 | | 122 | .0064 |
| Air | −148 | .0095 | | 212 | .0080 |
| | 32 | .0140 | | 302 | .0097 |
| | 212 | .0183 | Ethane | −94 | .0066 |
| | 392 | .0226 | | −29 | .0086 |
| | 572 | .0265 | | 32 | .0106 |
| Ammonia | −76 | .0095 | | 212 | .0175 |
| | 32 | .0128 | Ethyl acetate | 115 | .0072 |
| | 122 | .0157 | | 212 | .0096 |
| | 212 | .0185 | | 363 | .0141 |
| Benzene | 32 | .0052 | alcohol | 68 | .0089 |
| | 115 | .0073 | | 212 | .0124 |
| | 212 | .0103 | chloride | 32 | .0055 |
| | 363 | .0152 | | 212 | .0095 |
| | 413 | .0176 | | 363 | .0135 |
| Butane (n-) | 32 | .0078 | | 413 | .0152 |
| | 212 | .0135 | ether | 32 | .0077 |
| (iso-) | 32 | .0080 | | 115 | .0099 |
| | 212 | .0139 | | 212 | .0131 |
| Carbon dioxide | −58 | .0068 | | 363 | .0189 |
| | 32 | .0085 | | 413 | .0209 |
| | 212 | .0133 | Ethylene | −96 | .0064 |
| | 392 | .0181 | | 32 | .0101 |
| | 572 | .0228 | | 122 | .0131 |
| disulfide | 32 | .0040 | | 212 | .0161 |
| | 45 | .0042 | Heptane (n-) | 392 | .0112 |
| monoxide | −312 | .0041 | | 212 | .0103 |
| | −294 | .0046 | Hexane (n-) | 32 | .0072 |
| | 32 | .0135 | | 68 | .0080 |
| tetrachloride | 115 | .0041 | Hexene | 32 | .0061 |
| | 212 | .0052 | | 212 | .0109 |
| | 363 | .0065 | Hydrogen | −148 | .065 |
| Chlorine | 32 | .0043 | | −58 | .083 |
| | 132 | .0049 | | 32 | .100 |
| | | | | 122 | .115 |
| | | | | 212 | .129 |
| | | | | 572 | .178 |

**Table 11.4.**[4] *Cont'd.*

| Substance | $t$, °F | $k$ | Substance | $t$, °F | $k$ |
|---|---|---|---|---|---|
| Hydrogen and carbon dioxide | 32 | | Methylene chloride | 32 | .0039 |
| 0% H$_2$ | ..... | .0083 | | 115 | .0049 |
| 20% | ..... | .0165 | | 212 | .0063 |
| 40% | ..... | .0270 | | 413 | .0095 |
| 60% | ..... | .0410 | | | |
| 80% | ..... | .0620 | Nitric oxide | −94 | .0103 |
| 100% | ..... | .100 | | 32 | .0138 |
| Hydrogen and nitrogen | 32 | | Nitrogen | −148 | .0095 |
| 0% H$_2$ | ..... | .0133 | | 32 | .0140 |
| 20% | ..... | .0212 | | 122 | .0160 |
| 40% | ..... | .0313 | | 212 | .0180 |
| 60% | ..... | .0438 | Nitrous oxide | −98 | .0067 |
| 80% | ..... | .0635 | | 32 | .0087 |
| Hydrogen and nitrous oxide | 32 | | | 212 | .0128 |
| 0% H$_2$ | ..... | .0092 | Oxygen | −148 | .0095 |
| 20% | ..... | .0170 | | −58 | .0119 |
| 40% | ..... | .0270 | | 32 | .0142 |
| 60% | ..... | .0410 | | 122 | .0164 |
| 80% | ..... | .0650 | | 212 | .0185 |
| Hydrogen sulfide | 32 | .0076 | | | |
| | | | Pentane (*n*-) | 32 | .0074 |
| Mercury | 392 | .0197 | | 68 | .0083 |
| Methane | −148 | .0100 | (iso-) | 32 | .0072 |
| | −58 | .0145 | | 212 | .0127 |
| | 32 | .0175 | Propane | 32 | .0087 |
| | 122 | .0215 | | 212 | .0151 |
| Methyl alcohol | 32 | .0083 | | | |
| | 212 | .0128 | Sulfur dioxide | 32 | .0050 |
| acetate | 32 | .0059 | | 212 | .0069 |
| | 68 | .0068 | | | |
| Methyl chloride | 32 | .0053 | Water vapor,[19] zero pressure | 32 | .0092 |
| | 115 | .0072 | absolute | 200 | .0133 |
| | 212 | .0094 | | 400 | .0184 |
| | 363 | .0130 | | 600 | .0238 |
| | 413 | .0148 | | 800 | .0292 |
| | | | | 1000 | .0347 |

*a*The extreme temperature values given constitute the experimental range. For extrapolation to other temperatures, it is suggested that the data be plotted as log $k$ vs. log $T$, or that use be made of the assumption that the ratio $c_p u/k$ is practically independent of temperature (or of pressure, within moderate limits).

*b*The section on water vapor is from Ref. 19.

**Table 11.5.**[5] $C_p/C_v$ Ratios of Specific Heats of Gases at 1-atm Pressure

| Compound | Formula | Temperature, °C | Ratio of specific heats, $(\gamma) = C_p/C_v$ |
|---|---|---|---|
| Acetaldehyde | $C_2H_4O$ | 30 | 1.14 |
| Acetic acid | $C_2H_4O_2$ | 136 | 1.15 |
| Acetylene | $C_2H_2$ | 15 | 1.26 |
| | | −71 | 1.31 |
| Air | − | 925 | 1.36 |
| | | 17 | 1.403 |
| | | −78 | 1.408 |
| | | −118 | 1.415 |
| Ammonia | $NH_3$ | 15 | 1.310 |
| Argon | A | 15 | 1.668 |
| | | 0–100 | 1.67 |
| Benzene | $C_6H_6$ | 90 | 1.10 |
| Bromine | $Br_2$ | 20–350 | 1.32 |
| Carbon dioxide | $CO_2$ | 15 | 1.304 |
| | | −75 | 1.37 |
| disulfide | $CS_2$ | 100 | 1.21 |
| monoxide | CO | 15 | 1.404 |
| | | −180 | 1.41 |
| Chlorine | $Cl_2$ | 15 | 1.355 |
| Chloroform | $CHCl_3$ | 100 | 1.15 |
| Cyanogen | $(CN)_2$ | 15 | 1.256 |
| Cyclohexane | $C_6H_{12}$ | 80 | 1.08 |
| Dichlorodifluoromethane | $CCl_2F_2$ | 25 | 1.139 |
| Ethane | $C_2H_6$ | 100 | 1.19 |
| | | 15 | 1.22 |
| | | −82 | 1.28 |
| Ethyl alcohol | $C_2H_6O$ | 90 | 1.13 |
| ether | $C_4H_{10}O$ | 35 | 1.08 |
| | | 80 | 1.086 |
| Ethylene | $C_2H_4$ | 100 | 1.18 |
| | | 15 | 1.255 |
| | | −91 | 1.35 |
| Helium | He | −180 | 1.660 |
| Hexane (n-) | $C_6H_{14}$ | 80 | 1.08 |
| Hydrogen | $H_2$ | 15 | 1.410 |
| | | −76 | 1.453 |
| | | −181 | 1.597 |
| bromide | HBr | 20 | 1.42 |
| chloride | HCl | 15 | 1.41 |
| | | 100 | 1.40 |
| cyanide | HCN | 65 | 1.31 |
| | | 140 | 1.28 |
| | | 210 | 1.24 |

**Table 11.5.**[5]  *Cont'd.*

| Compound | Formula | Temperature, °C | Ratio of specific heats, $(\gamma) = C_p/C_v$ |
|---|---|---|---|
| Hydrogen (*cont'd*) | | | |
| iodide | HI | 20–100 | 1.40 |
| sulfide | $H_2S$ | 15 | 1.32 |
| | | −45 | 1.30 |
| | | −57 | 1.29 |
| Iodine | $I_2$ | 185 | 1.30 |
| Isobutane | $C_4H_{10}$ | 15 | 1.11 |
| Krypton | Kr | 19 | 1.68 |
| Mercury | Hg | 360 | 1.67 |
| Methane | $CH_4$ | 600 | 1.113 |
| | | 300 | 1.16 |
| | | 15 | 1.31 |
| | | −80 | 1.34 |
| | | −115 | 1.41 |
| Methyl acetate | $C_3H_6O_2$ | 15 | 1.14 |
| alcohol | $CH_4O$ | 77 | 1.203 |
| ether | $C_2H_6O$ | 6–30 | 1.11 |
| Methylal | $C_3H_8O_2$ | 13 | 1.06 |
| | | 40 | 1.09 |
| Neon | Ne | 19 | 1.64 |
| Nitric oxide | NO | 15 | 1.400 |
| | | −45 | 1.39 |
| | | −80 | 1.38 |
| Nitrogen | $N_2$ | 15 | 1.404 |
| | | −181 | 1.47 |
| Nitrous oxide | $N_2O$ | 100 | 1.28 |
| | | 15 | 1.303 |
| | | −30 | 1.31 |
| | | −70 | 1.34 |
| Oxygen | $O_2$ | 15 | 1.401 |
| | | −76 | 1.415 |
| | | −181 | 1.45 |
| Pentane (*n-*) | $C_5H_{12}$ | 86 | 1.086 |
| Phosphorus | P | 300 | 1.17 |
| Potassium | K | 850 | 1.77 |
| Sodium | Na | 750–920 | 1.68 |
| Sulfur dioxide | $SO_2$ | 15 | 1.29 |
| Xenon | Xe | 19 | 1.66 |

**Table 11.6.**[6]   Specific Heats of Organic Liquids[a]

| Compound | Formula | Temperature, °C | Sp. ht., cal/g °C |
|---|---|---|---|
| Acetal | $C_6H_{14}O_2$ | 0 | 0.467 |
| | | 19–99 | .520 |
| Acetic acid | $C_2H_4O_2$ | 26–95 | .522 |
| Acetone | $C_3H_6O$ | 3–22.6 | .514 |
| | | 0 | .506 |
| | | 24.2–49.4 | .538 |
| Acetonitrile | $C_2H_3N$ | 21–76 | .541 |
| Acetophenone | $C_8H_8O$ | 20–196 | .450 |
| Acetyl chloride | $C_2H_3ClO$ | 0 | .339 |
| Allyl acetate | $C_5H_8O_2$ | 0 | .430 |
| alcohol | $C_3H_6O$ | 0 | .386 |
| | | 21–96 | .665 |
| benzoate | $C_{10}H_{10}O$ | 20 | .388 |
| butyrate | $C_7H_{12}O_2$ | 20 | .451 |
| chloride | $C_3H_5Cl$ | 0 | .313 |
| chloroacetate | $C_5H_7ClO_2$ | 20 | .396 |
| dichloroacetate | $C_5H_6Cl_2O_2$ | 20 | .332 |
| isobutyrate | $C_7H_{12}O_2$ | 20 | .448 |
| propionate | $C_6H_{10}O_2$ | 20 | .451 |
| trichloroacetate | $C_5H_5Cl_3O_2$ | 20 | .288 |
| valerate | $C_8H_{14}O_2$ | 20 | .451 |
| Aminobenzoic acid (o-) | $C_7H_7NO_2$ | M.P. | .435 |
| (m-) | $C_7H_7NO_2$ | M.P. | .435 |
| (p-) | $C_7H_7NO_2$ | M.P. | .444 |
| Amyl alcohol (d-primary) | $C_5H_{12}O$ | 22–125 | .711 |
| (t-) | $C_5H_{12}O$ | 20–99 | .753 |
| Amylene | $C_5H_{10}$ | 0 | .282 |
| Anethole | $C_{10}H_{12}O$ | 23–233 | .511 |
| | | 22.48 | .551 |
| | | 24.59 | .564 |
| | | 25.23 | .612 |
| Aniline | $C_6H_7N$ | 8–82 | .512 |
| Benzaldehyde | $C_7H_6O$ | 22–172 | .428 |
| Benzene | $C_6H_6$ | 6–60 | .419 |
| | | 10 | .340 |
| | | 65 | .482 |
| Benzonitrile | $C_7H_5N$ | 22–186 | .441 |
| Benzophenone (β-) | $C_{13}H_{10}O$ | 3–40 | .382 |
| | | 0 | .346 |
| Benzyl alcohol | $C_7H_8O$ | 20–100 | .511 |
| | | 22–200 | .540 |
| chloride | $C_7H_7Cl$ | 0 | .323 |
| ethylene | $C_9H_{10}$ | 0 | .393 |
| Bromobenzene | $C_6H_5Br$ | 0 | .215 |
| | | 20–100 | .231 |
| | | 16.9–65 | .239 |
| Bromochlorobenzene (o-) | $C_6H_4BrCl$ | 0 | .215 |
| (m-) | $C_6H_4BrCl$ | 0 | .212 |
| Bromoiodobenzene (o-) | $C_6H_4BrI$ | 0 | .153 |
| | | 5–100 | .160 |
| | | 3.2–64.6 | .157 |
| | | 1.8–34 | .157 |
| (m-) | $C_6H_4BrI$ | 0 | .152 |
| | | 5–100 | .158 |
| | | 3.2–64.5 | .156 |
| | | 1.7–34.1 | .154 |
| | | 1.7–36.2 | .149 |

**Table 11.6.**[6]  *Cont'd.*

| Compound | Formula | Temperature, °C | Sp. ht., cal. g, °C |
|---|---|---|---|
| Bromophenol | $C_6H_5BrO$ | 18–77 | .316 |
| Butane (*n-*) | $C_4H_{10}$ | 0 | .549 |
| Butyl alcohol (*n-*) | $C_4H_{10}O$ | 21–115 | .687 |
| | | 30 | .582 |
| | | −76.2 | .443 |
| | | −33.3 | .453 |
| | | 2.3 | .526 |
| | | 19.2 | .563 |
| chloride (*n-*) | $C_4H_9Cl$ | 20 | .451 |
| formate (*n-*) | $C_5H_{10}O_2$ | 20 | .459 |
| propionate | $C_7H_{14}O_2$ | 20 | .459 |
| valerate | $C_9H_{18}O_2$ | 20 | .459 |
| Butyric acid (*n-*) | $C_4H_8O_2$ | 0 | .444 |
| | | 40 | .501 |
| | | 20–100 | .515 |
| Butyronitrile (*n-*) | $C_4H_7N$ | 21–113 | .547 |
| Caproic acid | $C_6H_{12}O_2$ | 29–105 | .531 |
| Capronitrile | $C_6H_{11}N$ | 18–156 | .541 |
| Carbon tetrachloride | $CCl_4$ | 0 | .198 |
| | | 20 | .201 |
| | | 30 | .200 |
| Carvacrol | $C_{10}H_{14}O$ | 24–233 | .575 |
| Chloral | $C_2HCl_3O$ | 17–53 | .250 |
| hydrate | $C_2H_5Cl_3O_2$ | 55–88 | .470 |
| Chlorobenzene | $C_6H_5Cl$ | 0 | .273 |
| | | 10 | .298 |
| | | 20 | .308 |
| Chlorobenzoic acid (*o-*) | $C_7H_5ClO_2$ | 0 | 0.390 |
| (*m-*) | $C_7H_5ClO_2$ | 0 | .265 |
| (*p-*) | $C_7H_5ClO_2$ | M.P. | .545 |
| Chloroform | $CHCl_3$ | 0 | .232 |
| | | 15 | .226 |
| | | 30 | .234 |
| Chlorophenol | $C_6H_5ClO$ | 0–20 | .399 |
| Chlorotoluene | $C_7H_7Cl$ | 0 | .315 |
| Cresol (*o-*) | $C_7H_8O$ | 0–20 | .497 |
| (*m-*) | $C_7H_8O$ | 21–197 | .551 |
| | | 0–20 | .477 |
| Cresyl methyl ether (*p-*) | $C_8H_{10}O$ | 0 | .404 |
| Crotonic acid | $C_4H_6O_2$ | 71.4 | .500 |
| Cyclohexanol | $C_6H_{12}O$ | 15–18 | .416 |
| Cyclohexanone | $C_6H_{10}O$ | 15–18 | .431 |
| *o-*Cymene | $C_{10}H_{14}$ | 0 | .398 |
| Decahydronaphthalene (cis-) | $C_{10}H_{18}$ | 15–18 | .393 |
| Decane | $C_{10}H_{22}$ b.p. = 159 | 21–154 | .588 |
| | $C_{10}H_{22}$ b.p. = 162 | 0–50 | .493 |
| | $C_{10}H_{22}$ b.p. = 172 | 0–50 | .500 |
| Decylene (γ-) | $C_{10}H_{20}$ | 0–50 | .467 |
| Diallyl oxalate | $C_8H_{10}O_4$ | 20 | .424 |
| succinate | $C_{10}H_{14}O_4$ | 20 | .450 |
| Diamylene | $C_{10}H_{20}$ | 20–130 | .543 |
| Dibromobenzene (*o-*) | $C_6H_4Br_2$ | 0 | .179 |
| (*m-*) | $C_6H_4Br_2$ | 0 | .175 |
| Dibutyl oxalate | $C_{10}H_{18}O_4$ | 20 | .439 |
| Dichlorodifluormethane | $CCl_2F_2$ | −43 | .21 |
| | | 21–106 | .349 |

**Table 11.6.**[(6)] *Cont'd.*

| Compound | Formula | Temperature, °C | Sp. ht., cal. g, °C |
|---|---|---|---|
| Dichloroacetic acid | $C_2H_2Cl_2O_2$ | 21–196 | .348 |
| Dichlorobenzene (o-) | $C_6H_4Cl_2$ | 0 | .269 |
| (m-) | $C_6H_4Cl_2$ | 0 | .269 |
| (p-) | $C_6H_4Cl_2$ | 53–99 | .297 |
| Diethylamine | $C_4H_{11}N$ | 22.5 | .516 |
| Diethyl carbonate | $C_5H_{10}O_3$ | 0 | .245 |
| | | 20–100 | .462 |
| ether (see Ether) | $C_4H_{10}O$ | 20.2–123 | .473 |
| ketone | $C_5H_{10}O$ | 20–98.5 | .555 |
| malate | $C_8H_{14}O_5$ | 24–186 | .473 |
| malonate | $C_7H_{12}O_4$ | 20 | .431 |
| oxalate | $C_6H_{10}O_4$ | 20 | .431 |
| succinate | $C_8H_{14}O_4$ | 20 | .450 |
| Dihydronaphthalene | $C_{10}H_{10}$ | 18–28 | .345 |
| Di-iodobenzene (m-) | $C_6H_4I_2$ | 34.2–99.6 | .139 |
| Di-isoamyl | $C_{10}H_{22}$ | 21.5–155 | .588 |
| oxalate | $C_{12}H_{22}O_4$ | 20 | .447 |
| Di-isobutylamine | $C_8H_{19}N$ | 22–130 | .569 |
| | | 0–20 | .416 |
| Dimethyl aniline | $C_8H_{11}N$ | 0 | .403 |
| naphthalene (β-) | $C_{12}H_{12}$ | 0 | .392 |
| pyrone | $C_7H_8O_2$ | 166 | .547 |
| Dinitrobenzene (m-) | $C_6H_4N_2O_4$ | M.P. | .404 |
| Diphenylamine | $C_{12}H_{11}N$ | 54 | .437 |
| | | 56 | .441 |
| | | 66 | .480 |
| Dipropyl ketone | $C_7H_{14}O$ | 20–140 | .550 |
| malonate | $C_9H_{16}O_4$ | 20 | .431 |
| oxalate (n-) | $C_8H_{14}O_4$ | 20 | .431 |
| succinate | $C_{10}H_{18}O_4$ | 20 | .450 |
| Dodecane | $C_{12}H_{26}$ | 14–20 | .505 |
| | | 0–50 | .498 |
| Dodecylene | $C_{12}H_{24}$ | 0–50 | .455 |
| Ether | $C_4H_{10}O$ | −100 | .511 |
| | | −50 | .515 |
| | | −5 | .525 |
| | | 0 | .521 |
| | | +30 | .545 |
| | | 80 | .687 |
| | | 120 | .800 |
| | | 140 | .819 |
| | | 180 | 1.037 |
| Ethyl acetate | $C_4H_8O_2$ | 20 | 0.457 |
| acetoacetate | $C_6H_{10}O_3$ | 0 | .428 |
| | | 20–100 | .475 |
| alcohol | $C_2H_6O$ (100%) | −20 | .505 |
| | | 0–98 | .680 |
| benzene | $C_8H_{10}$ | 0 | .392 |
| | | 30 | .407 |
| benzoate | $C_9H_{10}O_2$ | 20 | .387 |
| bromide | $C_2H_5Br$ | −100 | 0.194 |
| | | −20 | .206 |
| | | 5–10 | .216 |
| | | 10–15 | .213 |
| | | 15–20 | .214 |
| butyrate | $C_6H_{12}O_2$ | 20 | .457 |

**Table 11.6.**[(6)]  *Cont'd.*

| Compound | Formula | Temperature, °C | Sp. ht., cal. g, °C |
|---|---|---|---|
| Ethyl acetate (*cont'd*) | | | |
| chloride | $C_2H_5Cl$ | −28 to +4 | .426 |
| | | 0 | .367 |
| chloroacetate | $C_4H_7ClO_2$ | 9−138 | .416 |
| | | 20 | .397 |
| cresyl ether (*p*-) | $C_9H_{12}O$ | 0 | .427 |
| dichloroacetate | $C_4H_6Cl_2O_2$ | 20 | .328 |
| ether | $C_4H_{10}O$ | 0 | .521 |
| formate | $C_3H_6O_2$ | 14−49 | .508 |
| | | −20 to +14 | .454 |
| iodide | $C_2H_5I$ | −30 | .156 |
| | | 0 | .161 |
| | | 60 | .171 |
| isobutyrate | $C_6H_{12}O_2$ | 20 | .457 |
| propionate | $C_5H_{10}O_2$ | 20 | .457 |
| silicate | $C_8H_{20}SiO_4$ | 15−98 | .424 |
| sulfide | $C_4H_{10}S$ | 0 | .468 |
| | | 5−10 | .470 |
| | | 10−15 | .473 |
| | | 20−70 | .477 |
| trichloroacetate | $C_4H_5Cl_3O_2$ | 10−81 | .294 |
| | | 9−139 | .305 |
| | | 20 | .284 |
| valerate | $C_7H_{14}O_2$ | 20 | .457 |
| Ethylene bromide | $C_2H_4Br_2$ | 8−95 | .182 |
| | | 13−106 | .175 |
| | | 20 | .173 |
| chloride | $C_2H_4Cl_2$ | −30 | .278 |
| | | +20 | .299 |
| | | 30 | .304 |
| | | 50 | .313 |
| | | 60 | .318 |
| dichloroacetate | $C_6H_6Cl_4O_4$ | 0 | .321 |
| glycol | $C_2H_6O_2$ | −11.1 | .535 |
| | | 0 | .542 |
| | | +2.5 | .550 |
| | | 5.1 | .554 |
| | | 14.9 | .569 |
| | | 19.9 | .573 |
| Formamide | $CH_3NO$ | 19 | .549 |
| Formic acid | $CH_2O_2$ | 0 | .436 |
| | | 15.5 | .509 |
| | | 20−100 | .524 |
| Furfural | $C_5H_4O_2$ | 0 | .367 |
| | | 20−100 | .416 |
| Glycerol | $C_3H_8O_3$ | 15−50 | .576 |
| Glycol (ethylene) | $C_2H_6O_2$ | (see ethylene glycol) | |
| Heptaldehyde | $C_7H_{14}O$ | 0 | .364 |
| Heptane (*n*-) | $C_7H_{16}$ | 0−50 | .507 |
| | | 20 | .490 |
| | | 30 | .518 |
| Heptylene | $C_7H_{14}$ | 0−50 | .486 |
| Heptylic acid | $C_7H_{14}O_2$ | 9 | .556 |
| Hexadecane (*n*-) | $C_{16}H_{38}$ | 0−50 | .496 |
| Hexadiene (1,5-) | $C_6H_{10}$ | 0 | .405 |
| Hexahydrocresol (*o*-) | $C_7H_{14}O$ | 15−18 | .416 |
| (*m*-) | $C_7H_{14}O$ | 15−18 | .420 |
| (*p*-) | $C_7H_{14}O$ | 15−18 | .421 |

**Table 11.6.**[6]  *Cont'd.*

| Compound | Formula | Temperature, °C | Sp. ht., cal. g, °C |
|---|---|---|---|
| Hexane (*n*-) | $C_6H_{14}$ | 0–50 | .527 |
| | | 20–100 | .600 |
| Hexylene | $C_6H_{12}$ | 0–50 | .504 |
| Isoamyl acetate | $C_7H_{14}O_2$ | 20 | .459 |
| alcohol | $C_5H_{12}O$ | 0 | .502 |
| | | 20 | .535 |
| | | 30 | .570 |
| | | 47.9 | .662 |
| | | 10–117 | .693 |
| | | 21–130 | .695 |
| | | 75.5 | .688 |
| amine | $C_5H_{13}N$ | 22–91 | .614 |
| butyrate | $C_9H_{18}O_2$ | 20 | .459 |
| formate | $C_6H_{12}O_2$ | 16–65 | .509 |
| isobutyrate | $C_9H_{18}O_2$ | 20 | .459 |
| propionate | $C_8H_{16}O_2$ | 20 | .459 |
| succinate | $C_{14}H_{26}O_4$ | 0 | .449 |
| valerate | $C_{10}H_{20}O_2$ | 20 | .459 |
| Isobutane | $C_4H_{10}$ | 0 | .549 |
| Isobutyl acetate | $C_6H_{12}O_2$ | 20 | .459 |
| alcohol | $C_4H_{10}O$ | 21–109 | .716 |
| | | 30 | .603 |
| butyrate | $C_8H_{16}O_2$ | 20 | 0.459 |
| succinate | $C_{12}H_{22}O_4$ | 0 | .442 |
| Isobutyric acid | $C_4H_8O_2$ | 20 | .450 |
| Isoheptane | $C_7H_{16}$ | 0–50 | .501 |
| Isopentane | $C_5H_{12}$ | 0 | .512 |
| | | 8 | .527 |
| Isovaleric acid | $C_5H_{10}O_2$ | 20 | .463 |
| | | 23–93 | .590 |
| Lauric acid | $C_{12}H_{24}O_2$ | 40–100 | 0.572 |
| | | 57 | .515 |
| Mesitylene | $C_9H_{12}$ | 0 | .393 |
| Mesityl oxide | $C_6H_{10}O$ | 21–121 | .521 |
| Methyl acetate | $C_3H_6O_2$ | 15 | .468 |
| Methylal | $C_3H_8O_2$ | 15–41 | .521 |
| Methyl alcohol | $CH_4O$ | 5–10 | .590 |
| | | 15–20 | .601 |
| aniline | $C_7H_9N$ | 20–197 | .512 |
| benzoate | $C_8H_8O_2$ | 0 | .363 |
| butyl ketone | $C_6H_{12}O$ | 21–127 | .553 |
| butyrate (*n*-) | $C_5H_{10}O_2$ | 20 | .459 |
| chloroacetate | $C_3H_5ClO_2$ | 20 | .382 |
| cyclohexanone (*o*-) | $C_7H_{12}O$ | 15–18 | .436 |
| (*m*-) | $C_7H_{12}O$ | 15–18 | .441 |
| (*p*-) | $C_7H_{12}O$ | 15–18 | .441 |
| dichloroacetate | $C_3H_4Cl_2O_2$ | 20 | .311 |
| ethylketone | $C_4H_8O$ | 20–78 | .549 |
| ethylketoxime | $C_4H_9NO$ | 21.8–151.5 | .650 |
| formate | $C_2H_4O_2$ | 13–29 | .516 |
| hexyl ketone | $C_8H_{16}O$ | 22–168 | .552 |
| isobutyl ketone | $C_6H_{12}O$ | 20 | .459 |
| isopropyl ketone | $C_5H_{10}O$ | 20–91 | .525 |
| propionate | $C_4H_8O_2$ | 20 | .459 |
| trichloroacetate | $C_3H_3Cl_3O_2$ | 20 | .267 |
| valerate | $C_6H_{12}O_2$ | 20 | .459 |
| Methylene chloride | $CH_2Cl_2$ | 15–40 | .288 |

**Table 11.6.**[6]  *Cont'd.*

| Compound | Formula | Temperature, °C | Sp. ht., cal. g, °C |
|---|---|---|---|
| Myristic acid | $C_{14}H_{28}O_2$ | 56−100 | .539 |
| Naphthalene | $C_{10}H_8$ | 87.5 | .402 |
| Naphthylamine (α-) | $C_{10}H_9N$ | 53.2 | .475 |
|  |  | 94.2 | .476 |
| Nitrobenzene | $C_6H_5NO_2$ | 10 | .358 |
|  |  | 30 | .339 |
|  |  | 50 | .330 |
|  |  | 70 | .330 |
|  |  | 90 | .343 |
|  |  | 120 | .394 |
| Nitrobenzoic acid (p-) | $C_7H_5NO_4$ | M.P. | .449 |
| Nitromethane | $CH_3NO_2$ | 17 | .412 |
| Nitronaphthalene (α-) | $C_{10}H_7NO_2$ | 58.6 | .365 |
|  |  | 61.4 | .378 |
|  |  | 94.3 | .390 |
| Nonane | $C_9H_{20}$ | 0−50 | .503 |
| Nonylene | $C_9H_{18}$ | 0−50 | .485 |
| Octane (n-) | $C_8H_{18}$ | 0−50 | .505 |
|  |  | 20−123 | .578 |
| Octylene | $C_8H_{16}$ | 0−50 | .486 |
| Palmitic acid | $C_{16}H_{32}O_2$ | 65−104 | .653 |
| Paraldehyde | $C_6H_{12}O_3$ | 0 | .436 |
| Pentadecane | $C_{15}H_{32}$ | 0−50 | .497 |
| Pentadecylene | $C_{15}H_{30}$ | 0−50 | .471 |
| Phenetole | $C_8H_{10}O$ | 20 | .446 |
| Phenyl methyl ether | $C_7H_8O$ | 0 | .405 |
|  |  | 20−152 | .483 |
| Picoline (α-) | $C_6H_7N$ | 22−124 | .434 |
| Piperidine | $C_5H_{11}N$ | 20−98 | .523 |
| Propane | $C_3H_8$ | 0 | .576 |
| Propionaldehyde | $C_3H_6O$ | 0 | .522 |
| Propionic acid | $C_3H_6O_2$ | 0 | .444 |
|  |  | 20−137 | .560 |
| Propionitrile | $C_3H_5N$ | 0 | .508 |
|  |  | 19−95 | .538 |
| Propyl acetate (n-) | $C_5H_{10}O_2$ | 20 | .459 |
|   benzene | $C_9H_{12}$ | 0 | .400 |
|   benzoate | $C_{10}H_{12}O_2$ | 20 | .398 |
|   butyrate | $C_7H_{14}O_2$ | 20 | .459 |
|   chloroacetate | $C_5H_9ClO_2$ | 20 | .414 |
|   dichloroacetate | $C_5H_8Cl_2O_2$ | 20 | .341 |
|   formate (n-) | $C_4H_8O_2$ | 20 | .459 |
|   isobutyrate | $C_7H_{14}O_2$ | 20 | .459 |
|   phenyl ether | $C_9H_{12}O$ | 0 | .429 |
|   propionate | $C_6H_{12}O_2$ | 20 | .459 |
|   trichloroacetate | $C_5H_7Cl_3O_2$ | 20 | .297 |
|   valerate | $C_8H_{16}O_2$ | 20 | .459 |
| Pseudocumene | $C_9H_{12}$ | 20 | .414 |
| Pyridine | $C_5H_5N$ | 20 | .405 |
|  |  | 21−108 | .431 |
|  |  | 0−20 | .395 |
| Quinoline | $C_9H_7N$ | 0−20 | .352 |
| Salicylaldehyde | $C_7H_6O_2$ | 18 | .382 |
| Salol | $C_{13}H_{10}O_3$ | 44.1 | .391 |
| Stearic acid | $C_{18}H_{36}O_2$ | 75−137 | .550 |

**Table 11.6.**[6] *Cont'd.*

| Compound | Formula | Temperature, °C | Sp. ht., cal. g, °C |
|---|---|---|---|
| Tetrachloroethane | $C_2H_2Cl_4$ | 20 | .268 |
| Tetrachloroethylene | $C_2Cl_4$ | 20 | .216 |
| | | 24 | .211 |
| Tetradecane | $C_{14}H_{30}$ | 0–50 | 0.497 |
| Thymol (*m-*) | $C_{10}H_{14}O$ | 50 | .566 |
| | | 10 | .364 |
| Toluene | $C_7H_8$ | 85 | .534 |
| | | 12–99 | .440 |
| Toluidine (*o-*) | $C_7H_9N$ | 0 | .454 |
| | | 22–195 | .598 |
| | | 40.5 | .498 |
| (*p-*) | $C_7H_9N$ | 43 | .524 |
| | | 58 | .634 |
| | | 94 | .533 |
| Trichloroethane | $C_2H_3Cl_3$ | 20 | 0.266 |
| Trichloroethylene | $C_2Hcl_3$ | 20 | .223 |
| Tridecane | $C_{13}H_{28}$ | 0–50 | .499 |
| Tridecylene | $C_{13}H_{26}$ | 0–50 | .457 |
| Trinitrotoluene (2,4,6-) | $C_7H_5N_3O_6$ | . . . . . . . . | .335 |
| Undecane | $C_{11}H_{24}$ | 0–50 | .501 |
| Undecylene | $C_{11}H_{22}$ | 0–50 | .482 |
| Valeronitrile | $C_5H_9N$ | 23–121 | .520 |
| Xylene (*o-*) | $C_8H_{10}$ | 30 | .411 |
| (*m-*) | $C_8H_{10}$ | 0 | .383 |
| | | 9–40 | .400 |
| | | 16–35 | .387 |
| | | 30 | .401 |
| (*p-*) | $C_8H_{10}$ | 0 | .383 |
| | | 30 | .397 |
| | | 40.8 | .428 |
| dibromide (*o-*) | $C_8H_8Br_2$ | 15–40 | .183 |
| (*m-*) | $C_8H_8Br_2$ | 15–40 | .184 |
| (*p-*) | $C_8H_8Br_2$ | 15–40 | .180 |
| dichloride (*o-*) | $C_8H_8Cl_2$ | 15–40 | .283 |
| (*m-*) | $C_8H_8Cl_2$ | 15–40 | .295 |
| (*p-*) | $C_8H_8Cl_2$ | 15–40 | .282 |
| tetrachloride (*o-*) | $C_8H_6Cl_4$ | 15–40 | .240 |
| (*p-*) | $C_8H_6Cl_4$ | 15–40 | .242 |
| Xylyl ethyl ether (2,4-) | $C_{10}H_{14}O$ | 0 | .417 |

[a]See Figure 11.4 for specific heats of these liquids as a function of temperature.

**Table 11.7.**[7]   Specific Heats of Miscellaneous Materials

Specific Heats of Miscellaneous Liquids and Solids

| Material | Specific heat, cal/g °C |
|---|---|
| Alumina | 0.2 (100°C); 0.274 (1500°C) |
| Aluminum | 0.186 (100°C) |
| Asbestos | 0.25 |
| Asphalt | 0.22 |
| Bakelite | 0.3 to 0.4 |
| Brickwork | About 0.2 |
| Carbon | 0.168 (26° to 76°C) |
| | 0.314 (40° to 892°C) |
| | 0.387 (56° to 1450°C) |
| (gas retort) | 0.204 |
| (also see Graphite) | |
| Cellulose | 0.32 |
| Cement, Portland | 0.186 |
| Charcoal (wood) | 0.242 |
| Chrome brick | 0.17 |
| Clay | 0.224 |
| Coal | 0.26 to 0.37 |
| tar oils | 0.34 (15° to 90°C) |
| Coal tars | 0.35 (40°C); 0.45 (200°C) |
| Coke | 0.265 (21° to 400°C) |
| | 0.359 (21° to 800°C) |
| | 0.403 (21° to 1300°C) |
| Concrete | 0.156 (70° to 312°F); 0.219 (72° to 1472°F) |
| Cryolite | 0.253 (16° to 55°C) |
| Diamond | 0.147 |
| Fireclay brick | 0.198 (100°C); 0.298 (1500°C) |
| Fluorspar | 0.21 (30°C) |
| Gasoline | 0.53 |
| Glass (crown) | 0.16 to 0.20 |
| (flint) | 0.117 |
| (pyrex) | 0.20 |
| (silicate) | 0.188 to 0.204 (0 to 100°C) |
| | 0.24 to 0.26 (0 to 700°C) |
| wool | 0.157 |
| Granite | 0.20 (20° to 100°C) |
| Graphite | 0.165 (26° to 76°C); 0.390 (56° to 1450°C) |
| Gypsum | 0.259 (16° to 46°C) |
| Kerosene | 0.47 |
| Limestone | 0.217 |
| Litharge | 0.055 |
| Magnesia | 0.234 (100°C); 0.188 (1500°C) |
| Magnesite brick | 0.222 (100°C); 0.195 (1500°C) |
| Marble | 0.21 (18°C) |
| Pyrites (copper) | 0.131 (19° to 50°C) |
| (iron) | 0.136 (15° to 98°C) |
| Quartz | 0.17 (0°C); 0.28 (350°C) |
| Sand | 0.191 |
| Silica | 0.316 |
| Steel | 0.12 |
| Stone | About 0.2 |
| Turpentine | 0.42 (18°C) |
| Wood (oak) | 0.570 |
| Most woods vary between | 0.45 and 0.65 |

**Table 11.7.**[7] *Cont'd.*

Oils (animal, vegetable, mineral oils)

$$C_p = \frac{A}{(d_4^{15})^{1/2}} + B(t - 15), \text{ cal/g-°C}$$

| Oils | A | B |
|---|---|---|
| Castor | 0.500 | 0.0007 |
| Citron | (0.438 at 54°C) | |
| Fatty drying | 0.440 | 0.0007 |
|   nondrying | 0.450 | 0.0007 |
|   semidrying | 0.445 | 0.0007 |
|   oils (except castor) | 0.450 | 0.0007 |
| Naphthene base | 0.405 | 0.0009 |
| Olive | (0.47 at 7°C) | |
| Paraffin base | 0.425 | 0.0009 |
| Petroleum oils | 0.415 | 0.0009 |

| Porcelain | Average sp. heat between 20°C and | | | |
|---|---|---|---|---|
| | 100°C | 300°C | 500°C | 1100°C |
| Fired Berlin | 0.189 | 0.203 | 0.222 | 0.337 |
| Green Berlin | .185 | .197 | .228 | |
| Fired Berlin (glaze) | .179 | .189 | .199 | .245 |
| Green Berlin (glaze) | .170 | .183 | .208 | |
| Fired earthenware | .186 | .203 | .223 | .324 |
| Green earthenware | .181 | .192 | .215 | |

| | |
|---|---|
| Pyrex | 0.20 |
| Pyroxylin plastics | 0.34 to 0.38 |
| Rubber (vulcanized) | 0.415 |
| Silica brick | 0.202 (100°C); 0.195 (1500°C) |
| Silicon carbide brick | 0.202 (100°C) |
| Silk | 0.33 |
| Stoneware (common) | 0.185 to 0.191 (20° to 100°C) |
| Wool | 0.325 |
| Zirconium oxide | 0.11 (100°C); 0.179 (1500°C) |

**Table 11.8.**[8]  Specific Gravities and Molecular Weights of Liquids[a]

| Compound | Mol. wt. | s* | Compound | Mol. wt. | s* |
|---|---|---|---|---|---|
| Acetaldehyde | 44.1 | 0.78 | Ethyl iodide | 155.9 | 1.93 |
| Acetic acid, 100% | 60.1 | 1.05 | Ethyl glycol | 88.1 | 1.04 |
| Acetic acid, 70% | ...... | 1.07 | Formic acid | 46.0 | 1.22 |
| Acetic anhydride | 102.1 | 1.08 | Glycerol, 100% | 92.1 | 1.26 |
| Acetone | 58.1 | 0.79 | Glycerol, 50% | ...... | 1.13 |
| Allyl alcohol | 58.1 | 0.86 | n-Heptane | 100.2 | 0.68 |
| Ammonia, 100% | 17.0 | 0.61 | n-Hexane | 86.1 | 0.66 |
| Ammonia, 26% | ...... | 0.91 | Isopropyl alcohol | 60.1 | 0.79 |
| Amyl acetate | 130.2 | 0.88 | Mercury | 200.6 | 13.55 |
| Amyl alcohol | 88.2 | 0.81 | Methanol, 100% | 32.5 | 0.79 |
| Aniline | 93.1 | 1.02 | Methanol, 90% | ...... | 0.82 |
| Anisole | 108.1 | 0.99 | Methanol, 40% | ...... | 0.94 |
| Arsenic trichloride | 181.3 | 2.16 | Methyl acetate | 74.9 | 0.93 |
| Benzene | 78.1 | 0.88 | Methyl chloride | 50.5 | 0.92 |
| Brine, CaCl₂ 25% | ...... | 1.23 | Methyl ethyl ketone | 72.1 | 0.81 |
| Brine, NaCl 25% | ...... | 1.19 | Naphthalene | 128.1 | 1.14 |
| Bromotoluene, ortho | 171.0 | 1.42 | Nitric acid, 95% | ...... | 1.50 |
| Bromotoluene, meta | 171.0 | 1.41 | Nitric acid, 60% | ...... | 1.38 |
| Bromotoluene, para | 171.0 | 1.39 | Nitrobenzene | 123.1 | 1.20 |
| n-Butane | 58.1 | 0.60 | Nitrotoluene, ortho | 137.1 | 1.16 |
| i-Butane | 58.1 | 0.60 | Nitrotoluene, meta | 137.1 | 1.16 |
| Butyl acetate | 116.2 | 0.88 | Nitrotoluene, para | 137.1 | 1.29 |
| n-Butyl alcohol | 74.1 | 0.81 | n-Octane | 114.2 | 0.70 |
| i-Butyl alcohol | 74.1 | 0.82 | Octyl alcohol | 130.23 | 0.82 |
| n-Butyric acid | 88.1 | 0.96 | Pentachloroethane | 202.3 | 1.67 |
| i-Butyric acid | 88.1 | 0.96 | n-Pentane | 72.1 | 0.63 |
| Carbon dioxide | 44.0 | 1.29 | Phenol | 94.1 | 1.07 |
| Carbon disulfide | 76.1 | 1.26 | Phosphorus tribromide | 270.8 | 2.85 |
| Carbon tetrachloride | 153.8 | 1.60 | Phosphorus trichloride | 137.4 | 1.57 |
| Chlorobenzene | 112.6 | 1.11 | Propane | 44.1 | 0.59 |
| Chloroform | 119.4 | 1.49 | Propionic acid | 74.1 | 0.99 |
| Chlorosulfonic acid | 116.5 | 1.77 | n-Propyl alcohol | 60.1 | 0.80 |
| Chlorotoluene, ortho | 126.6 | 1.08 | n-Propyl bromide | 123.0 | 1.35 |
| Chlorotoluene, meta | 126.6 | 1.07 | n-Propyl chloride | 78.5 | 0.89 |
| Chlorotoluene, para | 126.6 | 1.07 | n-Propyl iodide | 170.0 | 1.75 |
| Cresol, meta | 108.1 | 1.03 | Sodium | 23.0 | 0.97 |
| Cyclohexanol | 100.2 | 0.96 | Sodium hydroxide, 50% | ...... | 1.53 |
| Dibromo methane | 187.9 | 2.09 | Stannic chloride | 260.5 | 2.23 |
| Dichloro ethane | 99.0 | 1.17 | Sulfur dicxide | 64.1 | 1.38 |
| Dichloro methane | 88.9 | 1.34 | Sulfuric acid, 100% | 98.1 | 1.83 |
| Diethyl oxalate | 146.1 | 1.08 | Sulfuric acid, 98% | ...... | 1.84 |
| Dimethyl oxalate | 118.1 | 1.42 | Sulfuric acid, 60% | ...... | 1.50 |
| Diphenyl | 154.2 | 0.99 | Sulfuryl chloride | 135.0 | 1.67 |
| Dipropyl oxalate | 174.1 | 1.02 | Tetra chloroethane | 167.9 | 1.60 |
| Ethyl acetate | 88.1 | 0.90 | Tetra chloroethylene | 165.9 | 1.63 |
| Ethyl alcohol, 100% | 46.1 | 0.79 | Titanium tetrachloride | 189.7 | 1.73 |
| Ethyl alcohol, 95% | ...... | 0.81 | Toluene | 92.1 | 0.87 |
| Ethyl alcohol, 40% | ...... | 0.94 | Trichloroethylene | 131.4 | 1.46 |
| Ethyl benzene | 106.1 | 0.87 | Vinyl acetate | 86.1 | 0.93 |
| Ethyl bromide | 108.9 | 1.43 | Water | 18.0 | 1.0 |
| Ethyl chloride | 64.5 | 0.92 | Xylene, ortho | 106.1 | 0.87 |
| Ethyl ether | 74.1 | 0.71 | Xylene, meta | 106.1 | 0.86 |
| Ethyl formate | 74.1 | 0.92 | Xylene, para | 106.1 | 0.86 |

[a]At approximately 68°F.  These values will be satisfactory for most engineering problems.

## Table 11.9.[9] Heats of Vaporization of Organic Compounds[a]

| Hydrocarbon compounds | Formula | Temperature, °C. | $\Delta H_v$, cal./g. |
|---|---|---|---|
| **Paraffins:** | | | |
| Methane | CH₄ | −161.6 | 121.87 |
| Ethane | C₂H₆ | −88.9 | 116.87 |
| Propane | C₃H₈ | 25 | 81.76 |
| | | −42.1 | 101.76 |
| n-Butane | C₄H₁₀ | 25 | 86.63 |
| | | −0.50 | 92.09 |
| 2-Methylpropane (isobutane) | C₄H₁₀ | 25 | 78.63 |
| | | −11.72 | 87.56 |
| n-Pentane | C₅H₁₂ | 25 | 87.54 |
| | | 36.08 | 85.38 |
| 2-Methylbutane (isopentane) | C₅H₁₂ | 25 | 81.47 |
| | | 27.86 | 80.97 |
| 2,2-Dimethylpropane (neopentane) | C₅H₁₂ | 25 | 72.15 |
| | | 9.45 | 75.37 |
| n-Hexane | C₆H₁₄ | 25 | 87.50 |
| | | 68.74 | 80.48 |
| 2-Methylpentane | C₆H₁₄ | 25 | 82.83 |
| | | 60.27 | 76.89 |
| 3-Methylpentane | C₆H₁₄ | 25 | 83.96 |
| | | 63.28 | 78.42 |
| 2,2-Dimethylbutane | C₆H₁₄ | 25 | 76.79 |
| | | 49.74 | 73.75 |
| 2,3-Dimethylbutane | C₆H₁₄ | 25 | 80.77 |
| | | 57.99 | 76.53 |
| n-Heptane | C₇H₁₆ | 25 | 87.18 |
| | | 98.43 | 76.45 |
| 2-Methylhexane | C₇H₁₆ | 25 | 83.02 |
| | | 90.05 | 73.4 |
| 3-Methylhexane | C₇H₁₆ | 25 | 83.68 |
| | | 91.95 | 74.1 |
| 3-Ethylpentane | C₇H₁₆ | 25 | 84.02 |
| | | 93.47 | 74.3 |
| 2,2-Dimethylpentane | C₇H₁₆ | 25 | 77.36 |
| | | 79.20 | 69.7 |
| 2,3-Dimethylpentane | C₇H₁₆ | 25 | 81.68 |
| | | 89.79 | 72.9 |
| 2,4-Dimethylpentane | C₇H₁₆ | 25 | 78.44 |
| | | 80.51 | 70.9 |
| 3,3-Dimethylpentane | C₇H₁₆ | 25 | 78.76 |
| | | 86.06 | 70.6 |
| 2,2,3-Trimethylbutane | C₇H₁₆ | 25 | 76.42 |
| | | 80.88 | 69.3 |
| n-Octane | C₈H₁₈ | 25 | 86.80 |
| | | 125.66 | 73.19 |
| 2-Methylheptane | C₈H₁₈ | 25 | 83.02 |
| | | 117.64 | 70.3 |
| 3-Methylheptane | C₈H₁₈ | 25 | 83.35 |
| | | 118.92 | 71.3 |
| 4-Methylheptane | C₈H₁₈ | 25 | 83.01 |
| | | 117.71 | 70.91 |
| 3-Ethylhexane | C₈H₁₈ | 25 | 82.95 |
| | | 118.53 | 71.7 |
| 2,2-Dimethylhexane | C₈H₁₈ | 25 | 78.02 |
| | | 106.84 | 67.7 |
| 2,3-Dimethylhexane | C₈H₁₈ | 25 | 81.17 |
| | | 115.60 | 70.2 |
| 2,4-Dimethylhexane | C₈H₁₈ | 25 | 79.02 |
| | | 109.43 | 68.5 |
| 2,5-Dimethylhexane | C₈H₁₈ | 25 | 79.21 |
| | | 109.10 | 68.6 |
| 3,3-Dimethylhexane | C₈H₁₈ | 25 | 78.54 |
| | | 111.97 | 68.5 |
| 3,4-Dimethylhexane | C₈H₁₈ | 25 | 81.55 |
| | | 117.72 | 70.2 |
| 2-Methyl-3-ethylpentane | C₈H₁₈ | 25 | 80.60 |
| | | 115.65 | 69.7 |
| 3-Methyl-3-ethylpentane | C₈H₁₈ | 25 | 79.49 |
| | | 118.26 | 69.3 |
| 2,2,3-Trimethylpentane | C₈H₁₈ | 25 | 77.24 |
| | | 109.84 | 67.3 |
| 2,2,4-Trimethylpentane | C₈H₁₈ | 25 | 73.50 |
| | | 99.24 | 64.87 |
| 2,3,3-Trimethylpentane | C₈H₁₈ | 25 | 77.87 |
| | | 114.76 | 68.1 |
| 2,3,4-Trimethylpentane | C₈H₁₈ | 25 | 78.90 |
| | | 113.47 | 68.37 |
| 2,2,3,3-Tetramethylbutane | C₈H₁₈ | 106.30 | 66.2 |

| Hydrocarbon compounds | Formula | Temperature, °C. | $\Delta H_v$, cal./g. |
|---|---|---|---|
| **Alkyl benzenes:** | | | |
| Benzene | C₆H₆ | 25 | 103.57 |
| | | 80.10 | 94.14 |
| Methylbenzene (toluene) | C₇H₈ | 25 | 98.55 |
| | | 110.62 | 86.8 |
| Ethylbenzene | C₈H₁₀ | 25 | 95.11 |
| | | 136.19 | 81.0 |
| 1,2-Dimethylbenzene (o-xylene) | C₈H₁₀ | 25 | 97.79 |
| | | 144.42 | 82.9 |
| 1,3-Dimethylbenzene (m-xylene) | C₈H₁₀ | 25 | 96.03 |
| | | 139.10 | 82.0 |
| 1,4-Dimethylbenzene (p-xylene) | C₈H₁₀ | 25 | 95.40 |
| | | 138.35 | 81.2 |
| n-Propylbenzene | C₉H₁₂ | 25 | 91.93* |
| | | 159.22 | 76.0 |
| Isopropylbenzene | C₉H₁₂ | 25 | 89.77 |
| | | 152.40 | 74.6 |
| 1-Methyl-2-ethylbenzene | C₉H₁₂ | 25 | 94.9 |
| | | 165.15 | 77.3 |
| 1-Methyl-3-ethylbenzene | C₉H₁₂ | 25 | 93.3 |
| | | 161.30 | 76.6 |
| 1-Methyl-4-ethylbenzene | C₉H₁₂ | 25 | 92.7 |
| | | 162.05 | 76.4 |
| 1,2,3-Trimethylbenzene | C₉H₁₂ | 25 | 97.56 |
| | | 176.15 | 79.6 |
| 1,2,4-Trimethylbenzene (pseudocumene) | C₉H₁₂ | 25 | 95.33 |
| | | 169.25 | 78.0 |
| 1,3,5-Trimethylbenzene (mesitylene) | C₉H₁₂ | 25 | 94.40 |
| | | 164.70 | 77.6 |
| **Alkyl cyclopentanes:** | | | |
| Cyclopentane | C₅H₁₀ | 25 | 97.1 |
| | | 49.26 | 93.1 |
| Methylcyclopentane | C₆H₁₂ | 25 | 89.83 |
| | | 71.81 | 83.2 |
| Ethylcyclopentane | C₇H₁₄ | 25 | 88.6 |
| | | 103.45 | 78.3 |
| 1,1-Dimethylcyclopentane | C₇H₁₄ | 25 | 82.5 |
| | | 87.5 | 74.6 |
| cis-1,2-Dimethylcyclopentane | C₇H₁₄ | 25 | 86.4 |
| | | 99.3 | 77.0 |
| trans-1,2-Dimethylcyclopentane | C₇H₁₄ | 25 | 83.9 |
| | | 91.9 | 75.5 |
| trans-1,3-Dimethylcyclopentane | C₇H₁₄ | 25 | 83.6 |
| | | 90.8 | 75.3 |
| **Alkyl cyclohexanes:** | | | |
| Cyclohexane | C₆H₁₂ | 25 | 93.81 |
| | | 80.74 | 85.6 |
| Methylcyclohexane | C₇H₁₄ | 25 | 86.07 |
| | | 100.94 | 76.9 |
| Ethylcyclohexane | C₈H₁₆ | 25 | 86.21 |
| | | 131.79 | 73.7 |
| 1,1-Dimethylcyclohexane | C₈H₁₆ | 25 | 80.9 |
| | | 119.50 | 70.7 |
| cis-1,2-Dimethylcyclohexane | C₈H₁₆ | 25 | 84.59 |
| | | 129.73 | 72.9 |
| trans-1,2-Dimethylcyclohexane | C₈H₁₆ | 25 | 81.70 |
| | | 123.42 | 71.1 |
| cis-1,3-Dimethylcyclohexane | C₈H₁₆ | 25 | 83.49 |
| | | 124.45 | 72.1 |
| trans-1,3-Dimethylcyclohexane | C₈H₁₆ | 25 | 81.42 |
| | | 120.09 | 70.9 |
| cis-1,4-Dimethylcyclohexane | C₈H₁₆ | 25 | 83.13 |
| | | 124.32 | 71.9 |
| trans-1,4-Dimethylcyclohexane | C₈H₁₆ | 25 | 80.67 |
| | | 119.35 | 70.4 |
| **Monoolefins:** | | | |
| Ethene (ethylene) | C₂H₄ | −103.71 | 115.39 |
| Propene (propylene) | C₃H₆ | −47.70 | 104.62 |
| 1-Butene | C₄H₈ | 25 | 86.8 |
| | | −6.25 | 93.36 |
| cis-2-Butene | C₄H₈ | 25 | 94.5 |
| | | 3.72 | 99.46 |
| trans-2-Butene | C₄H₈ | 25 | 91.8 |
| | | 0.88 | 96.94 |
| 2-Methylpropene (isobutene) | C₄H₈ | 25 | 87.7 |
| | | −6.90 | 94.22 |

| Non-hydrocarbon compounds | Formula | Temperature, °C. | $\Delta H_v$, cal./g. |
|---|---|---|---|
| Acetal | C₆H₁₄O₂ | 102.9 | 66.18 |
| Acetaldehyde | C₂H₄O | 21 | 136.17 |
| Acetic acid | C₂H₄O₂ | 118.3 | 96.75 |
| | | 140 | 94.37 |
| | | 220 | 81.23 |
| | | 321 | 0 |
| anhydride | C₄H₆O₃ | 137 | 92.2 |
| Acetone | C₃H₆O | 0 | 134.74 |
| | | 20 | 131.82 |
| | | 40 | 128.05 |
| | | 60 | 123.51 |
| | | 80 | 118.26 |
| | | 100 | 112.76 |
| | | 235 | 0 |

[a]The values for the hydrocarbons are from the tables of the American Petroleum Research Institute Research Project 44 at the National Bureau of Standards. The values for the nonhydrocarbon compounds were recalculated from data in *International Critical Tables*, Vol. 5.

**Table 11.9.**[(9)]  *Cont'd.*

| Non-hydrocarbon compounds | Formula | Temperature, °C. | ΔH$_v$, cal./g. | Non-hydrocarbon compounds | Formula | Temperature, °C. | ΔH$_v$, cal./g. |
|---|---|---|---|---|---|---|---|
| Acetonitrile | C$_2$H$_3$N | 80 | 173.68 | Ethyl nonylate | C$_{11}$H$_{22}$O$_2$ | 227 | 58.05 |
| Acetophenone | C$_8$H$_8$O | 203.7 | 77.16 | propionate | C$_5$H$_{10}$O$_2$ | 97.6 | 80.08 |
| Acetyl chloride | C$_2$H$_3$ClO | 51 | 78.84 | propyl ether | C$_6$H$_{12}$O | 60 | 82.66 |
| Air | | | 51.0 | valerate (n-) | C$_7$H$_{14}$O$_2$ | 98 | 77.16 |
| Allyl alcohol | C$_3$H$_6$O | 96 | 163.41 | | | | |
| Amyl alcohol (n-) | C$_5$H$_{11}$OH | 131 | 120.17 | Formic acid | CH$_2$O$_2$ | 101 | 119.93 |
| alcohol (t-) | C$_5$H$_{11}$OH | 102 | 105.83 | Furane | C$_4$H$_4$O | 31 | 95.32 |
| amine (n-) | C$_5$H$_{13}$N | 95 | 98.67 | Furfural | C$_5$H$_4$O$_2$ | 160.5 | 107.51 |
| bromide (n-) | C$_5$H$_{11}$Br | 129 | 48.26 | | | | |
| ether (n-) | C$_{10}$H$_{22}$O | 170 | 69.52 | Heptyl alcohol (n-) | C$_7$H$_{16}$O | 176 | 104.88 |
| iodide (n-) | C$_5$H$_{11}$I | 155 | 47.54 | Hexylmethyl ketone | C$_8$H$_{16}$O | 173 | 74.06 |
| methyl ketone (n-) | C$_7$H$_{14}$O | 149.2 | 82.66 | Hydrogen cyanide | HCN | 20 | 210.23 |
| Amylene | C$_5$H$_{10}$ | 12.5 | 75.01 | | | | |
| Anethole (p-) | C$_{10}$H$_{12}$O | 232 | 71.43 | Isoamyl acetate | C$_7$H$_{14}$O$_2$ | 143.6 | 69.04 |
| Aniline | C$_6$H$_7$N | 183 | 103.68 | alcohol | C$_5$H$_{12}$O | 130.2 | 119.78 |
| | | | | butyrate (n-) | C$_9$H$_{18}$O$_2$ | 169 | 61.88 |
| Benzaldehyde | C$_7$H$_6$O | 179 | 86.48 | formate | C$_6$H$_{12}$O$_2$ | 123 | 73.58 |
| Benzonitrile | C$_7$H$_5$N | 189 | 87.68 | isobutyrate | C$_9$H$_{18}$O$_2$ | 168 | 57.57 |
| Benzyl alcohol | C$_7$H$_8$O | 204.3 | 112.28 | propionate | C$_8$H$_{16}$O$_2$ | 161 | 65.22 |
| Butyl acetate (n-) | C$_6$H$_{12}$O$_2$ | 124 | 73.82 | valerate (n-) | C$_{10}$H$_{20}$O$_2$ | 187 | 56.14 |
| alcohol (n-) | C$_4$H$_{10}$O | 116.8 | 141.26 | Isobutyl acetate | C$_6$H$_{12}$O$_2$ | 115.3 | 73.75 |
| alcohol (s-) | C$_4$H$_{10}$O | 98.1 | 134.38 | alcohol | C$_4$H$_{10}$O | 106.9 | 138.08 |
| alcohol (t-) | C$_4$H$_{10}$O | 83 | 130.44 | butyrate (n-) | C$_8$H$_{16}$O$_2$ | 157 | 64.50 |
| formate | C$_5$H$_{10}$O$_2$ | 105.1 | 86.74 | formate | C$_5$H$_{10}$O$_2$ | 97 | 78.50 |
| methyl ketone (n-) | C$_6$H$_{12}$O | 127 | 82.42 | isovalerate | C$_9$H$_{18}$O$_2$ | 169 | 60.44 |
| propionate (n-) | C$_7$H$_{14}$O$_2$ | 144.9 | 71.74 | isobutyrate | C$_8$H$_{16}$O$_2$ | 148 | 63.31 |
| Butyric acid (n-) | C$_4$H$_8$O$_2$ | 163.5 | 113.96 | propionate | C$_7$H$_{14}$O$_2$ | 137 | 65.94 |
| Butyronitrile (n-) | C$_4$H$_7$N | 117.4 | 114.91 | valerate (n-) | C$_9$H$_{18}$O$_2$ | 169 | 57.81 |
| Bromobenzene | C$_6$H$_5$Br | 155.9 | 57.60 | Isobutyric acid | C$_4$H$_8$O$_2$ | 154 | 111.57 |
| | | | | Isopropyl alcohol | C$_3$H$_8$O | 82.3 | 159.35 |
| Capronitrile | C$_6$H$_{11}$N | 156 | 88.15 | methyl ketone | C$_5$H$_{10}$O | 92 | 89.83 |
| Carbon disulfide | CS$_2$ | 0 | 89.35 | Isovaleric acid | C$_5$H$_{10}$O$_2$ | 176.3 | 101.05 |
| | | 46.25 | 84.09 | | | | |
| | | 100 | 75.49 | Limonene | C$_{10}$H$_{16}$ | 165 | 69.52 |
| | | 140 | 67.37 | | | | |
| tetrachloride | CCl$_4$ | 0 | 52.06 | Mesityl oxide | C$_6$H$_{10}$O | 128 | 85.77 |
| | | 76.75 | 46.42 | Methyl acetate | C$_3$H$_6$O$_2$ | 0.0 | 113.96 |
| | | 200 | 32.73 | | | 56.3 | 98.09 |
| Carvacrol | C$_{10}$H$_{14}$O | 237 | 68.09 | Methylal | C$_3$H$_8$O$_2$ | 42 | 89.83 |
| Chloral | C$_2$HCl$_3$O | | 53.99 | Methyl alcohol | CH$_4$O | 0 | 284.29 |
| hydrate | C$_2$H$_3$Cl$_3$O$_2$ | 96 | 131.87 | | | 64.7 | 262.79 |
| Chlorobenzene | C$_6$H$_5$Cl | 130.6 | 77.59 | | | 100 | 241.29 |
| Chloroethyl alcohol (2-) | C$_2$H$_5$ClO | 126.7 | 122.94 | | | 160 | 193.51 |
| acetate (β-) | C$_4$H$_7$ClO$_2$ | 141.5 | 80.75 | | | 200 | 148.12 |
| Chloroform | CHCl$_3$ | 0 | 64.74 | | | 220 | 109.89 |
| | | 40 | 60.92 | | | 240 | 0 |
| | | 61.5 | 59.01 | amyl ketone (n-) | C$_7$H$_{14}$O | 149.2 | 82.66 |
| | | 100 | 55.19 | aniline | C$_7$H$_9$N | 194 | 95.56 |
| | | 260 | 0 | butyl ketone (n-) | C$_6$H$_{12}$O | 127 | 82.42 |
| Chlorotoluene (o-) | C$_7$H$_7$Cl | 158.1 | 72.63 | butyrate (n-) | C$_5$H$_{10}$O$_2$ | 102.6 | 79.79 |
| (p-) | C$_7$H$_7$Cl | 160.4 | 73.13 | chloride | CH$_3$Cl | −23.8 | 102.25 |
| Cresol (m-) | C$_7$H$_8$O | 202 | 1G0.58 | | | +15.0 | 96.04 |
| Cyanogen | (CN)$_2$ | 0 | 102.97 | | | 20.0 | 95.32 |
| chloride | CNCl | 13 | 134.98 | | | 25.0 | 94.60 |
| Cyclohexanol | C$_6$H$_{12}$O | 161.1 | 108.22 | | | 78.2 | 105.93 |
| Cyclohexyl chloride | C$_6$H$_{11}$Cl | 142.0 | 74.78 | ethyl ketone | C$_4$H$_8$O | 78.2 | 105.93 |
| | | | | ethyl ketoxime | C$_4$H$_9$NO | 182 | 115.87 |
| Dichloroacetic acid | C$_2$H$_2$Cl$_2$O$_2$ | 194.4 | 77.16 | formate | C$_2$H$_4$O$_2$ | 31.3 | 112.35 |
| Dichlorodifluoromethane | CCl$_2$F$_2$ | −29.8 | 40.40 | hexyl ketone | C$_8$H$_{16}$O | 173 | 74.06 |
| Diethylamine | C$_4$H$_{11}$N | 58 | 91.02 | iodide | CH$_3$I | 42 | 45.87 |
| carbonate | C$_5$H$_{10}$O$_3$ | 126 | 73.10 | isobutyrate | C$_5$H$_{10}$O$_2$ | 91.1 | 78.12 |
| ketone | C$_5$H$_{10}$O | 101 | 90.78 | isopropyl ketone | C$_5$H$_{10}$O | 92 | 89.83 |
| oxalate | C$_6$H$_{10}$O$_4$ | 185 | 67.61 | isovalerate | C$_6$H$_{12}$O$_2$ | 116 | 72.39 |
| Di-isobutylamine | C$_8$H$_{19}$N | 134 | 65.70 | phenyl ether | C$_7$H$_8$O | 153 | 81.46 |
| Dimethyl aniline | C$_8$H$_{11}$N | 193 | 80.75 | propionate | C$_4$H$_8$O$_2$ | 79.0 | 87.56 |
| carbonate | C$_3$H$_6$O$_3$ | 90 | 88.15 | valerate (n-) | C$_6$H$_{12}$O$_2$ | 116 | 70.00 |
| Dipropyl ketone | C$_7$H$_{14}$ | 143.5 | 75.73 | | | | |
| Dipropylamine (n-) | C$_6$H$_{15}$N | 108 | 75.73 | Naphthalene | C$_{10}$H$_8$ | 218 | 75.49 |
| | | | | Nitrobenzene | C$_6$H$_5$NO$_2$ | 210 | 79.08 |
| | | | | Nitromethane | CH$_3$NO$_2$ | 99.9 | 134.98 |
| Ethyl acetate | C$_4$H$_8$O$_2$ | 0.0 | 102.01 | | | | |
| alcohol | C$_2$H$_6$O | 78.3 | 204.26 | Octyl alcohol (n-) | C$_8$H$_{18}$O | 196 | 97.47 |
| Ethylamine | C$_2$H$_7$N | 15 | 145.97 | alcohol (dl-) (sec-) | C$_8$H$_{18}$O | 180 | 94.37 |
| Ethyl benzoate | C$_9$H$_{10}$O$_2$ | 213 | 64.50 | | | | |
| bromide | C$_2$H$_5$Br | 38.4 | 59.92 | Phenyl methyl ether | C$_7$H$_8$O | 153 | 81.46 |
| butyrate (n-) | C$_6$H$_{12}$O$_2$ | 118.9 | 74.68 | Picoline (α-) | C$_6$H$_7$N | 129 | 90.78 |
| caprylate | C$_{10}$H$_{20}$O$_2$ | 207 | 60.44 | Piperidine | C$_5$H$_{11}$N | 106 | 89.35 |
| chloride | C$_2$H$_5$Cl | 4.7 | 92.93 | Propionic acid | C$_3$H$_6$O$_2$ | 139.3 | 98.81 |
| | | 15.0 | 92.45 | Propionitrile | C$_3$H$_5$N | 97 | 134.26 |
| | | 20.0 | 92.22 | Propyl acetate (n-) | C$_5$H$_{10}$O$_2$ | 100.4 | 80.27 |
| | | 25.0 | 91.98 | alcohol (n-) | C$_3$H$_8$O | 97.2 | 164.36 |
| Ethylene bromide | C$_2$H$_4$Br$_2$ | 130.8 | 46.23 | butyrate (n-) | C$_7$H$_{14}$O$_2$ | 143.6 | 68.33 |
| chloride | C$_2$H$_4$Cl$_2$ | 0 | 85.29 | formate (n-) | C$_4$H$_8$O$_2$ | 80.0 | 88.13 |
| | | 82.3 | 77.33 | isobutyrate (n-) | C$_7$H$_{14}$O$_2$ | 134 | 63.79 |
| glycol | C$_2$H$_6$O$_2$ | 197 | 191.12 | isovalerate (n-) | C$_8$H$_{16}$O$_2$ | 156 | 64.50 |
| oxide | C$_2$H$_4$O | 13 | 138.56 | propionate (n-) | C$_6$H$_{12}$O$_2$ | 120.6 | 73.15 |
| Ethyl ether | C$_4$H$_{10}$O | 34.6 | 83.85 | Pyridine | C$_5$H$_5$N | 114.1 | 107.36 |
| formate | C$_3$H$_6$O$_2$ | 53.3 | 97.18 | | | | |
| iodide | C$_2$H$_5$I | 71.2 | 45.61 | Salicylaldehyde | C$_7$H$_6$O$_2$ | 196 | 74.78 |
| Ethylidine chloride | C$_2$H$_4$Cl$_2$ | 0.0 | 76.69 | | | | |
| | | 60 | 67.13 | Tetrachloroethane (1,1,2,2-) | C$_2$H$_2$Cl$_4$ | 145 | 55.07 |
| Ethyl isobutyl ether | C$_6$H$_{14}$O | 79.0 | 74.78 | Tetrachloroethylene | C$_2$Cl$_4$ | 120.7 | 50.05 |
| isobutyrate | C$_6$H$_{12}$O$_2$ | 109.2 | 72.05 | Toluidine (o-) | C$_7$H$_9$N | 198 | 95.08 |
| isovalerate | C$_7$H$_{14}$O$_2$ | 144 | 67.85 | Trichloroethylene | C$_2$HCl$_3$ | 85.7 | 57.24 |
| methyl ketone | C$_4$H$_8$O | 78.2 | 105.93 | | | | |
| methyl ketoxime | C$_4$H$_9$NO | 182 | 115.87 | Valeronitrile (n-) | C$_5$H$_9$N | 129 | 96.28 |

**Table 11.10.**[10]   Viscosities of Liquids: Coordinates for Use with Figure 11.1

| Liquid | X | Y | Liquid | X | Y |
|---|---|---|---|---|---|
| Acetaldehyde | 15.2 | 4.8 | Freon-113 | 12.5 | 11.4 |
| Acetic acid, 100% | 12.1 | 14.2 | Glycerol, 100% | 2.0 | 30.0 |
| Acetic acid, 70% | 9.5 | 17.0 | Glycerol, 50% | 6.9 | 19.6 |
| Acetic anhydride | 12.7 | 12.8 | Heptane | 14.1 | 8.4 |
| Acetone, 100% | 14.5 | 7.2 | Hexane | 14.7 | 7.0 |
| Acetone, 35% | 7.9 | 15.0 | Hydrochloric acid, 31.5% | 13.0 | 16.6 |
| Acetonitrile | 14.4 | 7.4 | Iodobenzene | 12.8 | 15.9 |
| Acrylic acid | 12.3 | 13.9 | Isobutyl alcohol | 7.1 | 18.0 |
| Allyl alcohol | 10.2 | 14.3 | Isobutyric acid | 12.2 | 14.4 |
| Allyl bromide | 14.4 | 9.6 | Isopropyl alcohol | 8.2 | 16.0 |
| Allyl iodide | 14.0 | 11.7 | Isopropyl bromide | 14.1 | 9.2 |
| Ammonia, 100% | 12.6 | 2.0 | Isopropyl chloride | 13.9 | 7.1 |
| Ammonia, 26% | 10.1 | 13.9 | Isopropyl iodide | 13.7 | 11.2 |
| Amyl acetate | 11.8 | 12.5 | Kerosene | 10.2 | 16.9 |
| Amyl alcohol | 7.5 | 18.4 | Linseed oil, raw | 7.5 | 27.2 |
| Aniline | 8.1 | 18.7 | Mercury | 18.4 | 16.4 |
| Anisole | 12.3 | 13.5 | Methanol, 100% | 12.4 | 10.5 |
| Arsenic trichloride | 13.9 | 14.5 | Methanol, 90% | 12.3 | 11.8 |
| Benzene | 12.5 | 10.9 | Methyl acetate | 14.2 | 8.2 |
| Brine, CaCl₂, 25% | 6.6 | 15.9 | Methyl acrylate | 13.0 | 9.5 |
| Brine, NaCl, 25% | 10.2 | 16.6 | Methyl i-butyrate | 12.3 | 9.7 |
| Bromine | 14.2 | 13.2 | Methyl n-butyrate | 13.2 | 10.3 |
| Bromotoluene | 20.0 | 15.9 | Methyl chloride | 15.0 | 3.8 |
| Butyl acetate | 12.3 | 11.0 | Methyl ethyl ketone | 13.9 | 8.6 |
| Butyl acrylate | 11.5 | 12.6 | Methyl formate | 14.2 | 7.5 |
| Butyl alcohol | 8.6 | 17.2 | Methyl iodide | 14.3 | 9.3 |
| Butyric acid | 12.1 | 15.3 | Methyl propionate | 13.5 | 9.0 |
| Carbon dioxide | 11.6 | 0.3 | Methyl propyl ketone | 14.3 | 9.5 |
| Carbon disulfide | 16.1 | 7.5 | Methyl sulfide | 15.3 | 6.4 |
| Carbon tetrachloride | 12.7 | 13.1 | Napthalene | 7.9 | 18.1 |
| Chlorobenzene | 12.3 | 12.4 | Nitric acid, 95% | 12.8 | 13.8 |
| Chloroform | 14.4 | 10.2 | Nitric acid, 60% | 10.8 | 17.0 |
| Chlorosulfonic acid | 11.2 | 18.1 | Nitrobenzene | 10.6 | 16.2 |
| Chlorotoluene, ortho | 13.0 | 13.3 | Nitrogen dioxide | 12.9 | 8.6 |
| Chlorotoluene, meta | 13.3 | 12.5 | Nitrotoluene | 11.0 | 17.0 |
| Chlorotoluene, para | 13.3 | 12.5 | Octane | 13.7 | 10.0 |
| Cresol, meta | 2.5 | 20.8 | Octyl alcohol | 6.6 | 21.1 |
| Cyclohexanol | 2.9 | 24.3 | Pentachloroethane | 10.9 | 17.3 |
| Cyclohexane | 9.8 | 12.9 | Pentane | 14.9 | 5.2 |
| Dibromomethane | 12.7 | 15.8 | Phenol | 6.9 | 20.8 |
| Dichloroethane | 13.2 | 12.2 | Phosphorus tribromide | 13.8 | 16.7 |
| Dichloromethane | 14.6 | 8.9 | Phosphorus trichloride | 16.2 | 10.9 |
| Diethyl ketone | 13.5 | 9.2 | Propionic acid | 12.8 | 13.8 |
| Diethyl oxalate | 11.0 | 16.4 | Propyl acetate | 13.1 | 10.3 |
| Diethylene glycol | 5.0 | 24.7 | Propyl alcohol | 9.1 | 16.5 |
| Diphenyl | 12.0 | 18.3 | Propyl bromide | 14.5 | 9.6 |
| Dipropyl ether | 13.2 | 8.6 | Propyl chloride | 14.4 | 7.5 |
| Dipropyl oxalate | 10.3 | 17.7 | Propyl formate | 13.1 | 9.7 |
| Ethyl acetate | 13.7 | 9.1 | Propyl iodide | 14.1 | 11.6 |
| Ethyl acrylate | 12.7 | 10.4 | Sodium | 16.4 | 13.9 |
| Ethyl alcohol, 100% | 10.5 | 13.8 | Sodium hydroxide, 50% | 3.2 | 25.8 |
| Ethyl alcohol, 95% | 9.8 | 14.3 | Stannic chloride | 13.5 | 12.8 |
| Ethyl alcohol, 40% | 6.5 | 16.6 | Succinonitrile | 10.1 | 20.8 |
| Ethyl benzene | 13.2 | 11.5 | Sulfur dioxide | 15.2 | 7.1 |
| Ethyl bromide | 14.5 | 8.1 | Sulfuric acid, 110% | 7.2 | 27.4 |
| 2-Ethyl butyl acrylate | 11.2 | 14.0 | Sulfuric acid, 100% | 8.0 | 25.1 |
| Ethyl chloride | 14.8 | 6.0 | Sulfuric acid, 98% | 7.0 | 24.8 |
| Ethyl ether | 14.5 | 5.3 | Sulfuric acid. 60% | 10.2 | 21.3 |
| Ethyl formate | 14.2 | 8.4 | Sulfuryl chloride | 15.2 | 12.4 |
| 2-Ethyl hexyl acrylate | 9.0 | 15.0 | Tetrachloroethane | 11.9 | 15.7 |
| Ethyl iodide | 14.7 | 10.3 | Thiophene | 13.2 | 11.0 |
| Ethyl propionate | 13.2 | 9.9 | Titanium tetrachloride | 14.4 | 12.3 |
| Ethyl propyl ether | 14.0 | 7.0 | Toluene | 13.7 | 10.4 |
| Ethyl sulfide | 13.8 | 8.9 | Trichloroethylene | 14.8 | 10.5 |
| Ethylene bromide | 11.9 | 15.7 | Triethylene glycol | 4.7 | 24.8 |
| Ethylene chloride | 12.7 | 12.2 | Turpentine | 11.5 | 14.9 |
| Ethylene glycol | 6.0 | 23.6 | Vinyl acetate | 14.0 | 8.8 |
| Ethylidene chloride | 14.1 | 8.7 | Vinyl toluene | 13.4 | 12.0 |
| Fluorobenzene | 13.7 | 10.4 | Water | 10.2 | 13.0 |
| Formic acid | 10.7 | 15.8 | Xylene, ortho | 13.5 | 12.1 |
| Freon-11 | 14.4 | 9.0 | Xylene, meta | 13.9 | 10.6 |
| Freon-12 | 16.8 | 15.6 | Xylene, para | 13.9 | 10.9 |
| Freon-21 | 15.7 | 7.5 | | | |
| Freon-22 | 17.2 | 4.7 | | | |

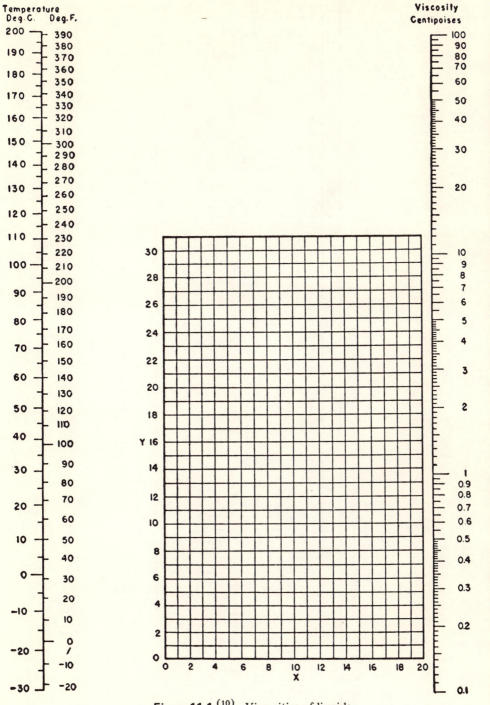

**Figure 11.1.**[(10)]   Viscosities of liquids.

**Table 11.11.**[11] Viscosities of Gases: Coordinates for Use with Figure 11.2

| No. | Gas | X | Y | No. | Gas | X | Y | No. | Gas | X | Y | No. | Gas | X | Y |
|---|---|---|---|---|---|---|---|---|---|---|---|---|---|---|---|
| 1 | Acetic acid | 7.7 | 14.3 | 15 | Chloroform | 8.9 | 15.7 | 29 | Freon-113 | 11.3 | 14.0 | 43 | Nitric oxide | 10.9 | 20.5 |
| 2 | Acetone | 8.9 | 13.0 | 16 | Cyanogen | 9.2 | 15.2 | 30 | Helium | 10.9 | 20.5 | 44 | Nitrogen | 10.6 | 20.0 |
| 3 | Acetylene | 9.8 | 14.9 | 17 | Cyclohexane | 9.2 | 12.0 | 31 | Hexane | 8.6 | 11.8 | 45 | Nitrosyl chloride | 8.0 | 17.6 |
| 4 | Air | 11.0 | 20.0 | 18 | Ethane | 9.1 | 14.5 | 32 | Hydrogen | 11.2 | 12.4 | 46 | Nitrous oxide | 8.8 | 19.0 |
| 5 | Ammonia | 8.4 | 16.0 | 19 | Ethyl acetate | 8.5 | 13.2 | 33 | $3H_2 + 1N_2$ | 11.2 | 17.2 | 47 | Oxygen | 11.0 | 21.3 |
| 6 | Argon | 10.5 | 22.4 | 20 | Ethyl alcohol | 9.2 | 14.2 | 34 | Hydrogen bromide | 8.8 | 20.9 | 48 | Pentane | 7.0 | 12.8 |
| 7 | Benzene | 8.5 | 13.2 | 21 | Ethyl chloride | 8.5 | 15.6 | 35 | Hydrogen chloride | 8.8 | 18.7 | 49 | Propane | 9.7 | 12.9 |
| 8 | Bromine | 8.9 | 19.2 | 22 | Ethyl ether | 8.9 | 13.0 | 36 | Hydrogen cyanide | 9.8 | 14.9 | 50 | Propyl alcohol | 8.4 | 13.4 |
| 9 | Butene | 9.2 | 13.7 | 23 | Ethylene | 9.5 | 15.1 | 37 | Hydrogen iodide | 9.0 | 21.3 | 51 | Propylene | 9.0 | 13.8 |
| 10 | Butylene | 8.9 | 13.0 | 24 | Fluorine | 7.3 | 23.8 | 38 | Hydrogen sulfide | 8.6 | 18.0 | 52 | Sulfur dioxide | 9.6 | 17.0 |
| 11 | Carbon dioxide | 9.5 | 18.7 | 25 | Freon-11 | 10.6 | 15.1 | 39 | Iodine | 9.0 | 18.4 | 53 | Toluene | 8.6 | 12.4 |
| 12 | Carbon disulfide | 8.0 | 16.0 | 26 | Freon-12 | 11.1 | 16.0 | 40 | Mercury | 5.3 | 22.9 | 54 | 2, 3, 3-Trimethylbutane | 9.5 | 10.5 |
| 13 | Carbon monoxide | 11.0 | 20.0 | 27 | Freon-21 | 10.8 | 15.3 | 41 | Methane | 9.9 | 15.5 | 55 | Water | 8.0 | 16.0 |
| 14 | Chlorine | 9.0 | 18.4 | 28 | Freon-22 | 10.1 | 17.0 | 42 | Methyl alcohol | 8.5 | 15.6 | 56 | Xenon | 9.3 | 23.0 |

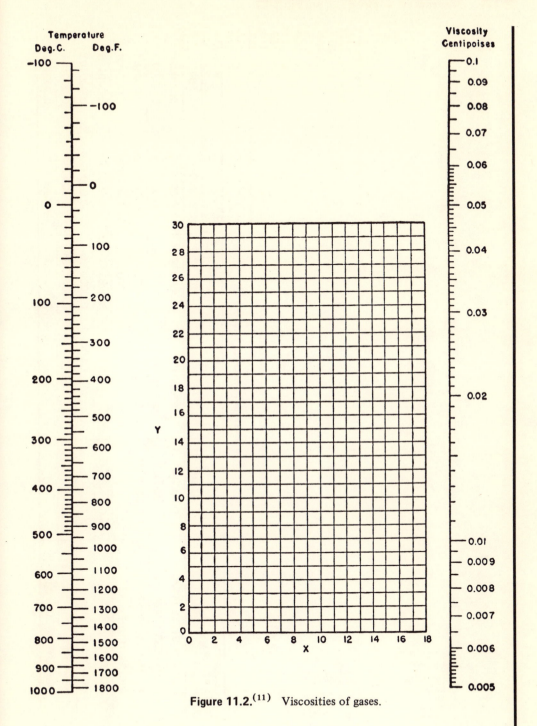

**Figure 11.2.**[(11)]  Viscosities of gases.

**Table 11.12.**[12] Coefficients of Linear Expansion—Approximate Values (in./ft-°F)

INSTANTANEOUS VALUES

| Material | Temperature—F | | | | |
|---|---|---|---|---|---|
| | −260 | −50 | 70 | 300 | 500 |
| CARBON STEEL (1020) | 0.000042 | 0.000070 | 0.000077 | 0.000086 | 0.000098 |
| STAINLESS STEEL (18-8) | 0.000067 | 0.000098 | 0.000102 | 0.000113 | 0.000119 |
| ALUMINUM (5052) | 0.000087 | 0.000138 | 0.000152 | 0.000170 | 0.000184 |
| NICKEL (PURE) | | | 0.000061 | 0.000096 | 0.000108 |
| MONEL | | | 0.000084 | 0.000107 | 0.000115 |
| INCONEL | | | 0.000070 | 0.000090 | 0.000101 |
| NI-O-NEL | | | 0.000073 | 0.000094 | 0.000101 |
| CARPENTER 20 | | | 0.000103 | 0.000107 | 0.000110 |

**MEAN VALUES**

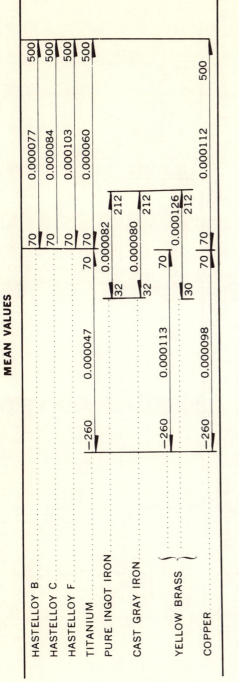
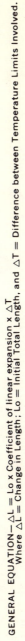

| | | | |
|---|---|---|---|
| HASTELLOY B | 70 | 0.000077 | 500 |
| HASTELLOY C | 70 | 0.000084 | 500 |
| HASTELLOY F | 70 | 0.000103 | 500 |
| TITANIUM | 70 | 0.000060 | 500 |
| | 70 | 0.000047 | |
| PURE INGOT IRON | 32 | 0.000082 | 212 |
| CAST GRAY IRON | 32 | 0.000080 | 212 |
| | −260 | 0.000113 | |
| YELLOW BRASS | 70 | 0.000126 | 212 |
| | 30 | | |
| COPPER | −260 | 0.000098 | |
| | 70 | 0.000112 | 500 |

GENERAL EQUATION—$\triangle L = L_o \times$ Coefficient of linear expansion $\times \triangle T$
Where $\triangle L =$ Change in Length; $L_o =$ Initial Total Length, and $\triangle T =$ Difference between Temperature Limits Involved.

**Table 11.13.**[13]  Thermodynamic Properties of Saturated Steam

| GAGE PRESSURE lbs. | TEMPERATURE deg. Fahr. | B. T. U. per lb. Heat of Liquid | B. T. U. per lb. Latent Heat of Evaporation | B. T. U. per lb. Total Heat of Steam | SPECIFIC VOLUME cu. ft. per lb. sat. vapor |
|---|---|---|---|---|---|
| 28 | 101 | 68 | 1037 | 1105 | 339 |
| 26 | 126 | 93 | 1023 | 1116 | 177 |
| 24 | 141 | 109 | 1014 | 1122 | 121 |
| 22 | 152 | 120 | 1007 | 1127 | 92 |
| 20 | 162 | 130 | 1001 | 1131 | 75 |
| 18 | 169 | 137 | 997 | 1134 | 63 |
| 16 | 176 | 144 | 993 | 1137 | 55 |
| 14 | 182 | 150 | 989 | 1139 | 48 |
| 12 | 187 | 155 | 986 | 1141 | 43 |
| 10 | 192 | 160 | 983 | 1143 | 39 |
| 8 | 197 | 165 | 980 | 1145 | 36 |
| 6 | 201 | 169 | 977 | 1146 | 33 |
| 4 | 205 | 173 | 975 | 1148 | 31 |
| 2 | 209 | 177 | 972 | 1149 | 29 |
| 0 | 212 | 180 | 970 | 1150 | 27 |
| 1 | 216 | 183 | 968 | 1151 | 25 |
| 2 | 219 | 187 | 965 | 1152 | 24 |
| 3 | 222 | 190 | 964 | 1154 | 22.5 |
| 4 | 224 | 193 | 962 | 1155 | 21.0 |
| 5 | 227 | 195 | 961 | 1156 | 20.0 |
| 6 | 230 | 198 | 959 | 1157 | 19.5 |
| 7 | 232 | 201 | 957 | 1158 | 18.5 |
| 8 | 235 | 203 | 956 | 1159 | 18.0 |
| 9 | 237 | 206 | 954 | 1160 | 17.0 |
| 10 | 240 | 208 | 952 | 1160 | 16.5 |
| 15 | 250 | 218 | 945 | 1163 | 14.0 |
| 20 | 259 | 227 | 940 | 1167 | 12.0 |
| 25 | 267 | 236 | 934 | 1170 | 10.5 |
| 30 | 274 | 243 | 929 | 1172 | 9.5 |
| 35 | 281 | 250 | 924 | 1174 | 8.5 |

Vacuum, In Hg (bracket covering gage pressure rows 28 through 0)

| | | | | | |
|---|---|---|---|---|---|
| 40 | 287 | 256 | 915 | 1177 | 8.0 |
| 45 | 292 | 262 | 915 | 1179 | 7.0 |
| 50 | 298 | 267 | 912 | 1180 | 6.7 |
| 55 | 303 | 272 | 908 | 1182 | 6.2 |
| 60 | 307 | 277 | 905 | | 5.8 |
| 65 | 312 | 282 | 901 | 1183 | 5.5 |
| 70 | 316 | 286 | 898 | 1184 | 5.2 |
| 75 | 320 | 290 | 895 | 1185 | 4.9 |
| 80 | 324 | 294 | 892 | 1186 | 4.7 |
| 85 | 328 | 298 | 889 | 1187 | 4.4 |
| 90 | 331 | 302 | 886 | 1188 | 4.2 |
| 95 | 335 | 306 | 883 | 1189 | 4.0 |
| 100 | 338 | 309 | 881 | 1190 | 3.9 |
| 110 | 344 | 316 | 876 | 1192 | 3.6 |
| 120 | 350 | 322 | 871 | 1193 | 3.3 |
| 125 | 353 | 325 | 868 | 1193 | 3.2 |
| 130 | 356 | 328 | 866 | 1194 | 3.1 |
| 140 | 361 | 334 | 861 | 1195 | 2.9 |
| 150 | 366 | 339 | 857 | 1196 | 2.7 |
| 160 | 371 | 344 | 853 | 1197 | 2.6 |
| 170 | 375 | 348 | 849 | 1197 | 2.5 |
| 180 | 380 | 353 | 845 | 1198 | 2.3 |
| 190 | 384 | 358 | 841 | 1198 | 2.2 |
| 200 | 388 | 362 | 837 | 1199 | 2.1 |
| 220 | 395 | 370 | 830 | 1200 | 2.0 |
| 240 | 403 | 378 | 823 | 1201 | 1.8 |
| 250 | 406 | 381 | 820 | 1201 | 1.75 |
| 260 | 409 | 385 | 817 | 1202 | 1.7 |
| 280 | 416 | 392 | 811 | 1203 | 1.6 |
| 300 | 422 | 399 | 805 | 1204 | 1.5 |
| 350 | 436 | 414 | 790 | 1204 | 1.3 |
| 400 | 448 | 428 | 776 | 1204 | 1.1 |
| 450 | 460 | 441 | 764 | 1205 | 1.0 |
| 500 | 470 | 453 | 751 | 1204 | 0.90 |
| 600 | 489 | 475 | 728 | 1203 | 0.75 |

Table 11.14.[14]  Physical Properties of Freon

| | | "FREON" 11 | "FREON" 12 | "FREON" 13 | "FREON" 13B1 | "FREON" 14 | "FREON" 21 | "FREON" 22 |
|---|---|---|---|---|---|---|---|---|
| Chemical Formula | | $CCl_3F$ | $CCl_2F_2$ | $CClF_3$ | $CBrF_3$ | $CF_4$ | $CHCl_2F$ | $CHClF_2$ |
| Molecular Weight | | 137.37 | 120.92 | 104.46 | 148.92 | 88.00 | 102.93 | 86.47 |
| Boiling Point at 1 atm | °C | 23.82 | −29.79 | −81.4 | −57.75 | −127.96 | 8.92 | −40.75 |
| | °F | 74.87 | −21.62 | −114.6 | −71.95 | −198.32 | 48.06 | −41.36 |
| Freezing Point | °C | −111 | −158 | −181' | −168 | −184² | −135 | −160 |
| | °F | −168 | −252 | −294 | −270 | −299 | −211 | −256 |
| Critical Temperature | °C | 198.0 | 112.0 | 28.9 | 67.0 | −45.67 | 178.5 | 96.0 |
| | °F | 388.4 | 233.6 | 83.9 | 152.6 | −50.2 | 353.3 | 204.8 |
| Critical Pressure | atm | 43.5 | 40.6 | 38.2 | 39.1 | 36.96 | 51.0 | 49.12 |
| | lbs/sq in abs | 639.5 | 596.9 | 561 | 575 | 543.2 | 750 | 721.9 |
| Critical Volume | cc/mol | 247 | 217 | 181 | 200 | 141 | 197 | 165 |
| | cu ft/lb | 0.0289 | 0.0287 | 0.0277 | 0.0215 | 0.0256 | 0.0307 | 0.0305 |
| Critical Density | g/cc | 0.554 | 0.558 | 0.578 | 0.745 | 0.626 | 0.522 | 0.525 |
| | lbs/cu ft | 34.6 | 34.8 | 36.1 | 46.5 | 39.06 | 32.6 | 32.76 |
| Density, Liquid at 25°C (77°F) | g/cc | 1.476 | 1.311 | 1.298 @ −30°C | 1.538 | 1.317 @ −80°C | 1.366 | 1.194 |
| | lbs/cu ft | 92.14 | 81.84 | 81.05 (−22°F) | 96.01 | 82.21 (−112°F) | 85.28 | 74.53 |
| Density, Sat'd Vapor at Boiling Point | g/l | 5.86 | 6.33 | 7.01 | 8.71 | 7.62 | 4.57 | 4.72 |
| | lbs/cu ft | 0.367 | 0.395 | 0.438 | 0.544 | 0.476 | 0.285 | 0.295 |
| Specific Heat, Liquid (Heat Capacity) at 25°C (77°F) | cal/(g) (°C) or Btu/(lb)(°F) | 0.208 | 0.232 | 0.247 @ −30°C (−22°F) | 0.208 | 0.294 @ −80°C (−112°F) | 0.256 | 0.300 |
| Specific Heat, Vapor, at Const Pressure (1 atm) at 25°C (77°F) | cal/(g)(°C) or Btu/(lb)(°F) | 0.142 @ 38°C (100°F) | 0.145 | 0.158 | 0.112 | 0.169 | 0.140 | 0.157 |
| Specific Heat Ratio at 25°C and 1 atm | $C_p/C_v$ | 1.137 @ 38°C (100°F) | 1.137 | 1.145 | 1.144 | 1.159 | 1.175 | 1.184 |
| Heat of Vaporization at Boiling Point | cal/g | 43.10 | 39.47 | 35.47 | 28.38 | 32.49 | 57.86 | 55.81 |
| | Btu/lb | 77.51 | 71.04 | 63.85 | 51.08 | 58.48 | 104.15 | 100.45 |
| Thermal Conductivity⁷ at 25°C (77°F) Btu/(hr) (ft) (°F) Liquid Vapor (1 atm) | | 0.0506 0.00451 | 0.0405 0.00557 | 0.0378 @ −30°C 0.00501 (−22°F) | 0.0234 0.00534 | 0.0361 @ −80°C 0.00463 (−112°F) | 0.0592 0.00506 | 0.0507 0.00609 |
| Viscosity⁷ at 25°C (77°F) Liquid Vapor (1 atm) | centipoise centipoise | 0.415 0.0107 | 0.214 0.0123 | 0.170 @ −30°C 0.0119 (−22°F) | 0.157 0.0154 | 0.23 @ −80°C 0.0116 (−112°F) | 0.313 0.0114 | 0.198 0.0127 |
| Surface Tension at 25°C (77°F) dynes/cm | | 18 | 9 | 14 @ −73°C (−100°F) | 4 | 4 @ −73°C (−100°F) | 18 | 8 |
| Refractive Index of Liquid at 25°C (77°F) | | 1.374 | 1.287 | 1.199 @ −73°C (−100°F) | 1.238 | 1.151 @ −73°C (−100°F) | 1.354 | 1.256 |
| Relative Dielectric Strengthᵉ at 1 atm and 25°C (77°F) (nitrogen=1) | | 3.71 | 2.46 | 1.65 | 1.83 | 1.06 | 1.85 | 1.27 |
| Dielectric Constant Liquid Vapor (1 atm)⁹ᵃ | | 2.28 @ 29°C 1.0036 @ 24°C⁹ᵇ | 2.13 @ 29°C 1.0032 (84°F) | 1.0024 @ 29°C (84°F) | | 1.0012 @ 24.5°C (76°F) | 5.34 @ 28°C 1.0070 @ 30°C | 6.11 @ 24°C 1.0071 @ 25.4°C |
| Solubility of "Freon" in Water at 1 atm and 25°C (77°F) | wt % | 0.11 | 0.028 | 0.009 | 0.03 | 0.0015 | 0.95 | 0.30 |
| Solubility of Water in "Freon" at 25°C (77°F) | wt % | 0.011 | 0.009 | | 0.0095 21°C (70°F) | | 0.13 | 0.13 |

## mily of Fluorocarbon Compounds

| REON" 23 | "FREON" 112 | "FREON" 113 | "FREON" 114 | FC 114B2 | "FREON" 115 | "FREON" 116 | "FREON" 500 | "FREON" 502 | "FREON" 503 |
|---|---|---|---|---|---|---|---|---|---|
| CHF₃ | CCl₂F-CCl₂F | CCl₂F-CClF₂ | CClF₂-CClF₂ | CBrF₂-CBrF₂ | CClF₂-CF₃ | CF₃-CF₃ | a | b | c |
| 70.01 | 203.84 | 187.38 | 170.93 | 259.85 | 154.47 | 138.01 | 99.31 | 111.64 | 87.28 |
| −82.03 / −115.66 | 92.8 / 199.0 | 47.57 / 117.63 | 3.77 / 38.78 | 47.26 / 117.06 | −39.1 / −38.4 | −78.2 / −108.8 | −33.5 / −28.3 | −45.42 / −49.76 | −87.9 / −126.2 |
| −155.2 / −247.4 | 26 / 79 | −35 / −31 | −94 / −137 | −110.5 / −166.8 | −106[10] / −159 | −100.6 / −149.1 | −159 / −254 | | |
| 25.9 / 78.6 | 278 / 532 | 214.1 / 417.4 | 145.7 / 294.3 | 214.5 / 418.1 | 80.0 / 175.9 | 19.7[4] / 67.5 | 105.5 / 221.9 | 82.2 / 179.9 | 19.5 / 67.1 |
| 47.7 / 701.4 | 34[3] / 500 | 33.7 / 495 | 32.2 / 473.2 | 34.4 / 506.1 | 30.8 / 453 | 29.4[4] / 432 | 43.67 / 641.9 | 40.2 / 591.0 | 43.0 / 632.2 |
| 133 / 0.0305 | 370[3] / 0.029 | 325 / 0.0278 | 293 / 0.0275 | 329 / 0.0203 | 259 / 0.0269 | 225 / 0.0262 | 200.0 / 0.03226 | 199 / 0.02857 | 155 / 0.0284 |
| 0.525 / 32.78 | 0.55[3] / 34 | 0.576 / 36.0 | 0.582 / 36.32 | 0.790 / 49.32 | 0.596 / 37.2 | 0.612 / 38.21 | 0.4966 / 31.0 | 0.561 / 35.0 | 0.564 / 35.21 |
| 0.670 / 41.82 | 1.634[6] @ 30°C (86°F) / 102.1 | 1.565 / 97.69 | 1.456 / 90.91 | 2.163 / 135.0 | 1.291 / 80.60 | 1.587[4] @ −73°C (−100°F) / 99.08 | 1.156 / 72.16 | 1.217 / 75.95 | 1.233 @ −30°C (−22°F) / 76.95 |
| 4.66 / 0.291 | 7.02[5] / 0.438 | 7.38 / 0.461 | 7.83 / 0.489 | | 8.37 / 0.522 | 9.01[4] / 0.562 | 5.278 / 0.3295 | 6.22 / 0.388 | 6.02 / 0.374 |
| 5 @ −30°C (−22°F) | | 0.218 | 0.243 | 0.166 | 0.285 | 0.232 @ −73°C (−100°F)[4] | 0.258 | 0.293 | 0.287 @ −30°C (−22°F) |
| 0.176 | | 0.161 @ 60°C (140°F) | 0.170 | | 0.164 | 0.182[11] @ 0 pressure | 0.175 | 0.164 | 0.16 |
| 1.191 / 0 pressure | | 1.080 @ 60°C (140°F) | 1.084 | | 1.091 | 1.085 (est) @ 0 pressure | 1.143 | 1.132 | 1.21 @ −34°C (−30°F) |
| 57.23 / 103.02 | 37 (est) / 67 | 35.07 / 63.12 | 32.51 / 58.53 | 25 (est) / 45 (est) | 30.11 / 54.20 | 27.97 / 50.35 | 48.04 / 86.47 | 41.21 / 74.18 | 42.86 / 77.15 |
| 9 / 50 @ −30°C (−22°F) | 0.040 | 0.0434 / 0.0044 (0.5 atm) | 0.0372 / 0.0060 | 0.027 | 0.0302 / 0.00724 | 0.045 / 0.0098 @ −73°C (−100°F) | 0.0432 | 0.0373 / 0.00670 | 0.0430 @ −30°C (−22°F) |
| 7 / 8 @ −30°C (−22°F) | 1.21[6] | 0.68 / 0.010 (0.1 atm) | 0.36 / 0.0112 | 0.72 | 0.193 / 0.0125 | 0.30 / 0.0148 | 0.192 / 0.0120 | 0.180 / 0.0126 | 0.144 @ −30°C (−22°F) |
| @ −73°C (−100°F) | 23 @ 30°C (86°F) | 17.3 | 12 | 18 | 5 | 16 @ −73°C (−100°F) | 8.4 | 5.9 | 6.1 @ −30°C (−22°F) |
| 5 @ −73°C (−100°F) | 1.413 | 1.354 | 1.288 | 1.367 | 1.214 | 1.206 @ −73°C (−100°F) | 1.273 | 1.234 | 1.209 @ −30°C (−22°F) |
| 1.04 | 5 (est) | 3.9 (0.44 atm) | 3.34 | 4.02 (0.44 atm) | 2.54 | 2.02 | | 1.3 | |
| 73 @ 25°C[9b] | 2.54 @ 25°C (77°F) | 2.41 @ 25°C (77°F) | 2.26 @ 25°C / 1.0043 @ 26.8°C | 2.34 @ 25°C (77°F) | 1.0035 @ 27.4°C | 1.0021 @ 23°C (73°F) | | 6.11 @ 25°C 1.0035 (0.5 atm) | |
| 0.10 | 0.012 (Sat'n Pres) | 0.017 (Sat'n Pres) | 0.013 | | 0.006 | | | | 0.042 |
| | | 0.011 | 0.009 | | | | 0.056 | 0.056 | |

a.  CCl₂F₂/CH₃CHF₂    (73.8/26.2% by wt.)
b.  CHClF₂/CClF₂CF₃   (48.8/51.2% by wt.)
c.  CHF₃/CClF₃        (40/60% by wt.)

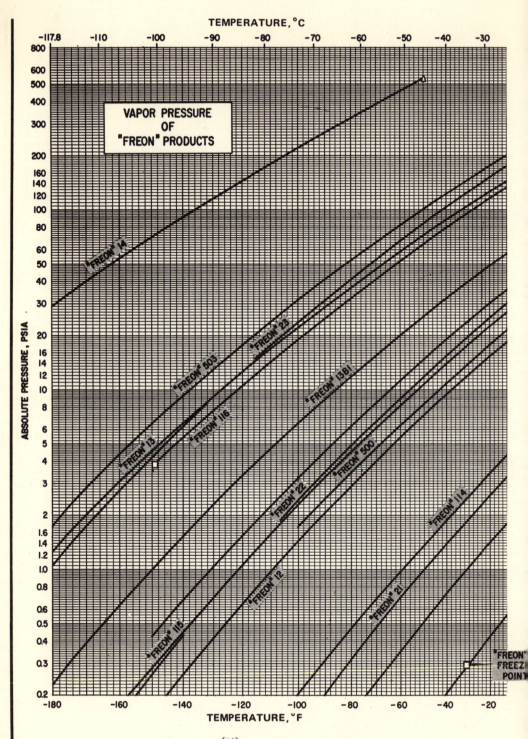

**Figure 11.3.**[14] Refrigerant properties.

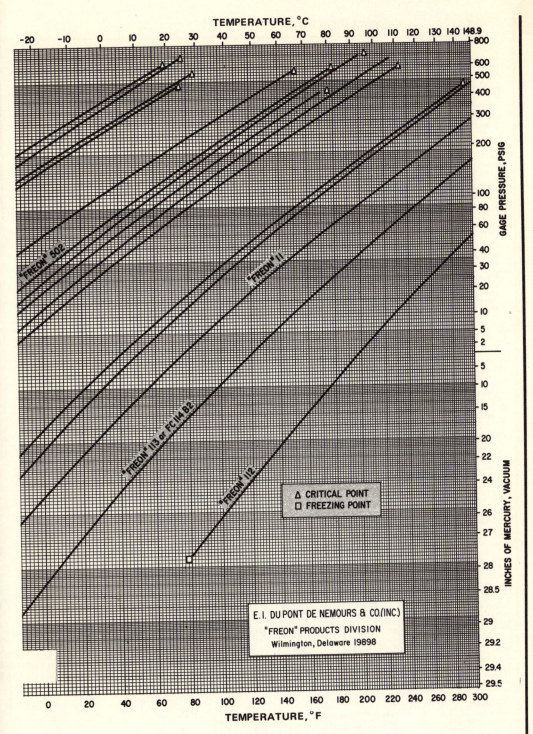

**Figure 11.3.**[14] *Cont'd.*

**Table 11.15.**[15]  Molecular Diffusivities ($D$. in cm$^2$/sec)

| Substance | Temp., °C. | Air | A | $H_2$ | $O_2$ | $N_2$ | $CO_2$ | $N_2O$ | $CH_4$ | $C_2H_6$ | $C_2H_4$ | $n\text{-}C_4H_{10}$ | $i\text{-}C_4H_{10}$ |
|---|---|---|---|---|---|---|---|---|---|---|---|---|---|
| Acetic acid | 0 | 0.1064 | | 0.416 | | | 0.0716 | | | | | | |
| Acetone | 0 | .109 | | .361 | | | | | | | | | |
| n-Amyl alcohol | 0 | .0589 | | .235 | | | | | | | | | |
| sec-Amyl alcohol | 30 | .072 | | | | | .0422 | | | | | | |
| Amyl butyrate | 0 | .040 | | | | | | | | | | | |
| Amyl formate | 0 | .0543 | | | | | | | | | | | |
| i-Amyl formate | 0 | .058 | | | | | | | | | | | |
| Amyl isobutyrate | 0 | .0419 | | .171 | | | | | | | | | |
| Amyl propionate | 0 | .046 | | .1914 | | | .0347 | | | | | | |
| Aniline | 0 | .0610 | | | | | | | | | | | |
| | 30 | .075 | | | | | | | | | | | |
| Anthracene | 0 | .0421 | | | | | | | | | | | |
| Argon | 20 | | | | | .194 | | | | | | | |
| Benzene | 0 | .077 | | .306 | 0.0797 | | .0528 | | | | | | |
| Benzidine | 0 | .0298 | | | | | | | | | | | |
| Benzyl chloride | 0 | .066 | | | | | | | | | | | |
| n-Butyl acetate | 0 | .058 | | | | | | | | | | | |
| i-Butyl acetate | 0 | .0612 | | .2364 | | | .0425 | | | | | | |
| n-Butyl alcohol | 0 | .0703 | | .2716 | | | .0476 | | | | | | |
| | 30 | .088 | | | | | | | | | | | |
| i-Butyl alcohol | 0 | .0727 | | .2771 | | | .0483 | | | | | | |
| Butyl amine | 0 | .0821 | | | | | | | | | | | |
| i-Butyl amine | 0 | .0853 | | | | | | | | | | | |
| i-Butyl butyrate | 0 | .0468 | | .185 | | | .0327 | | | | | | |
| i-Butyl formate | 0 | .0705 | | | | | | | | | | | |
| i-Butyl isobutyrate | 0 | .0457 | | .191 | | | .0364 | | | | | | |
| i-Butyl proprionate | 0 | .0529 | | .203 | | | .0366 | | | | | | |
| i-Butyl valerate | 0 | .0424 | | .173 | | | .0308 | | | | | | |
| Butyric acid | 0 | .067 | | .264 | | | .0476 | | | | | | |
| i-Butyric acid | 0 | .0679 | | .271 | | | .0471 | | | | | | |
| Cadmium | 0 | | | | | .17 | | | | | | | |
| Caproic acid | 0 | .050 | | | | | | | | | | | |
| i-Caproic acid | 0 | .0513 | | | | | | | | | | | |
| Carbon dioxide | 0 | .138 | | .550 | .139 | | | .096 | .153 | | | | |
| | 20 | | | | | .163 | | | | | | | |
| | 25 | | | | | | | .0996* | .002151† | | | | |
| | 500‡ | | | | .9 | | | | | | | | |
| Carbon disulfide | 0 | .0892 | | .369 | | | | | | | | | |
| Carbon monoxide | 0 | | | .651 | .185 | | .063 | | | | .116 | | |
| | 450‡ | | | | 1.0 | | .137 | | | | | | |
| Carbon tetrachloride | θ | | | .293 | 0.0636 | | | | | | | | |
| Chlorobenzene | 30 | .075 | | | | | | | | | | | |
| Chloroform | 0 | .091 | | | | | | | | | | | |
| Chloropicrin | 25 | .088 | | | | | | | | | | | |
| m-Chlorotoluene | 0 | .054 | | | | | | | | | | | |
| o-Chlorotoluene | 0 | .059 | | | | | | | | | | | |
| p-Chlorotoluene | 0 | .051 | | | | | | | | | | | |
| Cyanogen chloride | 0 | .111 | | | | | | | | | | | |
| Cyclohexane | 15 | | 0.0719 | .319 | .0744 | .0760 | | | | | | | |
| | 45 | .086 | | | | | | | | | | | |
| n-Decane | 90 | | | .306 | | .0841 | | | | | | | |
| Diethylamine | 0 | .0884 | | | | | | | | | | | |
| 2,3-Dimethyl butane | 15 | | .0657 | .301 | .0753 | .0751 | | | | | | | |
| Diphenyl | 0 | .0610 | | | | | | | | | | | |
| n-Dodecane | 126 | | | .308 | | .0813 | | | | | | | |
| Ethane | 0 | | | .459 | | | | | | | | | |
| Ethanol | 0 | | | .377 | | | .0686 | | | | | | |
| Ether (diethyl) | 0 | .0778 | | .298 | | | .0546 | | | | | | |
| Ethyl acetate | 0 | .0715 | | .273 | | | .0487 | | | | | | |
| | 30 | .089 | | | | | | | | | | | |
| Ethyl alcohol | 0 | .102 | | .375 | | | .0685 | | | | | | |
| Ethyl benzene | 0 | .0658 | | | | | | | | | | | |
| Ethyl n-butyrate | 0 | .0579 | | .224 | | | .0407 | | | | | | |
| Ethyl i-butyrate | 0 | .0591 | | .229 | | | .0413 | | | | | | |
| Ethylene | 0 | | | .486 | | | | | | | | | |
| Ethyl formate | 0 | .0840 | | .337 | | | .0573 | | | | | | |
| Ethyl propionate | 0 | .068 | | .236 | | | .0450 | | | | | | |
| Ethyl valerate | 0 | .0512 | | .205 | | | .0367 | | | | | | |
| Eugenol | 0 | .0377 | | | | | | | | | | | |
| Formic acid | 0 | .1308 | | .510 | | | .0874 | | | | | | |
| Helium | 0 | | .641 | | | | | | | | | | |
| | 20 | | | | | .705 | | | | | | | |
| n-Heptane | 38 | | | | | | | | | | | | |
| n-Hexane | 15 | | .0663 | .290 | .0753 | .0757 | | | .066§ | | | | |
| Hexyl alcohol | 0 | .0499 | | .200 | | | .0351 | | | | | | |
| Hydrogen | 0 | .611 | | | .697 | .674 | .550 | .535 | .625 | 0.459 | 0.486 | 0.272 | 0.277 |
| | 25 | | | | | | | | .646 | .537 | .726 | | |
| | 500 | | | 4.2 | | | | | | | | | |
| Hydrogen cyanide | 0 | .173 | | | | | | | | | | | |
| Hydrogen peroxide | 60 | .188 | | | | | | | | | | | |
| Iodine | 0 | .07 | | | | .070 | | | | | | | |
| Mercury | 0 | .112 | | .53 | | | | | | | | | |
| Mesitylene | 0 | .056 | | | | .13 | | | | | | | |
| Methane | 500 | | | | 1.1 | | | | | | | | |
| Methyl acetate | 0 | .084 | | .333 | | | .0567 | | | | | | |
| Methyl alcohol | 0 | .132 | | .506 | | | .0879 | | | | | | |
| Methyl butyrate | 0 | .0633 | | .242 | | | .0446 | | | | | | |
| Methyl i-butyrate | 0 | .0639 | | .257 | | | .0451 | | | | | | |
| Methyl cyclopentane | 15 | | .0731 | .318 | .0742 | .0758 | | | | | | | |
| Methyl formate | 0 | .0872 | | | | | | | | | | | |
| Methyl propionate | 0 | .0735 | | .295 | | | .0528 | | | | | | |

## Table 11.15.[15]  Cont'd.

| Substance | Temp., °C. | Air | A | $H_2$ | $O_2$ | $N_2$ | $CO_2$ | $N_2O$ | $CH_4$ | $C_2H_6$ | $C_2H_4$ | $n\text{-}C_4H_{10}$ | $i\text{-}C_4H_{10}$ |
|---|---|---|---|---|---|---|---|---|---|---|---|---|---|
| Methyl valerate | 0 | 0.0569 | | | | | | | | | | | |
| Naphthalene | 0 | .0513 | | | | | | | | | | | |
| Nitrogen | 0 | | | | 0.181 | | 0.165 | | | 0.148 | 0.163 | 0.0960 | 0.0908 |
|  | 25 | | | | | | | | | | | | |
| Nitrous oxide | 0 | | | 0.535 | | | .096 | | | | | | |
| n-Octane | 0 | .0505 | | | | | | | | | | | |
|  | 30 | | 0.0642 | .271 | .0705 | 0.0710 | | | | | | | |
| Oxygen | 0 | .178 | | .697 | | .181 | .139 | | | | | | |
| Phosgene | 0 | .095 | | | | | | | | | | | |
| Propionic acid | 0 | .0829 | | .330 | | | .0588 | | | | | | |
| Propyl acetate | 0 | .067 | | | | | .0577 | | | | | | |
| n-Propyl alcohol | 0 | .085 | | .315 | | | | | | | | | |
| i-Propyl alcohol | 0 | .0818 | | | | | | | | | | | |
|  | 30 | .101 | | | | | | | | | | | |
| n-Propyl benzene | 0 | .0481 | | | | | | | | | | | |
| i-Propyl benzene | 0 | .0489 | | | | | | | | | | | |
| n-Propyl bromide | 0 | .085 | | | | | | | | | | | |
| i-Propyl bromide | 0 | .0902 | | | | | .0364 | | | | | | |
| Propyl butyrate | 0 | .0530 | | .206 | | | | | | | | | |
| Propyl formate | 0 | .0712 | | .281 | | | .0490 | | | | | | |
| n-Propyl iodide | 0 | .079 | | | | | | | | | | | |
| i-Propyl iodide | 0 | .0802 | | .212 | | | .0388 | | | | | | |
| n-Propyl isobutyrate | 0 | .0549 | | | | | | | | | | | |
| i-Propyl isobutyrate | 0 | .059 | | | | | | | | | | | |
| Propyl propionate | 0 | .057 | | .212 | | | .0395 | | | | | | |
| Propyl valerate | 0 | .0466 | | .189 | | | .0341 | | | | | | |
| Safrol | 0 | .0434 | | | | | | | | | | | |
| i-Safrol | 0 | .0455 | | .418 | | | | | | | | | |
| Sulfur hexafluoride | 25 | | | | | | | | | | | | |
| Toluene | 0 | .076 | .071 | | | | | | | | | | |
|  | 30 | .088 | | | | | | | | | | | |
| Trimethyl carbinol | 0 | .087 | | | | | | | | | | | |
| 2,2,4-Trimethyl pentane | 30 | | .0618 | .288 | .0688 | .0705 | | | | | | | |
| 2,2,3-Trimethyl heptane | 90 | | | .270 | | .0684 | | | | | | | |
| n-Valeric acid | 0 | .050 | | | | | .0376 | | | | | | |
| i-Valeric acid | 0 | .0544 | | .212 | | | .138 | | | | | | |
| Water | 0 | .220 | | .75 | | | | | | | | | |
|  | 450 | | | | 1.3 | | | | | | | | |

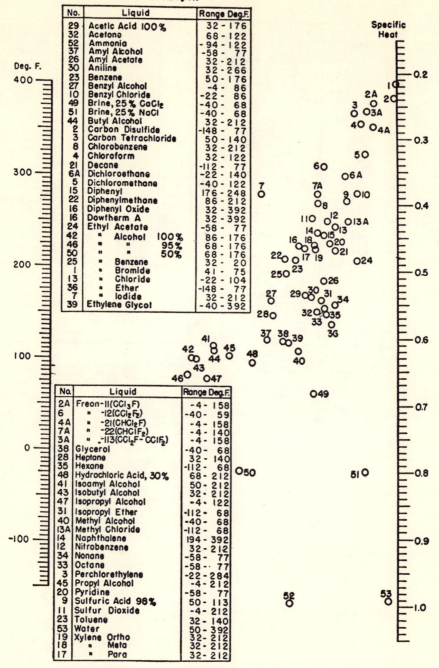

Specific heat = Btu/(lb.)(Deg. F.)

| No. | Liquid | Range Deg.F. |
|-----|--------|--------------|
| 29 | Acetic Acid 100% | 32 - 176 |
| 32 | Acetone | 68 - 122 |
| 52 | Ammonia | -94 - 122 |
| 37 | Amyl Alcohol | -58 - 77 |
| 26 | Amyl Acetate | 32 - 212 |
| 30 | Aniline | 32 - 266 |
| 23 | Benzene | 50 - 176 |
| 27 | Benzyl Alcohol | -4 - 86 |
| 10 | Benzyl Chloride | -22 - 86 |
| 49 | Brine, 25% CaCl₂ | -40 - 68 |
| 51 | Brine, 25% NaCl | -40 - 68 |
| 44 | Butyl Alcohol | 32 - 212 |
| 2 | Carbon Disulfide | -148 - 77 |
| 3 | Carbon Tetrachloride | 50 - 140 |
| 8 | Chlorobenzene | 32 - 212 |
| 4 | Chloroform | 32 - 122 |
| 21 | Decane | -112 - 77 |
| 6A | Dichloroethane | -22 - 140 |
| 5 | Dichloromethane | -40 - 122 |
| 15 | Diphenyl | 176 - 248 |
| 22 | Diphenylmethane | 86 - 212 |
| 16 | Diphenyl Oxide | 32 - 392 |
| 16 | Dowtherm A | 32 - 392 |
| 24 | Ethyl Acetate | -58 - 77 |
| 42 | " Alcohol 100% | 86 - 176 |
| 46 | " " 95% | 68 - 176 |
| 50 | " " 50% | 68 - 176 |
| 25 | " Benzene | 32 - 20 |
| 1 | " Bromide | 41 - 75 |
| 13 | " Chloride | -22 - 104 |
| 36 | " Ether | -148 - 77 |
| 7 | " Iodide | 32 - 212 |
| 39 | Ethylene Glycol | -40 - 392 |

| No. | Liquid | Range Deg.F. |
|-----|--------|--------------|
| 2A | Freon-11(CCl₃F) | -4 - 158 |
| 6 | " -12(CCl₂F₂) | -40 - 59 |
| 4A | " -21(CHCl₂F) | -4 - 158 |
| 7A | " -22(CHClF₂) | -4 - 140 |
| 3A | " -113(CCl₂F-CClF₂) | -4 - 158 |
| 38 | Glycerol | -40 - 68 |
| 28 | Heptane | 32 - 140 |
| 35 | Hexane | -112 - 68 |
| 48 | Hydrochloric Acid, 30% | 68 - 212 |
| 41 | Isoamyl Alcohol | 50 - 212 |
| 43 | Isobutyl Alcohol | 32 - 212 |
| 47 | Isopropyl Alcohol | -4 - 122 |
| 31 | Isopropyl Ether | -112 - 68 |
| 40 | Methyl Alcohol | -40 - 68 |
| 13A | Methyl Chloride | -112 - 68 |
| 14 | Naphthalene | 194 - 392 |
| 12 | Nitrobenzene | 32 - 212 |
| 34 | Nonane | -58 - 77 |
| 33 | Octane | -58 - 77 |
| 3 | Perchlorethylene | -22 - 284 |
| 45 | Propyl Alcohol | -4 - 212 |
| 20 | Pyridine | -58 - 77 |
| 9 | Sulfuric Acid 98% | 50 - 113 |
| 11 | Sulfur Dioxide | -4 - 212 |
| 23 | Toluene | 32 - 140 |
| 53 | Water | 50 - 392 |
| 19 | Xylene Ortho | 32 - 212 |
| 18 | " Meta | 32 - 212 |
| 17 | " Para | 32 - 212 |

**Figure 11.4.**[16]  Specific heats of liquids.

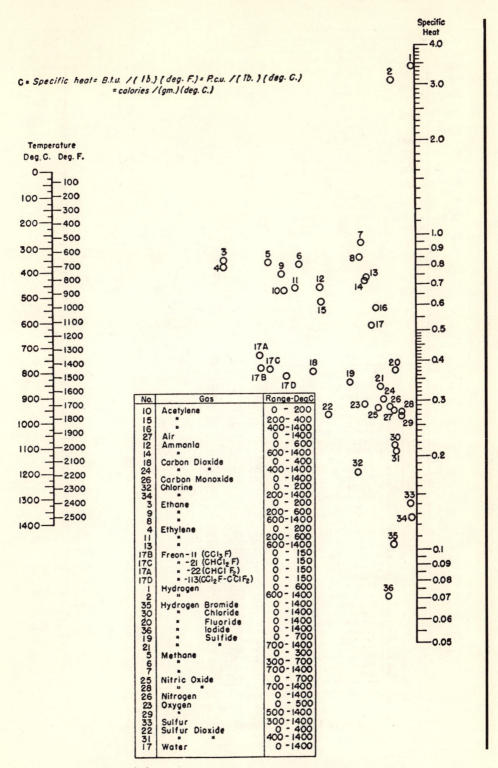

**Figure 11.5.**[17]   Specific heats ($C_p$) of gases at 1-atm pressure.

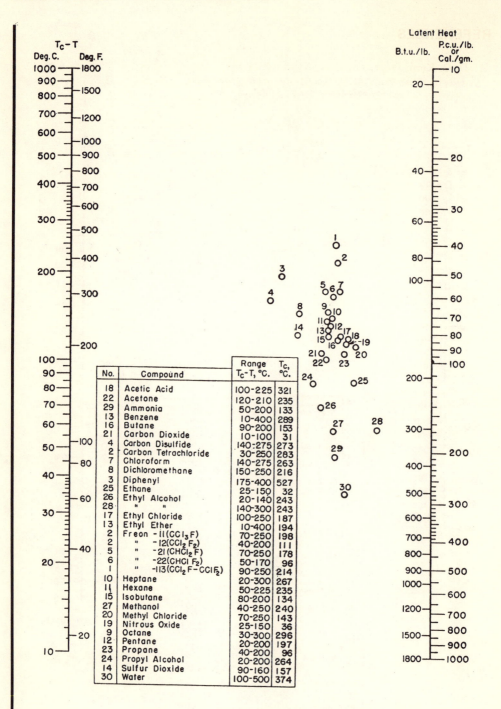

**Figure 11.6.**[18] Latent heats of vaporization.

# REFERENCES

1. D. Q. Kern, *Process Heat Transfer*, McGraw-Hill, New York, pp. 795–798. (Originally published in L. S. Maihi, *Mechanical Engineers' Handbook*, McGraw-Hill, New York, 1941.)
2. Ref. 1, p. 799.
3. R. H. Perry and C. H. Chilton, *Chemical Engineers' Handbook*, 5th ed., McGraw-Hill, New York, 1973, p. 3-214.
4. R. H. Perry, C. H. Chilton, and S. D. Kirkpatrick, *Chemical Engineers' Handbook*, 4th ed., McGraw-Hill, New York, 1963, p. 3-206.
5. Ref. 3, p. 3-134.
6. Ref. 3, pp. 3-126–3-128.
7. Ref. 3, p. 3-136.
8. Ref. 1, p. 808.
9. Ref. 3, pp. 3-115–3-116.
10. Ref. 3, pp. 3-212–3-213.
11. Ref. 3, pp. 3-210–3-211.
12. Platecoil Product Data Manual No. 5-63, Platecoil Division, Tranter Manufacturing, Inc., Lansing, Michigan, p. 57.
13. Ref. 12, p. 76.
14. Freon Product Information Report No. B-2, E. I. DuPont de Nemours & Co., Wilmington, Del., 1971, pp. 9–10.
15. Ref. 4, pp. 14-22–14-23.
16. Ref. 1, p. 804.
17. Ref. 3, p. 3-130.
18. Ref. 4, p. 3-114.
19. Keenan, J. H., and Keyes, F. G., Thermodynamic Properties of Steam, 33rd Printing, Wiley, New York, 1961, p. 76.

# 12 Dimensions and Properties of Piping

**Table 12.1.**[1]  Dimensions and Properties of Steel Pipe

| Nominal pipe size (IPS) (in.) | Outside diameter (in.) | Schedule No.[a] | Wall thickness (in.) | Inside diameter (in.) | External surface (ft² per ft length) | Weight (lb/ft) | Internal area[b] (ft²) | Working pressure PSIA |
|---|---|---|---|---|---|---|---|---|
| ⅛ | 0.405 | 40 (s) | 0.068 | 0.269 | 0.106 | 0.244 | 0.00039 | 314 |
|   | 0.405 | 80 (x) | 0.095 | 0.215 | 0.106 | 0.314 | 0.00025 | 1084 |
| ¼ | 0.540 | 40 (s) | 0.088 | 0.364 | 0.14` | 0.424 | 0.00072 | 649 |
|   | 0.540 | 80 (x) | 0.119 | 0.302 | 0.14. | 0.535 | 0.00050 | 1353 |
| ⅜ | 0.675 | 40 (s) | 0.091 | 0.493 | 0.177 | 0.567 | 0.00133 | 574 |
|   | 0.675 | 80 (x) | 0.126 | 0.423 | 0.177 | 0.738 | 0.00098 | 1191 |
| ½ | 0.840 | 40 (s) | 0.109 | 0.622 | 0.220 | 0.850 | 0.00211 | 697 |
|   | 0.840 | 80 (x) | 0.147 | 0.546 | 0.220 | 1.00 | 0.00163 | 1266 |
|   | 0.840 | .. xx | 0.294 | 0.252 | 0.220 | 1.71 | 0.00035 | 3824 |
| ¾ | 1.050 | 40 (s) | 0.113 | 0.824 | 0.275 | 1.13 | 0.00370 | 604 |
|   | 1.050 | 80 (x) | 0.154 | 0.742 | 0.275 | 1.47 | 0.00300 | 1078 |
|   | 1.050 | .. xx | 0.308 | 0.434 | 0.275 | 2.44 | 0.00103 | 3134 |
| 1 | 1.315 | 40 (s) | 0.133 | 1.049 | 0.344 | 1.68 | 0.00600 | 651 |
|   | 1.315 | 80 (x) | 0.179 | 0.957 | 0.344 | 2.17 | 0.00500 | 1083 |
|   | 1.315 | .. xx | 0.358 | 0.599 | 0.344 | 3.66 | 0.00196 | 2963 |
| 1¼ | 1.660 | 40 (s) | 0.140 | 1.380 | 0.435 | 2.27 | 0.01039 | 440 |
|   | 1.660 | 80 (x) | 0.191 | 1.278 | 0.435 | 3.00 | 0.00891 | 805 |
|   | 1.660 | .. xx | 0.382 | 0.896 | 0.435 | 5.21 | 0.00438 | 2318 |
| 1½ | 1.900 | 40 (s) | 0.145 | 1.610 | 0.497 | 2.72 | 0.01414 | 417 |
|   | 1.900 | 80 (x) | 0.200 | 1.500 | 0.497 | 3.65 | 0.01227 | 756 |
|   | 1.900 | .. xx | 0.400 | 1.100 | 0.497 | 6.41 | 0.00660 | 2122 |
| 2 | 2.375 | 40 (s) | 0.154 | 2.067 | 0.622 | 3.65 | 0.02330 | 376 |
|   | 2.375 | 80 (x) | 0.218 | 1.939 | 0.622 | 5.02 | 0.02051 | 690 |
|   | 2.375 | .. xx | 0.436 | 1.503 | 0.622 | 9.03 | 0.01232 | 1861 |
| 2½ | 2.875 | 40 (s) | 0.203 | 2.469 | 0.753 | 5.79 | 0.03325 | 505 |
|   | 2.875 | 80 (x) | 0.276 | 2.323 | 0.753 | 7.66 | 0.02943 | 806 |
|   | 2.875 | .. xx | 0.552 | 1.771 | 0.753 | 13.7 | 0.01711 | 2048 |
| 3 | 3.500 | 40 (s) | 0.216 | 3.068 | 0.916 | 7.57 | 0.05134 | 454 |
|   | 3.500 | 80 (x) | 0.300 | 2.900 | 0.916 | 10.3 | 0.04587 | 734 |
|   | 3.500 | .. xx | 0.600 | 2.300 | 0.916 | 18.5 | 0.02885 | 1829 |
| 3½ | 4.000 | 40 (s) | 0.226 | 3.548 | 1.05 | 9.11 | 0.06866 | 425 |
|   | 4.000 | 80 (x) | 0.318 | 3.364 | 1.05 | 12.5 | 0.06172 | 692 |
|   | 4.000 | .. xx | 0.636 | 2.728 | 1.05 | 22.9 | 0.04059 | 1699 |
| 4 | 4.500 | 40 (s) | 0.237 | 4.026 | 1.18 | 10.8 | 0.08840 | 403 |
|   | 4.500 | 80 (s) | 0.337 | 3.826 | 1.18 | 14.9 | 0.07984 | 663 |
|   | 4.500 | .. xx | 0.674 | 3.152 | 1.18 | 27.5 | 0.05419 | 1602 |
| 5 | 5.563 | 40 (s) | 0.258 | 5.047 | 1.46 | 14.6 | 0.13898 | 498 |
|   | 5.563 | 80 (x) | 0.375 | 4.813 | 1.46 | 20.8 | 0.12635 | 825 |
|   | 5.563 | .. xx | 0.750 | 4.063 | 1.46 | 38.6 | 0.09004 | 1951 |

**Table 12.1.**[1]  *Cont'd.*

| Nominal pipe size (IPS) (in.) | Outside diameter (in.) | Schedule No.[a] | Wall thickness (in.) | Inside diameter (in.) | External surface (ft² per ft length) | Weight (lb/ft) | Internal area[b] (ft²) | Working pressure PSIA |
|---|---|---|---|---|---|---|---|---|
| 6 | 6.625 | 40 (s) | 0.280 | 6.065 | 1.73 | 18.0 | 0.20063 | 467 |
|  | 6.625 | 80 (x) | 0.432 | 5.761 | 1.73 | 28.6 | 0.18102 | 825 |
|  | 6.625 | .. xx | 0.864 | 4.897 | 1.73 | 53.1 | 0.13079 | 1912 |
| 8 | 8.625 | 30 (s) | 0.277 | 8.071 | 2.26 | 24.7 | 0.35529 | 351 |
|  | 8.625 | 40 (s) | 0.322 | 7.981 | 2.26 | 28.6 | 0.34741 | 431 |
|  | 8.625 | 80 (x) | 0.500 | 7.625 | 2.26 | 43.4 | 0.31711 | 753 |
|  | 8.625 | .. xx | 0.875 | 6.875 | 2.26 | 72.4 | 0.25779 | 1460 |
| 10 | 10.750 | .. (s) | 0.279 | 10.192 | 2.81 | 31.2 | 0.56656 | 285 |
|  | 10.750 | 30 (s) | 0.307 | 10.136 | 2.81 | 34.2 | 0.56035 | 324 |
|  | 10.750 | 40 (s) | 0.365 | 10.020 | 2.81 | 40.5 | 0.54760 | 405 |
|  | 10.750 | 60 (x) | 0.500 | 9.750 | 2.81 | 54.7 | 0.51849 | 600 |
| 12 | 12.750 | 30 (s) | 0.330 | 12.090 | 3.34 | 43.8 | 0.79722 | 299 |
|  | 12.750 | .. (s) | 0.375 | 12.000 | 3.34 | 49.6 | 0.78540 | 352 |
|  | 12.750 | .. (x) | 0.500 | 11.750 | 3.34 | 65.4 | 0.75301 | 503 |
| 14 | 14.000 | 30 (s) | 0.375 | 13.250 | 3.67 | 54.6 | 0.95754 | 458 |
|  | 14.000 | .. (x) | 0.500 | 13.000 | 3.67 | 72.1 | 0.92175 | 653 |
| 16 | 16.000 | 30 (s) | 0.375 | 15.250 | 4.18 | 62.4 | 1.26843 | 400 |
|  | 16.000 | 40 (x) | 0.500 | 15.000 | 4.18 | 82.8 | 1.22718 | 570 |
| 18 | 18.000 | .. (s) | 0.375 | 17.250 | 4.71 | 70.6 | 1.62295 | 355 |
|  | 18.000 | .. (x) | 0.500 | 17.000 | 4.71 | 93.5 | 1.57625 | 506 |
| 20 | 20.000 | 20 (s) | 0.375 | 19.250 | 5.23 | 78.6 | 2.02110 | 319 |
|  | 20.000 | 30 (s) | 0.500 | 19.000 | 5.23 | 104.2 | 1.96895 | 454 |
| 24 | 24.000 | 20 | 0.375 | 23.250 | 6.29 | 94.6 | 2.94831 | 265 |
|  | 24.000 | .. (s) | 0.500 | 23.000 | 6.29 | 125.5 | 2.88525 | 378 |

[a]The letters "s," "x," and "xx" in the column of schedule numbers indicate standard, extra strong, and double extra strong pipe, respectively.

[b]The values shown in ft² for the internal area also represent the volume in cubic feet per foot of pipe length.

**Table 12.2.**[2]  Dimensional Data, Plastic-Lined Pipe[a,b]

| Pipe size | Lined pipe o.d. A | Linear thickness B | | Approx. wt/ft (lb) w/o flange | Minimum available pipe length[c] | |
|---|---|---|---|---|---|---|
| | | SL, PPL, KL | TFE | | Gasket joint | Molded raised face |
| 1 | 1.315 | 0.150 | 0.125 | 2.4 | 2.188 | 2.438 |
| 1.5 | 1.900 | 0.160 | 0.125 | 3.4 | 2.438 | 2.688 |
| 2 | 2.375 | 0.172 | 0.125 | 4.4 | 2.813 | 3.063 |
| 2.5[d] | 2.875 | 0.175 | — | 7.0 | 3.063 | 3.313 |
| 3 | 3.500 | 0.175 | 0.125 | 8.5 | 3.188 | 3.438 |
| 4 | 4.500 | 0.207 | 0.125 | 12.2 | 3.313 | 3.563 |
| 6 | 6.625 | 0.218 | 0.140 | 22.5 | 3.688 | 3.938 |
| 8 | 8.625 | 0.218 | 0.150 | 28.6 | 3.938 | 4.188 |

[a]All dimensions are nominal and in inches.
[b]SL, Saran lined; PPL, Polypropylene lined; KL, Kynar lined; TFE, polytetrafluoroethylene resin lined.
[c]Based on minimum clearance between flange bolts.
[d]Size not available with liner of Kynar or TFE resin.

# REFERENCES

1. Platecoil Catalog No. 5-63, Platecoil Division, Tranter Manufacturing, Inc., Lansing, Michigan.
2. Dow Plastic Lined Pipe and Fittings Catalog, Dow Chemical USA, Midland, Michigan, 1972, p. 11.

# 13    Pump Sizing

## PERFORMANCE

Use the procedure outlined in Figure 13.1 to predict pump requirements. See Section 6.1 to obtain equivalent lengths of fittings and to calculate line loss per foot of pipe due to friction.

## HORSEPOWER

Brake horsepower (bhp) is the actual horsepower delivered to the pump shaft. Hydraulic horsepower (whp) is the liquid horsepower delivered by the pump.

$$\text{whp} = \frac{Q \times H \times SG}{3,960} \tag{13.1}$$

$$\text{bhp} = \frac{\text{whp}}{\text{pump efficiency}} \tag{13.2}$$

where $Q$ is the capacity (gpm); $H$ is the head (ft); $SG$ is the specific gravity.

## NET POSITIVE SECTION HEAD AVAILABLE (NPSH)[2]

NPSH available is the total suction head in feet of liquid absolute, determined at the suction nozzle and corrected to datum, less the vapor pressure of the liquid in feet absolute.

$$\text{NPSH}_a = H_p \pm H_s - H_f - H_{vp} \tag{13.3}$$

where $H_p$ is the absolute pressure on the surface of the liquid where the pump takes suction in feet of liquid; $H_s$ is the static elevation of the liquid above the centerline of the pump in feet of liquid (If the liquid line is below the pump centerline, $H_s$ is minus); $H_f$ is the friction and entrance head losses in the suction piping in feet of liquid; $H_{vp}$ is the absolute vapor pressure of the liquid at the pumping temperature in feet of liquid; $\text{NPSH}_a$ is the net positive suction head available.

### PUMP CALCULATION SHEET

| PLANT | NO. UNITS | JOB NUMBER |
|---|---|---|
| LOCATION | BLDG. NO. | PUMP NUMBER |
| BY | | DATE |

**CONDITIONS (left)**

- Liquid Pumped _____
- T S | Pumping Temperature _____ °F
- Specific Gravity at T _____
- Viscosity at T _____ cps–cts
- Vapor Pressure at T _____ psia
- $F_N$ Normal Flow _____ lb./hr.
- $F_M$ Max. Flow _____ lb./hr.
- $Q_N$ Normal gpm = $F_N/500S$ _____ gpm
- $Q_M$ Max. gpm = $F_M/500S$ _____ gpm
- Corrosion or Erosion Factors:

**CONDITIONS (right)**

- G | PSI per ft. = .433S
- $C_Q$ | Viscosity Capacity Correction _____
- $C_E$ | Viscosity Efficiency Correction _____
- $C_H$ | Viscosity Head Correction _____
- $\dfrac{\text{Control Valve Drop}}{\text{Design Dynamic Drop}} = \dfrac{\Delta_c}{\Delta}$
- $\dfrac{\text{Safety Head}}{\text{Design Pump Head}} = \dfrac{\Delta s}{(P_{ND}+\Delta_c-P_s)_D}$

**PIPING (See Ft. for Fitting Data)**

| Pipe I.D. | SUCTION, $d_S$ = | | | Discharge, $d_D$ = | | | Remarks |
|---|---|---|---|---|---|---|---|
| Fitting Type | Equiv. Length, $L_e$ | No. $N_F$ | $L_e \times N_F$ | Equiv. Length, $L_e$ | No. $N_F$ | $L_e \times N_F$ | |
| 90° elbow | | | | | | | |
| 45° elbow | | | | | | | |
| Tee-run | | | | | | | |
| Tee-branch | | | | | | | |
| | | | | | | | |
| | | | | | | | |
| | | | | | | | |
| | | | | | | | |
| | | | | | | | |
| Total Fitting Equiv. Length = | | | | = | | | |
| Actual Pipe Length = | | | | = | | | |
| Equiv. + Actual, $L_S$ = | | | | $L_D$ = | | | |
| Net Static Head, $h_S$ = | | | | $h_D$ = | | | |

**SUCTION**

| | | Normal | Design | Units |
|---|---|---|---|---|
| $V_s$ | Velocity, $0.408\,Q/d_s^2$ | | | FPS |
| $f_s$ | Friction Loss | | | PSI/ft |
| | Line Loss = $f_S L_S$ | – | – | PSI |
| | Entrance = $V_s^2 G/129$ | – | – | PSI |
| | Misc. | – | – | PSI |
| | Equip. Press. (min.) | | | PSIA |
| | Static Head = $h_S G$ | | | PSI |
| $P_s$ | Suction Head | | | PSIA |
| | Velocity Head = $V_s^2 G/64$ | | | PSI |
| | Vapor Press. at T | – | – | PSIA |
| $P_{HS}$ | Net Suction Head | | | PSI |
| | NPSH = $P_{HS}/G$ | | | Ft. |

**DISCHARGE**

| | | Normal | Design | Units |
|---|---|---|---|---|
| $V_D$ | Velocity, $0.408\,Q/d_D^2$ | | | FPS |
| $f_D$ | Friction Loss | | | PSI/ft |
| | Line Loss = $f_D L_D$ | | | PSI |
| | Orifice | | | |
| | Exchangers | | | |
| | Misc. | | | |
| $\Delta$ | Dynamic Drop | | | PSI |
| | Equip. Press. (max.) | | | PSIA |
| | Static Head = $h_D G$ | | | PSI |
| | Expansion = $V_D^2 G/64$ | | | PSI |
| $P_{ND}$ | Net Discharge Head | | | PSIA |
| $\Delta c$ | Control Valve | | | PSI |
| $\Delta s$ | Safety Head | | | PSI |
| $P_D$ | Discharge Head | | | PSI |
| $P$ | Pump Head = $P_D - P_S$ | | | PSI |
| $h$ | Differential Head = $P/G$ | | | Ft. |

**HP**

| | | Normal | Design | Units |
|---|---|---|---|---|
| whp | Hyd. HP = $Qhs/3960$ | | | HP |
| bhp | Brake HP = whp/pump eff. | | | HP |

**Figure 13.1.** Pump calculation sheet.

The available NPSH must always be equal to or greater than the NPSH required by the pump. The required NPSH can be determined from the pump performance curve, furnished by the vendor.

Special caution should be exercised with liquids containing dissolved gasses. NPSH problems can be avoided by multiplying the required NPSH value by at least a factor of 2.[4]

## AFFINITY LAWS[1]

1. For impeller diameter held constant:

$$\frac{Q_1}{Q_2} = \frac{N_1}{N_2} \tag{13.4}$$

$$\frac{H_1}{H_2} = \left(\frac{N_1}{N_2}\right)^2 \tag{13.5}$$

$$\frac{(bhp)_1}{(bhp)_2} = \left(\frac{N_1}{N_2}\right)^3 \tag{13.6}$$

2. For pump speed (rpm) held constant:

$$\frac{Q_1}{Q_2} = \frac{D_1}{D_2} \tag{13.7}$$

$$\frac{H_1}{H_2} = \left(\frac{D_1}{D_2}\right)^2 \tag{13.8}$$

$$\frac{(bhp)_1}{(bhp)_2} = \left(\frac{D_1}{D_2}\right)^3 \tag{13.9}$$

where bhp is the brake horsepower; $D$ is the impeller diameter; $H$ is the total head; $Q$ is the capacity (gpm); $N$ is the pump speed (rpm).

## VISCOSITY CORRECTION FACTORS[3]

The performance of centrifugal pumps is affected when handling viscous liquids. A marked increase in brake horsepower, a reduction in head, and some reduction in capacity occur with moderate and high viscosities. Figures 13.2 and 13.3 provide corrections for various viscosities. However, these charts are to be used only within the following limitations.

1. Do not extrapolate.
2. Use only for pumps of conventional design, in their normal operating range, with open or closed impellers.

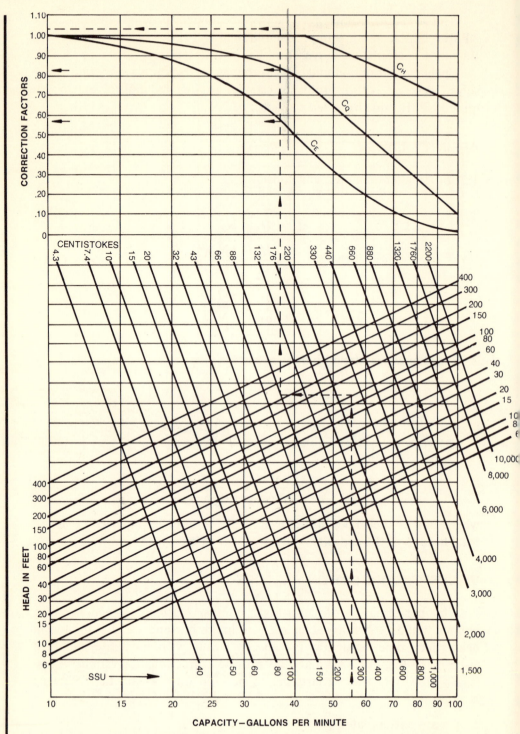

**Figure 13.2.**[3]  Viscosity correction chart (10–100 gpm).  cSt = cP/*SG*.

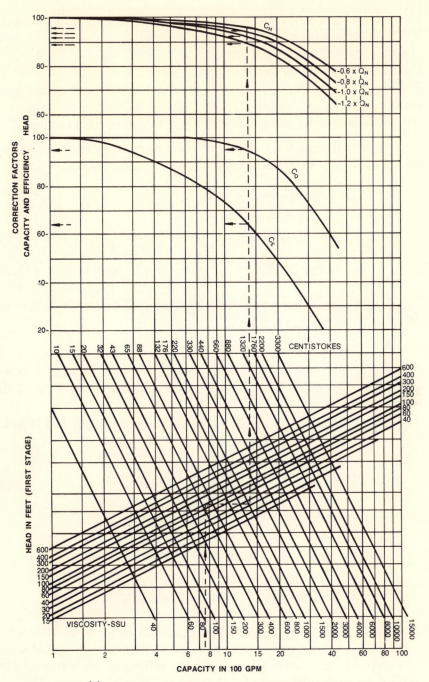

**Figure 13.3.**[3]   Viscosity correction curve (100–10,000 gpm).   cSt = cP/*SG*.

3. Do not use for mixed flow or axial flow pumps or for pumps of special hydraulic design for either viscous or nonuniform liquids.
4. Use only where NPSH is adequate.
5. Use only on Newtonian liquids.
6. The curves should be applied only for the maximum efficiency point for the pump.
7. Where more accurate information is required, contact the vendor.

The following equations are used for determining the viscous performance where the water performance of the pump is known:

$$Q_{vis} = C_q \times Q_w \qquad (13.10)$$

$$H_{vis} = C_h \times H_w \qquad (13.11)$$

$$E_{vis} = C_e \times E_w \qquad (13.12)$$

$$bhp_{vis} = \frac{Q_{vis} \times H_{vis} \times SG}{3960 \times E_{vis}} \qquad (13.13)$$

$C_q$, $C_h$, and $C_e$ are determined from Figures 13.2 and 13.3, which are based on water performance.

### Example*

Select a pump to deliver 750 gpm at 100-ft total head of a liquid having a viscosity of 1,000 SSU and a specific gravity of 0.90 at the pumping temperature.

Enter the chart (Figure 13.3) with 750 gpm, go up to 100-ft head, over to 1,000 SSU, and then up to the correction factors.

$$C_q = 0.95$$

$$C_h = 0.92 \text{ (for } 1.0\ Q_{nw})$$

$$C_e = 0.635$$

$$Q_w = \frac{750}{0.95} = 790 \text{ gpm}$$

$$H_w = \frac{100}{0.92} = 108.8 \cong 109\text{-ft head}$$

Select a pump for a water capacity of 790 gpm at 109-ft head. The selection should be at or close to the maximum efficiency point for water performance. If the pump selected has an efficiency on water of 81% at 790 gpm, then the efficiency for the viscous liquid will be as follows:

$$E_{vis} = 0.635 \times 81\% = 51.5\%$$

*After Ref. 3, with permission.

The brake horsepower for pumping the viscous liquid will be:

$$\text{bhp}_{\text{vis}} = \frac{750 \times 100 \times 0.90}{3960 \times 0.515} = 33.1 \text{ hp}$$

## NOMENCLATURE

bhp   Brake horsepower

$\text{bhp}_{\text{vis}}$   Viscous brake horsepower—the horsepower required by the pump for the viscous conditions

$C_e$   Efficiency correction factor

$C_h$   Head correction factor

$C_q$   Capacity correction factor

$D$   Impeller diameter

$E_{\text{vis}}$   Viscous efficiency (%)—the efficiency when pumping a viscous liquid.

$E_w$   Water efficiency, per cent

$H$   Total head (ft)

$H_{\text{vis}}$   Viscous head (ft)—the head when pumping a viscous liquid.

$H_w$   Water head (ft)

$N$   Pump speed (rpm)

NPSH   Net positive suction head

$Q$   Capacity (gpm)

$Q_{\text{vis}}$   Viscous capacity (gpm)—the capacity when pumping a viscous liquid.

$Q_w$   Water capacity (gpm)

$SG$   Specific gravity

whp   Hydraulic horsepower

## REFERENCES

1. Goulds Pump Manual, Goulds Pumps, Inc., Seneca Falls, New York, 1976–1977.
2. D. R. Rankin, It's easy to determine NPSH, *Petroleum Refiner*, *32*, 129–132 (1953).
3. *Hydraulic Institute Standards*, 13th ed., Hydraulic Institute, 1230 Keith Building, Cleveland, Ohio (1975).
4. I. Taylor, What NPSH for dissolved gas?, *Hydrocarbon Processing*, Vol. 46, No. 8, August (1967).

## SELECTED READING

*Engineering Manual*, Marlow Pumps, Fluid Handling Division, International Telephone and Telegraph, Midland Park, N.J.
*Goulds Pump Manual*, 1976–77 Edition (or later if available), Goulds Pumps, Inc., Seneca Falls, N.Y.

F. A. Holland and F. S. Chapman, Centrifugal pumps, *Chem. Eng.* July 4 (1966).

*Hydraulic Institute Standards*, 13th ed., Hydraulic Institute, 1230 Keith Building, Cleveland, Ohio (1975).

J. K. Jacobs, How to select and specify process pumps, *Hydrocarbon Processing*, Vol. 44, No. 6, June (1965).

C. Jackson, How to prevent pump cavitation, *Hydrocarbon Processing*, May (1973).

R. Kern, Use nomographs to quickly size pump piping and components, *Hydrocarbon Processing*, March (1973).

*Pump Engineering Manual*, The Duriron Company, Dayton, Ohio, 1968.

D. R. Rankin, What is NPSH?, *Petroleum Refiner*, Vol. 32, No. 5, May (1953).

D. R. Rankin, It's easy to determine NPSH, *Petroleum Refiner*, Vol. 32, No. 6, June (1953).

I. Taylor, What NPSH for dissolved gas?, *Hydrocarbon Processing*, Vol. 46, No. 8, August (1967).

# 14 Safety Relief Valves and Rupture Disks

The following material is intended for use *only with single phase systems*. It should not be used for systems that are simultaneously discharging vapors and liquids. Such systems involve complex two-phase flow calculations, which are beyond the scope of this work. In fact, the entire area of emergency relief systems is undergoing study which will result in changes in the calculational procedures. This work is being done by the Design Institute for Emergency Relief Systems (DIERS) sponsored by the American Institute of Chemical Engineers.

## BASIS FOR DESIGN

Perhaps the most important factor to consider in sizing a relief system is the basis for the design. Factors to consider are:

1. Exposure to fire.
2. Inlet and or outlet block valves closed. This can apply to long liquid filled lines or process equipment—compressors, heat exchangers, pumps, columns, etc. On heating, by either ambient or process conditions, tremendous pressure can develop.
3. Control valve failure.
4. Runaway chemical reaction.
5. Failure of a heat exchanger tube.
6. Loss of coolant.
7. Water hammer, due to a rapidly closing valve.
8. Positive displacement equipment.
9. Human error.

These factors, and others, must be considered singly and in combination, to arrive at a basis for design.

## SAFETY RELIEF VALVES

Figure 14.1 shows the standard parts of a conventional safety relief valve. Figure 14.2 is for a balanced bellows safety relief valve. In the

**243**

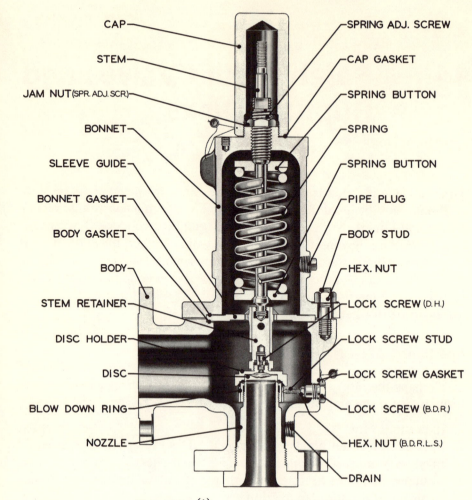

**Figure 14.1.**[1]   Conventional relief valve.

conventional valve, the discharge pressure is affected by the backpressure, or the pressure on the discharge side of the valve. If backpressure is constant, no serious problems will arise since the spring can be adjusted to compensate. For discharges into a header, where the backpressure can vary, the spring-adjustment solution is not acceptable. For this case, the balanced bellows valve should be used.

The bellows is so proportioned that the area open to the atmosphere is exactly equal to the area of the disc exposed to the process pressure thus, theoretically, eliminating any effect of backpressure on the set point. Actually, because of practical limitations, this is only approximately true—but the variation can generally be tolerated if the backpressure does not exceed 40% of the set pressure. With conventional valves the customary limit of variable backpressure in 10% of the set pressure.[2]

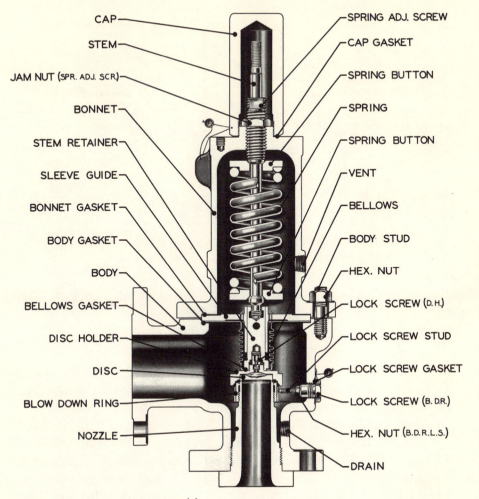

**Figure 14.2.**[1]   Balanced bellows relief valve.

An explanation of the physical principles that apply to each type of valve is given in Figure 14.3.

## GAS AND VAPORS[3]

### Conventional Valves

The general equation for sizing conventional relief valves for gas and vapor flow is[3]

$$A = \frac{W}{CKK_b P}\left(\frac{TZ}{M}\right)^{1/2} \tag{14.1}$$

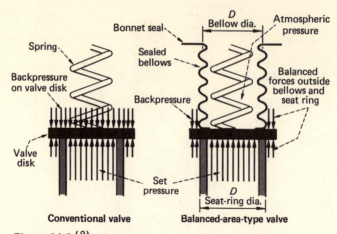

**Figure 14.3.**[9]   Comparison of a balanced bellows valve and a conventional valve.

Set pressure plus overpressure values ($P$) recommended by ASME code are:[3]

$1.1 P_s$ for unified pressure vessels
$1.2 P_s$ for vessels in case of fire
$1.3 P_s$ for piping

$K$ is the valve discharge coefficient which is determined by the manufacturer. Typically, for a nozzle, $K = 0.975$.

$K_b$ can be determined from Table 14.1.

A partial listing of data for $k$ (specific heat ratio) is given in Table ~~12.7~~ 14.5 and Figure 12.5. If $k$ is unknown, $k = 1$ will be conservative. $C$ is given in Figure 14.4.

Use Table 14.2 to select a relief valve with a standard orifice size. The valve selected must have an orifice at least as large as the area calculated.

**Table 14.1.**[3]   $K_b$ Factor for Conventional Valves in Gas or Vapor Service

| Absolute backpressure, % | Constant backpressure factor, $K_b$ |
|---|---|
| 55 or less | 1.0 |
| 60 | 0.995 |
| 65 | 0.975 |
| 70 | 0.945 |
| 75 | 0.90 |
| 80 | 0.845 |

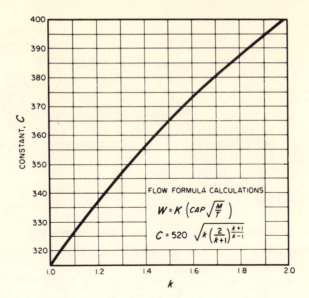

| $k$ | Constant $C$ | $k$ | Constant $C$ | $k$ | Constant $C$ |
|---|---|---|---|---|---|
| 1.00 | 315 | 1.26 | 343 | 1.52 | 366 |
| 1.02 | 318 | 1.28 | 345 | 1.54 | 368 |
| 1.04 | 320 | 1.30 | 347 | 1.56 | 369 |
| 1.06 | 322 | 1.32 | 349 | 1.58 | 371 |
| 1.08 | 324 | 1.34 | 351 | 1.60 | 372 |
| 1.10 | 327 | 1.36 | 352 | 1.62 | 374 |
| 1.12 | 329 | 1.38 | 354 | 1.64 | 376 |
| 1.14 | 331 | 1.40 | 356 | 1.66 | 377 |
| 1.16 | 333 | 1.42 | 358 | 1.68 | 379 |
| 1.18 | 335 | 1.44 | 359 | 1.70 | 380 |
| 1.20 | 337 | 1.46 | 361 | 2.00 | 400 |
| 1.22 | 339 | 1.48 | 363 | 2.20 | 412 |
| 1.24 | 341 | 1.50 | 364 | | |

**Figure 14.4.**[4]  $C$ vs. specific heat ratio $k$.

**Table 14.2.**[5]  Orifice Sizes for Relief Valves

| Orifice letter | D | E | F | G | H | J |
|---|---|---|---|---|---|---|
| area (in.$^2$) | 0.110 | 0.196 | 0.307 | 0.503 | 0.785 | 1.287 |
| Orifice letter | K | L | M | N | P | Q |
| area (in.$^2$) | 1.838 | 2.853 | 3.600 | 4.340 | 6.380 | 11.05 |
| Orifice letter | R or S[a] | | T[a] | | T or U[a] | |
| area (in.$^2$) | 16.00 | | 19.64 | | 26.00 | |

[a]Code letters and orifice areas not consistent for these large orifices between various manufacturers.

### Balanced Valves[3]

The same general equation (14.1) is used for sizing balanced valves for vapor or gas service. However, $K_v$ is substituted for $K_b$.[3]

$$A = \frac{W}{CKK_vP} \left(\frac{TZ}{M}\right)^{1/2} \tag{14.2}$$

$K_v$ is obtained from Figure 14.5 (10% overpressure) or Figure 14.6 (20% overpressure).

## STEAM SERVICE

Since the physical properties of steam are known, the sizing equation can be simplified:[3]

$$A = \frac{W_s}{51.5\,KPK_b K_{sh}} \tag{14.3}$$

Values for $K_{sh}$ are given in Table 14.3.

## LIQUIDS

The capacity of safety relief valves handling liquids with a water-like viscosity is given by:[6]

$$A = \frac{V_L}{24.3\,K_a[(\Delta P)/SG)]^{1/2}} \tag{14.4}$$

$K_a$ is the effective area factor for a partially open valve. It can be determined from Figure 14.7. Figure 14.7 was computed from the $K_p$ curves supplied by typical (Farris, Kunkle) valve manufacturers.

As an example, select a conventional valve to release 1000 gpm of water to a reservoir at 25 psig. Set pressure is 100 psig, and the maximum allowable working pressure is 105 psig. ASME code allows a 10% pressure accumulation. Piping pressure drop will be assumed negligible.

Inlet pressure = (1.1)(105) = 115.5 psig.

$\Delta P$ = 115.5 - 25 = 90.5 psi

$P_d$ = 100 - 25 = 75 psi

$P_o$ = the pressure to begin opening the valve with no backpressure (100 - 25 = 75) plus the additional pressure to just completely open the valve (normally 25% of the process set pressure—0.25 × 100 = 25).

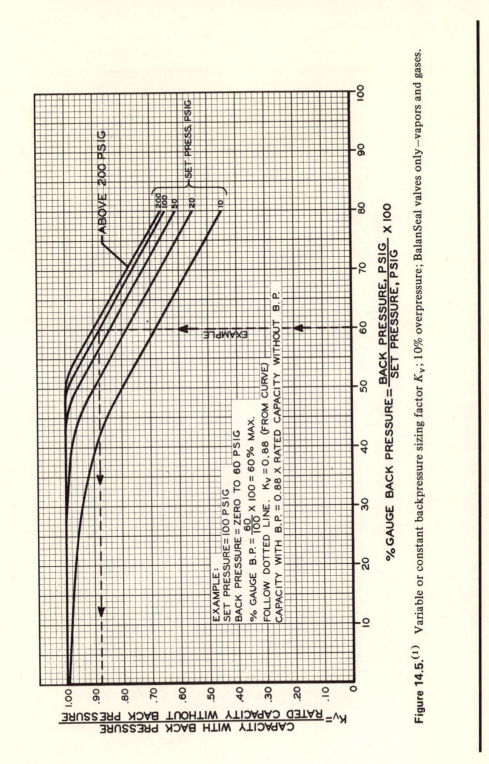

**Figure 14.5.**[1]  Variable or constant backpressure sizing factor $K_v$; 10% overpressure; BalanSeal valves only—vapors and gases.

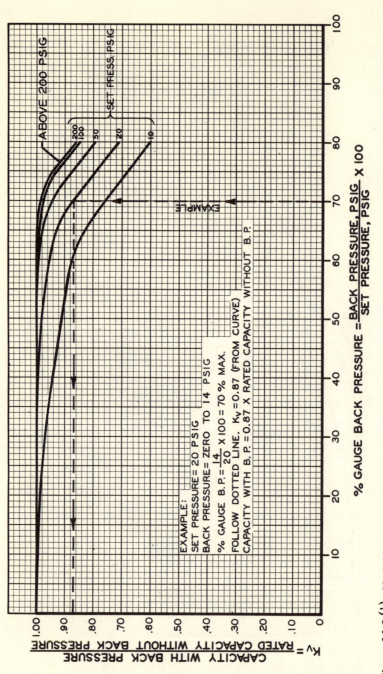

**Figure 14.6.**[(1)]  Variable or constant backpressure sizing factor $K_v$; 20% overpressure; BalanSeal valves only—vapors and gases.

TOTAL STEAM TEMPERATURE IN DEGREES FAHRENHEIT

| SET PRESSURE p.s.i.g. | SATURATED STEAM TEMP. °F. | 280 | 300 | 320 | 340 | 360 | 380 | 400 | 420 | 440 | 460 | 480 | 500 | 520 | 540 | 560 | 580 | 600 | 620 | 640 | 660 | 680 | 700 | 720 | 740 | 760 | 780 | 800 | 820 | 840 | 860 | 880 | 900 | 920 | 940 | 960 | 980 | 1000 |
|---|---|---|---|---|---|---|---|---|---|---|---|---|---|---|---|---|---|---|---|---|---|---|---|---|---|---|---|---|---|---|---|---|---|---|---|---|---|---|
| 15 | 250 | 1.00 | 1.00 | 1.00 | .99 | .99 | .98 | .98 | .97 | .96 | .95 | .94 | .93 | .92 | .91 | .90 | .89 | .88 | .87 | .86 | .86 | .85 | .84 | .83 | .83 | .82 | .81 | .81 | .80 | .79 | .79 | .78 | .78 | .77 | .76 | .76 | .75 | .75 |
| 20 | 259 | 1.00 | 1.00 | 1.00 | .99 | .99 | .98 | .98 | .97 | .96 | .95 | .94 | .93 | .92 | .91 | .90 | .89 | .88 | .87 | .86 | .86 | .85 | .84 | .83 | .83 | .82 | .81 | .81 | .80 | .79 | .79 | .78 | .78 | .77 | .76 | .76 | .75 | .75 |
| 40 | 287 | — | 1.00 | 1.00 | 1.00 | .99 | .98 | .98 | .97 | .96 | .95 | .94 | .93 | .92 | .91 | .90 | .89 | .88 | .87 | .86 | .86 | .85 | .84 | .83 | .83 | .82 | .81 | .81 | .80 | .80 | .79 | .78 | .78 | .77 | .77 | .76 | .75 | .75 |
| 60 | 308 | — | — | 1.00 | 1.00 | .99 | .99 | .98 | .97 | .96 | .95 | .94 | .93 | .92 | .91 | .90 | .89 | .89 | .88 | .87 | .86 | .85 | .84 | .84 | .83 | .82 | .82 | .81 | .80 | .80 | .79 | .78 | .78 | .77 | .77 | .76 | .76 | .75 |
| 80 | 324 | — | — | 1.00 | 1.00 | 1.00 | .99 | .99 | .98 | .97 | .96 | .94 | .93 | .92 | .91 | .90 | .89 | .89 | .88 | .87 | .86 | .85 | .84 | .84 | .83 | .82 | .82 | .81 | .80 | .80 | .79 | .78 | .78 | .77 | .77 | .76 | .76 | .75 |
| 100 | 338 | — | — | — | 1.00 | 1.00 | 1.00 | .99 | .98 | .97 | .96 | .95 | .94 | .93 | .92 | .91 | .90 | .89 | .88 | .87 | .86 | .85 | .85 | .84 | .83 | .82 | .82 | .81 | .80 | .80 | .79 | .78 | .78 | .77 | .77 | .76 | .76 | .75 |
| 120 | 350 | — | — | — | — | 1.00 | 1.00 | .99 | .98 | .97 | .96 | .95 | .94 | .93 | .92 | .91 | .90 | .89 | .88 | .87 | .86 | .85 | .85 | .84 | .83 | .82 | .82 | .81 | .80 | .80 | .79 | .78 | .78 | .77 | .77 | .76 | .76 | .75 |
| 140 | 361 | — | — | — | — | — | 1.00 | 1.00 | .98 | .98 | .96 | .95 | .94 | .93 | .92 | .91 | .90 | .89 | .88 | .87 | .86 | .86 | .85 | .84 | .83 | .82 | .82 | .81 | .81 | .80 | .79 | .79 | .78 | .77 | .77 | .76 | .76 | .75 |
| 160 | 371 | — | — | — | — | — | 1.00 | 1.00 | .99 | .98 | .97 | .95 | .94 | .93 | .92 | .91 | .90 | .89 | .88 | .87 | .86 | .86 | .85 | .84 | .83 | .82 | .82 | .81 | .81 | .80 | .79 | .79 | .78 | .77 | .77 | .76 | .76 | .75 |
| 180 | 380 | — | — | — | — | — | — | 1.00 | .99 | .98 | .97 | .96 | .95 | .93 | .92 | .91 | .90 | .89 | .89 | .87 | .86 | .86 | .85 | .84 | .83 | .82 | .82 | .81 | .81 | .80 | .79 | .79 | .78 | .78 | .77 | .76 | .76 | .75 |
| 200 | 388 | — | — | — | — | — | — | 1.00 | .99 | .97 | .96 | .95 | .93 | .93 | .92 | .91 | .90 | .89 | .88 | .87 | .86 | .86 | .85 | .84 | .83 | .83 | .82 | .81 | .81 | .80 | .79 | .79 | .78 | .77 | .77 | .76 | .75 | .75 |
| 220 | 395 | — | — | — | — | — | — | 1.00 | .99 | .98 | .96 | .95 | .94 | .93 | .92 | .91 | .90 | .89 | .88 | .87 | .87 | .86 | .85 | .84 | .84 | .83 | .82 | .81 | .81 | .80 | .80 | .79 | .78 | .78 | .77 | .76 | .76 | .76 |
| 240 | 403 | — | — | — | — | — | — | — | 1.00 | .98 | .97 | .96 | .95 | .94 | .93 | .92 | .91 | .90 | .89 | .88 | .87 | .86 | .86 | .85 | .84 | .83 | .83 | .82 | .81 | .81 | .80 | .79 | .79 | .78 | .77 | .77 | .76 | .76 |
| 260 | 409 | — | — | — | — | — | — | — | 1.00 | .98 | .97 | .96 | .95 | .94 | .93 | .92 | .91 | .90 | .89 | .88 | .87 | .86 | .86 | .85 | .84 | .83 | .83 | .82 | .81 | .81 | .80 | .79 | .79 | .78 | .77 | .77 | .76 | .76 |
| 280 | 416 | — | — | — | — | — | — | — | 1.00 | .99 | .97 | .96 | .95 | .94 | .93 | .92 | .91 | .90 | .89 | .88 | .87 | .86 | .86 | .85 | .84 | .83 | .83 | .82 | .81 | .81 | .80 | .79 | .79 | .78 | .77 | .77 | .76 | .76 |
| 300 | 422 | — | — | — | — | — | — | — | — | 1.00 | .98 | .96 | .95 | .94 | .93 | .92 | .91 | .90 | .89 | .88 | .87 | .86 | .86 | .85 | .84 | .83 | .82 | .82 | .81 | .80 | .80 | .79 | .78 | .78 | .78 | .76 | .76 | .76 |
| 350 | 436 | — | — | — | — | — | — | — | — | 1.00 | .99 | .99 | .97 | .96 | .94 | .93 | .92 | .91 | .90 | .89 | .88 | .87 | .86 | .86 | .85 | .84 | .83 | .82 | .82 | .81 | .80 | .80 | .80 | .78 | .78 | .77 | .77 | .76 |
| 400 | 448 | — | — | — | — | — | — | — | — | — | 1.00 | .99 | .98 | .96 | .95 | .94 | .92 | .91 | .90 | .89 | .88 | .87 | .86 | .86 | .85 | .84 | .84 | .82 | .82 | .82 | .81 | .80 | .80 | .79 | .78 | .77 | .77 | .76 |
| 450 | 460 | — | — | — | — | — | — | — | — | — | — | 1.00 | .99 | .97 | .95 | .94 | .93 | .92 | .91 | .89 | .88 | .88 | .87 | .86 | .85 | .84 | .84 | .83 | .82 | .82 | .81 | .81 | .80 | .79 | .79 | .77 | .77 | .77 |
| 500 | 470 | — | — | — | — | — | — | — | — | — | — | 1.00 | .99 | .98 | .96 | .95 | .93 | .92 | .91 | .90 | .89 | .88 | .87 | .86 | .85 | .84 | .84 | .83 | .82 | .82 | .81 | .81 | .80 | .79 | .79 | .78 | .78 | .77 |
| 550 | 480 | — | — | — | — | — | — | — | — | — | — | — | 1.00 | .99 | .97 | .95 | .94 | .92 | .91 | .90 | .89 | .88 | .87 | .86 | .85 | .84 | .83 | .83 | .82 | .82 | .81 | .80 | .80 | .79 | .78 | .78 | .77 | .76 |
| 600 | 489 | — | — | — | — | — | — | — | — | — | — | — | 1.00 | .99 | .98 | .96 | .94 | .93 | .92 | .90 | .89 | .88 | .88 | .86 | .86 | .85 | .84 | .83 | .82 | .82 | .81 | .81 | .81 | .80 | .79 | .78 | .78 | .77 |
| 650 | 497 | — | — | — | — | — | — | — | — | — | — | — | 1.00 | 1.00 | .98 | .97 | .95 | .94 | .92 | .91 | .90 | .89 | .89 | .88 | .86 | .86 | .85 | .84 | .82 | .82 | .81 | .81 | .80 | .79 | .79 | .78 | .78 | .77 |
| 700 | 506 | — | — | — | — | — | — | — | — | — | — | — | — | 1.00 | 1.00 | .97 | .95 | .94 | .93 | .91 | .90 | .89 | .89 | .88 | .87 | .86 | .85 | .84 | .83 | .83 | .81 | .81 | .80 | .79 | .79 | .78 | .78 | .77 |
| 750 | 513 | — | — | — | — | — | — | — | — | — | — | — | — | 1.00 | 1.00 | .98 | .96 | .95 | .93 | .92 | .91 | .90 | .89 | .88 | .87 | .86 | .86 | .84 | .83 | .83 | .82 | .82 | .81 | .79 | .79 | .78 | .78 | .77 |
| 800 | 520 | — | — | — | — | — | — | — | — | — | — | — | — | — | 1.00 | .99 | .97 | .95 | .94 | .92 | .91 | .90 | .88 | .87 | .86 | .85 | .86 | .84 | .83 | .83 | .81 | .80 | .80 | .79 | .78 | .78 | .77 | .76 |
| 850 | 527 | — | — | — | — | — | — | — | — | — | — | — | — | — | 1.00 | .99 | .98 | .96 | .95 | .93 | .92 | .91 | .89 | .88 | .87 | .86 | .86 | .85 | .83 | .83 | .82 | .82 | .81 | .79 | .79 | .78 | .78 | .77 |
| 900 | 533 | — | — | — | — | — | — | — | — | — | — | — | — | — | 1.00 | 1.00 | .99 | .97 | .95 | .94 | .93 | .92 | .90 | .89 | .88 | .87 | .87 | .85 | .83 | .83 | .82 | .82 | .81 | .80 | .80 | .79 | .78 | .78 |
| 950 | 540 | — | — | — | — | — | — | — | — | — | — | — | — | — | — | 1.00 | .99 | .97 | .96 | .95 | .93 | .93 | .91 | .90 | .89 | .88 | .88 | .86 | .84 | .84 | .83 | .82 | .82 | .80 | .80 | .79 | .78 | .78 |
| 1000 | 546 | — | — | — | — | — | — | — | — | — | — | — | — | — | — | 1.00 | .99 | .98 | .96 | .95 | .94 | .93 | .92 | .90 | .89 | .89 | .88 | .86 | .85 | .84 | .83 | .82 | .82 | .80 | .80 | .79 | .78 | .78 |
| 1050 | 552 | — | — | — | — | — | — | — | — | — | — | — | — | — | — | 1.00 | 1.00 | .99 | .97 | .95 | .93 | .92 | .90 | .89 | .88 | .86 | .86 | .85 | .84 | .83 | .82 | .81 | .80 | .80 | .79 | .78 | .77 | .77 |
| 1100 | 558 | — | — | — | — | — | — | — | — | — | — | — | — | — | — | 1.00 | 1.00 | .99 | .98 | .96 | .94 | .92 | .91 | .89 | .88 | .87 | .87 | .85 | .84 | .83 | .82 | .81 | .81 | .80 | .80 | .79 | .78 | .78 |
| 1150 | 563 | — | — | — | — | — | — | — | — | — | — | — | — | — | — | — | 1.00 | .99 | .98 | .97 | .95 | .93 | .92 | .90 | .89 | .88 | .87 | .86 | .84 | .83 | .82 | .82 | .81 | .80 | .80 | .79 | .78 | .78 |
| 1200 | 569 | — | — | — | — | — | — | — | — | — | — | — | — | — | — | — | 1.00 | 1.00 | .99 | .97 | .95 | .94 | .92 | .90 | .90 | .88 | .87 | .86 | .85 | .83 | .83 | .82 | .81 | .81 | .80 | .79 | .78 | .78 |
| 1250 | 574 | — | — | — | — | — | — | — | — | — | — | — | — | — | — | — | 1.00 | 1.00 | 1.00 | .97 | .95 | .94 | .93 | .91 | .90 | .88 | .88 | .87 | .85 | .84 | .83 | .82 | .82 | .81 | .80 | .79 | .78 | .78 |
| 1300 | 579 | — | — | — | — | — | — | — | — | — | — | — | — | — | — | — | 1.00 | 1.00 | 1.00 | .98 | .96 | .94 | .92 | .91 | .89 | .88 | .88 | .86 | .85 | .84 | .83 | .82 | .82 | .80 | .80 | .79 | .78 | .77 |
| 1350 | 584 | — | — | — | — | — | — | — | — | — | — | — | — | — | — | — | — | 1.00 | 1.00 | .99 | .96 | .94 | .93 | .92 | .90 | .89 | .88 | .86 | .85 | .84 | .83 | .82 | .81 | .81 | .80 | .79 | .79 | .78 |
| 1400 | 588 | — | — | — | — | — | — | — | — | — | — | — | — | — | — | — | — | 1.00 | 1.00 | .99 | .96 | .95 | .94 | .92 | .90 | .89 | .89 | .86 | .84 | .84 | .83 | .82 | .81 | .81 | .80 | .79 | .78 | .78 |
| 1450 | 593 | — | — | — | — | — | — | — | — | — | — | — | — | — | — | — | — | 1.00 | 1.00 | .99 | .98 | .95 | .94 | .92 | .91 | .89 | .89 | .87 | .85 | .84 | .83 | .82 | .82 | .81 | .80 | .79 | .78 | .78 |
| 1500 | 597 | — | — | — | — | — | — | — | — | — | — | — | — | — | — | — | — | 1.00 | 1.00 | 1.00 | .98 | .96 | .94 | .92 | .91 | .90 | .89 | .87 | .86 | .84 | .84 | .83 | .82 | .81 | .80 | .79 | .78 | .78 |

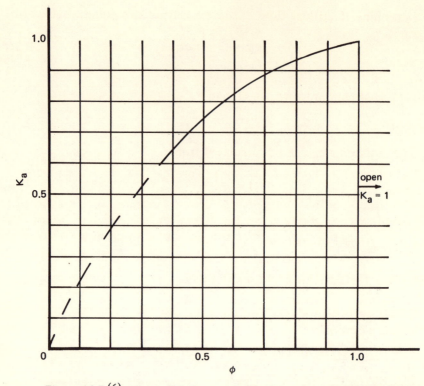

**Figure 14.7.**[6]  $\phi$ vs. effective area factor $K_a$, computed from $K_p$ curves (for springs that allow valve to be just fully open at 25% overpressure). $\phi = (\Delta P - P_d)/(P_o - P_d)$ (standard valve); $\phi = (P - P_s)/(P_o - P_s)$ (balanced valve).

$$P_o = 75 + 25 = 100 \text{ psig}$$

$$\phi = \frac{\Delta P - P_d}{P_o - P_d} = \frac{90.5 - 75}{100 - 75} = 0.62$$

$$K_a = 0.84$$

$$A = \frac{1000}{24.3\,(0.84)\,(90.5)^{1/2}} = 5.15 \text{ in.}^2$$

## RUPTURE DISKS

### Vapors and Gases[5]

For critical flow, the area is calculated from

$$W = (0.81)\,CAP(M/ZT)^{1/2} \tag{14.5}$$

Use the values of $C$ found in Figure 14.4. If $k$ is unknown, use $k = 1.0$ and $C = 315$. Compressibility, $Z$, can be taken as 1.0 for moderate pressures.

Determine if critical flow exists by solving the equation for critical pressure,

$$P_c = P \left( \frac{2}{k + 1} \right)^{k/(k-1)} \tag{14.6}$$

If the downstream pressure is less than $P_c$, critical flow will exist.

## Liquids[5]

The required area can be calculated from

$$A = \frac{144 W'}{c(2gh)^{1/2}} \tag{14.7}$$

$c = .61$ for $N_{Re} > 200$; $c = .50$ for $200 > N_{Re} > 100$; $c = .40$ for $100 > N_{Re} > 50$; $h$ = set pressure in feet of liquid; $W'$ = liquid flow (ft³/sec).

## VENTING REQUIREMENTS FOR FIRE

1. Heat absorbed by a vessel during a fire can be determined from the following equations:[7]

$$Q = 20,000 A_R \qquad \text{for } A_R < 200 \tag{14.8}$$
$$Q = 199,300 A_R^{0.566} \qquad \text{for } 200 < A_R < 1000 \tag{14.9}$$
$$Q = 963,400 A_R^{0.338} \qquad \text{for } > 1000 \tag{14.10}$$

where $A_R$ is the wetted area of the vessel, which is 55% of total exposed area of a sphere, 75% of total exposed area of a horizontal tank, and the first 30 ft above grade of exposed shell area of a vertical tank.

## Protection Factors

| Protection factor, $F$ | Condition |
|---|---|
| 0.5 | For drainage (see NFPA No. 30, paragraph 2172). |
| 0.3 | For approved water spray (see NFPA Nos. 15 and 30, paragraph 2172). |
| 0.3 | For approved insulation [see NFPA No. 30, paragraph 2157(a)]. |
| 0.15 | For approved water spray with approved insulation. |

## Example

A vessel is protected with a water spray deluge system. The wetted area is 575 ft.² $F = 0.3$.

$$Q = (1.993 \times 10^{+5})(575)^{0.566}(0.3)$$
$$Q = (7.27 \times 10^{+6})(0.3) = 2.18 \times 10^{+6} \text{ Btu/hr}$$

2. The required relief capacity is calculated from:

$$W = Q/\Delta H_v \qquad (14.11)$$

3. If the latent heat is known for only one temperature, the Watson correlation may be used to estimate its value at other temperatures.[8]

$$\Delta H_{v2} = \Delta H_{v1} \left( \frac{1 - T_{r2}}{1 - T_{r1}} \right)^{0.38} \qquad (14.12)$$

where $\Delta H_{v2}$ is the latent heat of vaporization (Btu/lb); $T$ is the temperature (°R; $T_r = T/T_c$); $T_c$ is the critical temperature (°R).

## TOTAL HEAT INPUT

The total heat of the system must include not only heat from fire, but heat input from the other sources listed under Basis for Design.

## VENT PIPING

For calculating the pressure drop in the discharge piping, use Section 6.1. In general, the piping should be as direct and as vertical as possible. Horizontal runs and elbows should be avoided. Vent lines must never have any pockets. Valves should never be installed between the relieving device and the vessel it is protecting. If in doubt about size, use the larger size. Keep friction drop at a minimum.

## NOMENCLATURE

| | |
|---|---|
| $A$ | Nozzle throat area, or orifice flow area (in.$^2$) |
| $A_R$ | Wetted area of a vessel exposed to fire (ft$^2$) |
| $C$ | Constant depending on ratio of specific heats, $C_p/C_v$ (see Figure 14.4) |
| $c$ | Orifice coefficients for liquids |
| $g$ | Acceleration of gravity (32.0 ft/sec$^2$) |
| $h$ | Height of liquid (ft) |
| $\Delta H_v$ | Latent heat of vaporization (Btu/lb) |
| $K$ | Coefficient of discharge for vapors or gases |
| $K_a$ | Effective area factor for a partially open valve |
| $K_b$ | Gas or vapor flow correction factor for constant backpressure |
| $K_{sh}$ | Sizing factor for superheat on steam (=1.0 for saturated steam) |

$K_v$    Gas or vapor correction factor for variable backpressure (for balanced valves only)

$k$    $C_p/C_v$, ratio of specific heats

$M$    Molecular weight of vapor, gas, or liquid

$N_{Re}$    Reynold's number

$P$    Absolute pressure in system at time of relief from overpressure (psia)

$P_c$    Critical pressure (psia)

$P_d$    Differential set pressure (psi)

$P_o$    Pressure to just fully open valve with no backpressure (psig). Usual convention is to design the valve to be just fully open at 25% above the process set pressure.

$P_s$    Set pressure (same as $P_d$ for a balanced valve) (psig)

$\Delta P$    Actual dynamic pressure differential across safety valve, inlet pressure minus back pressure or downstream pressure (psi)

$Q$    Total heat absorption, Btu/hr

$SG$    Specific gravity of liquid, referenced to water at the same temperature

$T$    Absolute temperature (degrees Rankine) ($^\circ R = ^\circ F + 460$)

$T_r$    Reduced temperature, see Equation (14.12)

$V_L$    Liquid capacity (gpm)

$W$    Flowrate (lb/hr)

$W_s$    Steam flow (lb/hr)

$W'$    Liquid flow (ft$^3$/sec)

$Z$    Compressibility factor, deviation of actual gas from perfect gas law. Usually $Z = 1.0$ at low pressure.

$\phi$    Valve factor (see Figure 14.7)

## REFERENCES

1. Teledyne Farris Company, Catalog No. FE336, 1975, pp. 2.2, 2.3, 3.5.
2. J. S. Rearick, How to design pressure relief systems, *Hydrocarbon Processing*, August, 104–105 (1969).
3. R. Kern, Pressure-relief valves for process plants, *Chem. Eng.* February 28, 191 (1977).
4. ASME Code, Section VIII, Division 1 Pressure Vessels, The American Society of Mechanical Engineers, 345 E. 47th Street, New York, New York, 1974, p. 328.
5. E. E. Ludwig, *Applied Process Design for Chemical and Petrochemical Plants*, Vol. 1, Gulf Publishing Company, Houston, Texas, 1964, pp. 215–238.
6. J. E. Huff, Dow Chemical, U.S.A., private communication.
7. *National Fire Code*, Vol. 2, Number 30, National Fire Protection Association, 470 Atlantic Avenue, Boston, Massachusetts, 1976, pp. 30-1–30-113.
8. R. C. Reid, and T. K. Sherwood, *The Properties of Gases and Liquids*, 2nd ed., McGraw-Hill Book Company, 1966, p. 148.
9. M. Isaacs, Pressure-relief systems, *Chem. Eng.* February 22, 113–114 (1971).

## SELECTED READING

F. E. Anderson, Pressure-relieving devices, *Chem. Eng.* May 24 (1976).

ASME Code, Section VIII, Division 1 Pressure Vessels, The American Society of Mechanical Engineers, 345 E. 47th Street, New York, N.Y., 1974, p. 328.

J. Conison, How to design a pressure relief system, *Chem. Eng.* July 25 (1960).

J. E. Huff, A General Approach to the Sizing of Emergency Pressure Relief Systems, presented at the European Federation of Chemical Engineering, 2nd International Symposium on Loss Prevention and Safety Promotion in the Process Industries, Heidelberg, Federal Republic of Germany, September, 1977.

M. Isaacs, Pressure-relief systems, *Chem. Eng.* February 22 (1971).

R. Kern, Pressure-relief valves for process plants, *Chem. Eng.* February 28 (1977).

E. E. Ludwig, Applied process design for chemical and petrochemical plants, Volume 1, Gulf Publishing Company, Houston, Texas, 1964.

National Fire Code, Volume 2, Number 30, National Fire Protection Association, 470 Atlantic Avenue, Boston, Mass, 1976, pp. 30-1 to 30-113.

J. S. Rearick, How to design pressure relief systems (Part 1), *Hydrocarbon Processing*, August (1969).

J. S. Rearick, How to design pressure relief systems (Part 2), *Hydrocarbon Processing*, September (1969).

J. E. Righom, Spring-loaded relief valves, *Chem. Eng.* February 10 (1958).

Teledyne Farris Company, Catalog No. FE336, 1975.

# 15 Steam Ejectors for Vacuum Service

## GENERAL CHARACTERISTICS[1]

High-pressure motive steam passes through the nozzle throat, expands, and leaves it at a high velocity (which entrains the gas or vapor in the suction chamber) and enters the diffuser. In the diffuser, the velocity head is converted to pressure. See Figure 15.1. The entrained gas is thus compressed from the low absolute pressure at the suction to a higher pressure at the discharge. The ratio of discharge pressure over suction pressure (ratio of compression) is about 6 or 10 to 1, per stage. As with other types of gas compressors, the compression stages are operated in series to obtain a greater overall compression, or lower suction pressure.

Ejectors can be built for higher compression ratios (absolute discharge pressure/absolute suction pressure) which should not exceed about 10:1 for reasonable steam economy. For higher compression ratios (lower absolute suction pressures) ejectors are staged in series and an intercondenser provided between stages to condense the operating steam of the

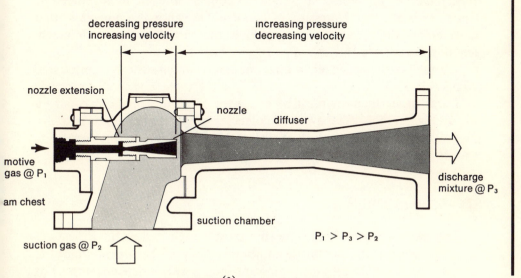

decreasing pressure
increasing velocity

increasing pressure
decreasing velocity

nozzle extension

nozzle

diffuser

motive
gas @ $P_1$

discharge
mixture @ $P_3$

am chest

suction chamber

$P_1 > P_3 > P_2$

suction gas @ $P_2$

**Figure 15.1.**[2] Typical steam ejector.

**257**

preceding stage, plus any entrained suction vapor which is condensible at the higher intercondenser pressure. This reduces the load on the following stage.

When a condenser is used between ejector stages, it is known as an intercondenser, and may be either direct contact (barometric) or surface type. An intercondenser reduces the size and steam consumption of the next ejector stage by condensing out the vapors with cooling water, and draining them to atmosphere through a 34-ft barometric leg.

If there are condensible vapors in the mixture entering an ejector which can be condensed out with available cooling water, a precooler or precondenser should be installed ahead of the ejector.

The capacity of an ejector is more nearly proportional to weight than volume. Therefore, all capacity curves are given in pounds per hour of equivalent air (29 molecular weight), and correction factors are given for other molecular weights (Figure 15.2) and for elevated suction temperatures (Figure 15.3).

The capacity of an ejector for any given suction conditions is proportional to the diffuser throat area. If more capacity is desired, a larger diffuser and nozzle must be substituted.

The amount of condensable vapor carried by the noncondensable gas can be estimated from

$$W_v = \frac{W_n M_v P_v}{M_n P_n} \tag{15.1}$$

where $W_v$, $W_n$ are the weights of condensable and noncondensable vapors (lb/hr); $M_v$, $M_n$ are molecular weights of condensable and noncondensable vapors; $P_v$ is the vapor pressure of condensable vapor at temperature that vapor enters the ejector (mm Hg); $P_n$ is the partial pressure of noncondensable vapor (mm Hg); $P_n = P_t - P_v$.

Air released from water in direct contact condensers can be estimated from Table 15.1.

Gases other than air must be converted to equivalent lb/hr of air by use of Figure 15.2. The average molecular weight of the vapor is multiplied by $M$ to get the equivalent (lb/hr) of air. A temperature correction is made using Figure 15.3.

## STEAM PRESSURE[3]

The motive steam design pressure must be selected as the lowest expected pressure at the ejector steam nozzle (Table 15.2). Operation of the unit will be unstable for steam pressures below the design pressure. For

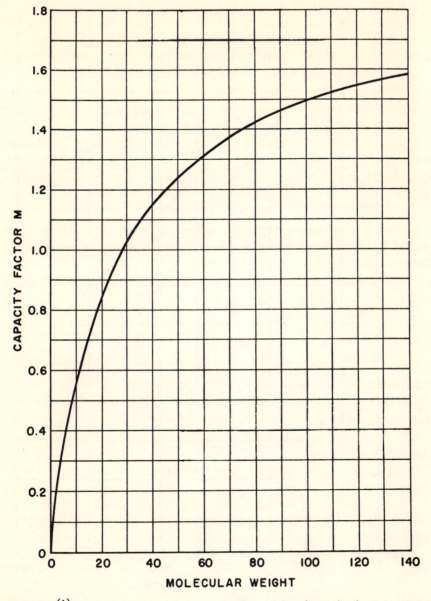

**Figure 15.2.**[(1)]  Capacity correction for molecular weight of entrained gas or vapor.

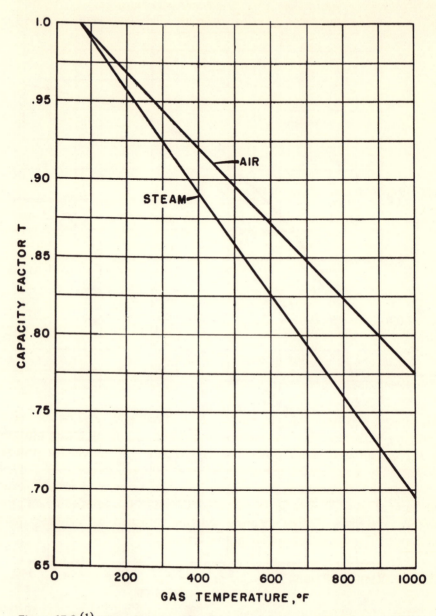

**Figure 15.3.**[1]  Capacity correction for temperature of entrained air or steam.

**Table 15.1**[3]  Air Released from Water under Vacuum

| Inlet water temperature (°F) | Pounds air released per hour per 1,000 gpm water |
|:---:|:---:|
| 40 | 16.8 |
| 50 | 14.9 |
| 60 | 13.2 |
| 70 | 11.8 |
| 80 | 10.7 |
| 90 | 9.7 |
| 100 | 8.8 |

**Table 15.2.**[3]  Pressure Range for Ejectors

| Number of stages | Minimum practical absolute pressure (mm Hg) | Range operating suction pressure (mm Hg) | Closed test pressure (mm Hg) |
|:---:|:---:|:---:|:---:|
| 1 | 50 | >75, (3 in.) | 37–50 |
| 2 | 5 | 10–100 | 5 |
| 3 | 2 | 1–25 | 1 |
| 4 | 0.2 | 0.25–3 | 0.05–0.1 |
| 5 | 0.03 | 0.03–0.3 | 0.005–0.01 |
| 6 | 0.003 | | |
| 7 | 0.001–0.0005 | | |

this reason, it is important to have a pressure reducing valve on the steam upstream of the ejector. The valve will maintain a constant steam pressure to the ejector and insure optimum performance. Steam ejector systems are normally designed for a steam pressure of 100–105 psig.

Figure 15.4 illustrates the consequences of using an incorrect steam pressure. Figure 15.4 is for an ejector designed to use an inlet steam pressure of 110 psig. Other ejectors may be designed for different inlet pressures. The manufacturer should be consulted.

## STEAM CONSUMPTION[3]

Steam consumption can be estimated using Figure 15.5. If surface condensers are used instead of barometric condensers, the steam consumption will be somewhat less than that shown in Figure 15.5. This is because air is released from the water used in the barometric condensers, putting more load on the ejectors.

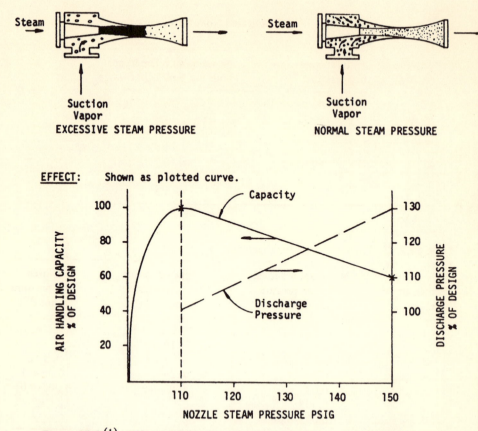

Figure 15.4.[1]  Effect of using incorrect steam pressure on steam ejectors.

## VAPOR CAPACITY[3]

The total capacity of the system is the sum of the following.

1. Air leakage.
2. Noncondensable gases from the process.
3. Condensable vapors being carried by the noncondensable gases.
4. Air released from water in direct contact condensers.

Air leakage is difficult to estimate and depends on the system. Moderately tight processes of up to 500 ft³ total volume can experience up to 10 lb/hr of leakage. Larger systems, or those with excessive fittings, may have 20 lb/hr of air leakage. Table 15.3 presents data for estimating the air leakage through various types of fittings, valves, etc.

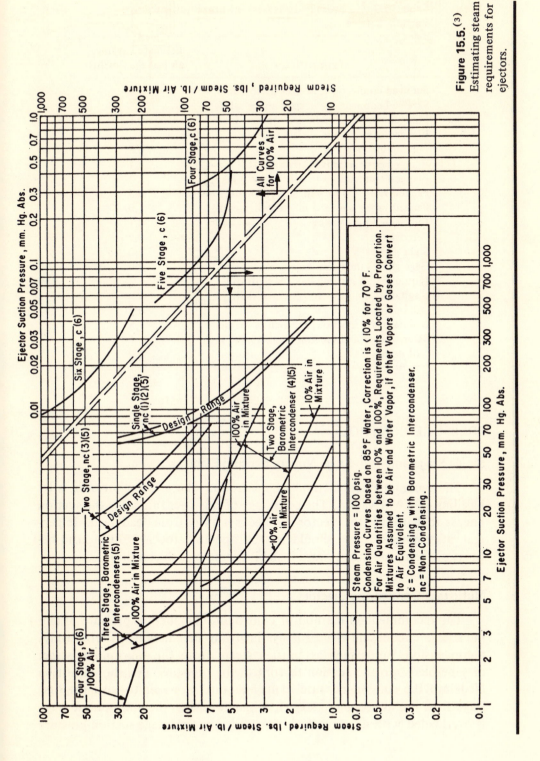

**Figure 15.5.**[3] Estimating steam requirements for ejectors.

**Table 15.3.**[5]   Estimated Air Leakage into Equipment in Vacuum Service

| Type fitting | Estimated average air leakage (lbs/hr) |
|---|---|
| Screwed connections in sizes up to 2 in. | 0.1 |
| Screwed connections in sizes above 2 in. | 0.2 |
| Flanged connections in sizes up to 6 in. | 0.5 |
| Flanged connections in sizes 6–24 in. including manholes | 0.8 |
| Flanged connections in sizes 2–6 ft | 1.1 |
| Flanged connections in sizes above 6 ft | 2.0 |
| Packed valves up to $\frac{1}{2}$-in. stem diameter | 0.5 |
| Packed valves above $\frac{1}{2}$-in. stem diameter | 1.0 |
| Lubricated plug valves | 0.1 |
| Petcocks | 0.2 |
| Sight glasses | 1.0 |
| Gage glasses including gage cocks | 2.0 |
| Liquid sealed stuffing box for shaft of agitators, pumps, etc., per inch shaft diameter | 0.3 |
| Ordinary stuffing box, per inch of diameter | 1.5 |
| Safety valves and vacuum breakers, per inch of nominal size | 1.0 |

## PRESSURE CONTROL[1]

### Single-Stage and Two-Stage Noncondensing Ejectors

The recommended method of controlling the suction pressure is by bleeding in air or vapor. The advantage of this over throttling the suction line, is that it loads the ejector and keeps it away from dead head. In general, atmospheric stages should not be operated too near dead head because of their tendency toward instability under such conditions.

### Multistage Condensing Ejectors

The recommended method of controlling the suction pressure of any multistage condensing ejector is by throttling the suction. Alternatively, a condensible vapor such as exhaust steam from the atmospheric stage, may be bled into the suction to control the pressure. Air should not be bled into the suction because it may require more air to load the first stage than the atmospheric stage can handle. When the atmospheric stage is overloaded with air, a marked increase in the first stage suction pressure

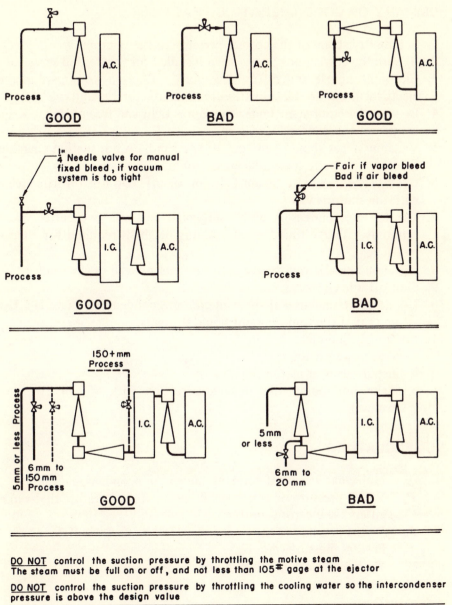

**Figure 15.6.**[1]  Pressure control of ejectors.

occurs, preventing the best of pressure controllers from maintaining a steady pressure.

Controlling pressures below 5 mm absolute is not recommended. Figure 15.6 depicts good and bad control practices for steam jet systems.

## SUMMARY OF GOOD OPERATING PRACTICES

1. Supply dry steam of the correct pressure to the ejectors.
2. Be sure the correct nozzle has been installed for the desired service.
3. Clean the nozzle and diffuser regularly. Replace worn parts if the ejector will not dead head within the manufacturer's limits.
4. Be sure the equipment under vacuum is tight and free of leaks. Keep condenser tubes clean.
5. Vacuum stages should run cool to the hand. A hot suction chamber on an atmospheric stage indicates trouble.
6. Operate ejectors with steam fully on or off. Do not throttle steam with the shut off valve.
7. Use of intercondensers with multistage jets will give better steam economy. Thirty four feet of building height is required for a barometric leg.
8. With condensable vapors, use a precondenser ahead of the ejectors.
9. Shut off idle ejectors.
10. Use an aftercondenser if the vent contains condensable material. Use a trap tank if the vent contains liquid droplets.
11. For a seal leg piping, *do not:*
    a. use screwed fittings,
    b. install valves or meters, etc., or
    c. make horizontal runs (change direction with 45° or 30° runs instead).

## NOMENCLATURE

$M_v$, $M_n$  Molecular weight of condensable and noncondensable vapor
$P_v$  Vapor pressure of condensable vapor at temperature that vapor enters the ejector (mm Hg)
$P_n$  Partial pressure of the noncondensable vapor (mm Hg)
$P_t$  Process pressure (mm Hg)
$W_v$, $W_n$  Weight of condensable and noncondensable vapor (lb/hr)

## REFERENCES

1. Dow Chemical Company, Steam Ejector Manual, private communication.
2. Ejector Guide Book, Worthington Corporation, Harrison, New Jersey, 1960, p. 4.
3. E. E. Ludwig, *Applied Process Design for Chemical and Petrochemical Plants*, Vol. 1, Gulf Publishing Co., Houston, Texas, 1964, pp. 182–214.
4. R. H. Perry, C. H. Chilton, and S. D. Kirkpatrick, *Chemical Engineers Handbook*, 4th ed., McGraw-Hill, New York, 1963, pp. 6-29–6-32.
5. D. H. Jackson, Selection and use of ejectors, *Chem. Eng. Prog. 44*, 347 (1948).

## SELECTED READING

Ejector Guide Book, Worthington Corporation, Harrison, N. H., 1975.

Frumerman, Steam jet ejectors, *Chem. Eng.* June (1956).

D. H. Jackson, Selection and use of ejectors, *Chem. Eng. Prog.* Vol. 44 (1948).

Ernest E. Ludwig, *Applied Process Design for Chemical and Petrochemical Plants*, Vol. 1, Gulf Publishing Co., Houston, Texas, 1964.

R. H. Perry and C. H. Chilton, *Chemical Engineers Handbook*, 5th ed., McGraw-Hill, New York, N.Y., 1974.

# 16    Tank Capacity

## VERTICAL TANKS

1. Use Tables 16.1 or 16.2 to determine the vertical volume, not including the head(s).
2. Find the volume of the head(s) from Table 16.3.

## HORIZONTAL TANKS (PARTIALLY FILLED)

1. Use Table 16.4 to determine the volume of liquid in the tank, not including the heads. In Table 16.4 $H$ is the liquid depth and $D$ is the tank inside diameter.[1]
2. Use Table 16.5 along with Table 16.3 to determine the volume of liquid in each of the heads.

**Table 16.1.**[1]    Volume of Cylinders, 15- to 148-in. Diameter

| Diam., in. | Gal/in. | Diam., in. | Gal/in. | Diam., in. | Gal/in. | Diam., in. | Gal/in. |
|------|------|------|------|------|------|------|------|
| 15.0 | 0.765 | 28.0 | 2.666 | 52 | 9.19 | 96 | 31.34 |
| 15.5 | 0.817 | 28.5 | 2.762 | 53 | 9.55 | 98 | 32.66 |
| 16.0 | 0.871 | 29.0 | 2.860 | 54 | 9.92 | 100 | 34.00 |
| 16.5 | 0.926 | 29.5 | 2.960 | 55 | 10.29 | 102 | 35.38 |
| 17.0 | 0.983 | 30.0 | 3.060 | 56 | 10.66 | 104 | 36.78 |
| | | | | 57 | 11.05 | 106 | 38.20 |
| 17.5 | 1.041 | 31 | 3.268 | 58 | 11.44 | 108 | 39.66 |
| 18.0 | 1.102 | 32 | 3.482 | 59 | 11.84 | 110 | 41.14 |
| 18.5 | 1.164 | 33 | 3.703 | 60 | 12.24 | 112 | 42.65 |
| 19.0 | 1.227 | 34 | 3.931 | 62 | 13.07 | 114 | 44.19 |
| 19.5 | 1.293 | 35 | 4.165 | 64 | 13.93 | 116 | 45.75 |
| 20.0 | 1.360 | 36 | 4.407 | 66 | 14.81 | 118 | 47.34 |
| 20.5 | 1.429 | 37 | 4.655 | 68 | 15.72 | 120 | 48.96 |
| 21.0 | 1.499 | 38 | 4.910 | 70 | 16.66 | 122 | 50.6 |
| 21.5 | 1.572 | 39 | 5.17 | 72 | 17.63 | 124 | 52.3 |
| 22.0 | 1.646 | 40 | 5.44 | 74 | 18.62 | 126 | 54.0 |
| 22.5 | 1.721 | 41 | 5.72 | 76 | 19.64 | 128 | 55.7 |
| 23.0 | 1.799 | 42 | 6.00 | 78 | 20.69 | 130 | 57.5 |
| 23.5 | 1.878 | 43 | 6.29 | 80 | 21.76 | 132 | 59.2 |
| 24.0 | 1.958 | 44 | 6.58 | 82 | 22.86 | 134 | 61.1 |
| 24.5 | 2.041 | 45 | 6.89 | | | | |
| | | | | 84 | 23.99 | 136 | 62.9 |
| 25.0 | 2.125 | 46 | 7.20 | 86 | 25.15 | 138 | 64.8 |
| 25.5 | 2.211 | 47 | 7.51 | 88 | 26.33 | 140 | 66.6 |
| 26.0 | 2.299 | 48 | 7.83 | | | | |
| 26.5 | 2.388 | 49 | 8.16 | 90 | 27.54 | 142 | 68.6 |
| 27.0 | 2.479 | 50 | 8.50 | 92 | 28.78 | 144 | 70.5 |
| 27.5 | 2.571 | 51 | 8.84 | 94 | 30.04 | 146 | 72.5 |
| | | | | | | 148 | 74.5 |

**Table 16.2.**[1]   Volume of Cylinders, 10- to 98-ft Diameter

| Diam. ft | in. | Gal/ft | Diam. ft | Gal/ft |
|---|---|---|---|---|
| 10 | 0 | 588 | 30 | 5,288 |
| 10 | 3 | 617 | 31 | 5,650 |
| 10 | 6 | 648 | 32 | 6,020 |
| 10 | 9 | 679 | 33 | 6,400 |
| 11 | 0 | 711 | 34 | 6,790 |
| 11 | 3 | 744 | 35 | 7,200 |
| 11 | 6 | 777 | 36 | 7,610 |
| 11 | 9 | 811 | 37 | 8,040 |
| 12 | 0 | 846 | 38 | 8,480 |
| 12 | 3 | 882 | 39 | 8,940 |
| 12 | 6 | 918 | 40 | 9,400 |
| 12 | 9 | 955 | 41 | 9,880 |
| 13 | 0 | 993 | 42 | 10,360 |
| 13 | 3 | 1,031 | 43 | 10,860 |
| 13 | 6 | 1,071 | 44 | 11,370 |
| 13 | 9 | 1,111 | 45 | 11,900 |
| 14 | 0 | 1,152 | 46 | 12,430 |
| 14 | 3 | 1,193 | 47 | 12,980 |
| 14 | 6 | 1,235 | 48 | 13,540 |
| 14 | 9 | 1,278 | 49 | 14,110 |
| 15 | 0 | 1,322 | 50 | 14,690 |
| 15 | 6 | 1,411 | 51 | 15,280 |
| 16 | 0 | 1,504 | 52 | 15,890 |
| 16 | 6 | 1,599 | 53 | 16,500 |
| 17 | 0 | 1,698 | 54 | 17,130 |
| 17 | 6 | 1,799 | 55 | 17,770 |
| 18 | 0 | 1,904 | 56 | 18,420 |
| 18 | 6 | 2,011 | 57 | 19,090 |
| 19 | 0 | 2,121 | 58 | 19,760 |
| 19 | 6 | 2,234 | 59 | 20,450 |
| 20 | 0 | 2,350 | 60 | 21,150 |
| 20 | 6 | 2,469 | 62 | 22,580 |
| 21 | 0 | 2,591 | 64 | 24,060 |
| 21 | 6 | 2,716 | 66 | 25,590 |
| 22 | 0 | 2,844 | 68 | 27,170 |
| 22 | 6 | 2,974 | 70 | 28,790 |
| 23 | 0 | 3,108 | 72 | 30,460 |
| 23 | 6 | 3,244 | 74 | 32,170 |
| 24 | 0 | 3,384 | 76 | 33,390 |
| 24 | 6 | 3,526 | 78 | 35,740 |
| 25 | 0 | 3,672 | 80 | 37,600 |
| 25 | 6 | 3,820 | 82 | 39,500 |
| 26 | 0 | 3,972 | 84 | 41,450 |
| 26 | 6 | 4,126 | 86 | 43,450 |
| 27 | 0 | 4,283 | 88 | 45,500 |
| 27 | 6 | 4,443 | 90 | 47,590 |
| 28 | 0 | 4,606 | 92 | 49,730 |
| 28 | 6 | 4,772 | 94 | 51,910 |
| 29 | 0 | 4,941 | 96 | 54,140 |
| 29 | 6 | 5,113 | 98 | 56,420 |

**Table 16.3.**[2]  Approximate Volume of Heads (U.S. Gallons); Includes I.C.R. Area; No Straight Flange

| I.D. | ASME F&D | Std. F&D | Dished Only | Elliptical | Hemispherical | Brighton 80-10 |
|------|----------|----------|-------------|------------|---------------|----------------|
| 12" | .576 | .543 | .420 | .980 | 2.000 | .813 |
| 18" | 1.840 | 1.730 | 1.400 | 3.300 | 6.600 | 2.776 |
| 24" | 4.500 | 3.880 | 3.200 | 7.800 | 15.700 | 6.580 |
| 30" | 8.970 | 7.240 | 6.300 | 15.300 | 30.600 | 13.000 |
| 36" | 15.700 | 12.300 | 10.900 | 26.400 | 52.900 | 22.000 |
| 42" | 25.900 | 19.300 | 17.300 | 42.000 | 84.000 | 35.000 |
| 48" | 37.300 | 28.400 | 25.800 | 62.700 | 125.300 | 53.000 |
| 54" | 53.100 | 40.200 | 36.700 | 89.200 | 178.500 | 75.000 |
| 60" | 72.900 | 54.600 | 50.400 | 122.400 | 244.800 | 103.000 |
| 66" | 98.200 | 72.400 | 67.000 | 162.900 | 325.800 | 137.000 |
| 72" | 127.000 | 95.600 | 87.00 | 211.500 | 423.000 | 178.000 |
| 78" | 162.000 | 120.000 | 110.700 | 268.900 | 537.800 | 226.000 |
| 84" | 203.000 | 150.000 | 138.200 | 335.900 | 671.700 | 282.000 |
| 90" | 251.000 | 180.000 | 170.000 | 413.100 | 826.200 | 347.000 |
| 96" | 308.000 | 214.000 | 206.300 | 501.400 | 1002.700 | 421.000 |
| 102" | 370.000 | 257.000 | 247.000 | 601.000 | 1203.000 | 505.000 |
| 108" | 439.000 | 311.000 | 294.000 | 714.000 | 1428.000 | 600.000 |
| 114" | 516.000 | 360.000 | 346.000 | 840.000 | 1679.000 | 705.000 |
| 120" | 604.000 | 428.000 | 403.000 | 979.000 | 1958.000 | 823.000 |
| 126" | 701.000 | 494.000 | 467.000 | 1134.000 | 2267.000 | 952.000 |
| 132" | 813.000 | 569.000 | 526.000 | 1303.000 | 2607.000 | 1094.000 |
| 138" | 926.000 | 682.000 | 613.000 | 1489.000 | 2978.000 | 1251.000 |
| 144" | 1026.000 | 739.000 | 696.000 | 1692.000 | 3384.000 | 1421.000 |

**Table 16.4.**[1] Volume of Partially Filled Horizontal Cylinders

| $H/D$ | Fraction of volume | $H/D$ | Fraction of volume |
|-------|-------------------|-------|-------------------|
| 0.01 | 0.00169 | 0.51 | 0.51273 |
| .02 | .00477 | .52 | .52546 |
| .03 | .00874 | .53 | .53818 |
| .04 | .01342 | .54 | .55088 |
| .05 | .01869 | .55 | .56356 |
| .06 | .02450 | .56 | .57621 |
| .07 | .03077 | .57 | .58884 |
| .08 | .03748 | .58 | .60142 |
| .09 | .04458 | .59 | .61397 |
| .10 | .05204 | .60 | .62647 |
| .11 | .05985 | .61 | .63892 |
| .12 | .06797 | .62 | .65131 |
| .13 | .07639 | .63 | .66364 |
| .14 | .08509 | .64 | .67590 |
| .15 | .09406 | .65 | .68808 |
| .16 | .10327 | .66 | .70019 |
| .17 | .11273 | .67 | .71221 |
| .18 | .12240 | .68 | .72413 |
| .19 | .13229 | .69 | .73652 |
| .20 | .14238 | .70 | .74769 |
| .21 | .15266 | .71 | .75930 |
| .22 | .16312 | .72 | .77079 |
| .23 | .17375 | .73 | .78216 |
| .24 | .18455 | .74 | .79340 |
| .25 | .19550 | .75 | .80450 |
| .26 | .20660 | .76 | .81545 |
| .27 | .21784 | .77 | .82625 |
| .28 | .22921 | .78 | .83688 |
| .29 | .24070 | .79 | .84734 |
| .30 | .25231 | .80 | .85762 |
| .31 | .26348 | .81 | .86771 |
| .32 | .27587 | .82 | .87760 |
| .33 | .28779 | .83 | .88727 |
| .34 | .29981 | .84 | .89673 |
| .35 | .31192 | .85 | .90594 |
| .36 | .32410 | .86 | .91491 |
| .37 | .33636 | .87 | .92361 |
| .38 | .34869 | .88 | .93203 |
| .39 | .36108 | .89 | .94015 |
| .40 | .37353 | .90 | .94796 |
| .41 | .38603 | .91 | .95542 |
| .42 | .39858 | .92 | .96252 |
| .43 | .41116 | .93 | .96923 |
| .44 | .42379 | .94 | .97550 |
| .45 | .43644 | .95 | .98131 |
| .46 | .44912 | .96 | .98658 |
| .47 | .46182 | .97 | .99126 |
| .48 | .47454 | .98 | .99523 |
| .49 | .48727 | .99 | .99831 |
| .50 | .50000 | 1.00 | 1.00000 |

**Table 16.5.**[1]   Volume of Partially Filled Heads on Horizontal Tanks

| $H/D_i$ | Fraction of volume | $H/D_i$ | Fraction of volume |
|---------|--------------------|---------|--------------------|
| 0.02 | 0.0012 | 0.52 | 0.530 |
| .04 | .0047 | .54 | .560 |
| .06 | .0104 | .56 | .590 |
| .08 | .0182 | .58 | .619 |
| .10 | .0280 | .60 | .648 |
| .12 | .0397 | .62 | .677 |
| .14 | .0533 | .64 | .705 |
| .16 | .0686 | .66 | .732 |
| .18 | .0855 | .68 | .758 |
| .20 | .1040 | .70 | .784 |
| .22 | .1239 | .72 | .8087 |
| .24 | .1451 | .74 | .8324 |
| .26 | .1676 | .76 | .8549 |
| .28 | .1913 | .78 | .8761 |
| .30 | .216 | .80 | .8960 |
| .32 | .242 | .82 | .9145 |
| .34 | .268 | .84 | .9314 |
| .36 | .295 | .86 | .9467 |
| .38 | .323 | .88 | .9603 |
| .40 | .352 | .90 | .9720 |
| .42 | .381 | .92 | .9818 |
| .44 | .410 | .94 | .9896 |
| .46 | .440 | .96 | .9953 |
| .48 | .470 | .98 | .9988 |
| .50 | .500 | 1.00 | 1.0000 |

# REFERENCES

1. R. H. Perry and C. H. Chilton, *Chemical Engineers' Handbook*, 5th ed., McGraw-Hill, New York, 1973 pp. 6-86–6-88.
2. Engineering Data TE-12, Brighton Corporation, Cincinatti, Ohio, 1969, p. 33.

# 17 Dimensions and Properties of Steel Tubing

Table 17.1.[1]  Dimensions and Properties of Steel Tubing

| Out-side diam-eter (in.) | B.W.G. gage | Wall thick-ness (in.) | Inside diam-eter (in.) | External surface (ft² per ft length) | Weight lb/ft | Internal area[a] (ft²) |
|---|---|---|---|---|---|---|
| ¼ | 22 | 0.028 | 0.194 | 0.0655 | 0.067 | 0.00021 |
| ¼ | 24 | 0.022 | 0.206 | 0.0655 | 0.054 | 0.00023 |
| ¼ | 26 | 0.018 | 0.214 | 0.0655 | 0.045 | 0.00025 |
| ⅜ | 18 | 0.049 | 0.277 | 0.0982 | 0.173 | 0.00042 |
| ⅜ | 20 | 0.035 | 0.305 | 0.0982 | 0.139 | 0.00051 |
| ⅜ | 22 | 0.028 | 0.319 | 0.0982 | 0.105 | 0.00056 |
| ⅜ | 24 | 0.022 | 0.331 | 0.0982 | 0.083 | 0.00060 |
| ½ | 16 | 0.065 | 0.370 | 0.1309 | 0.3020 | 0.00075 |
| ½ | 18 | 0.049 | 0.402 | 0.1309 | 0.2360 | 0.00088 |
| ½ | 20 | 0.035 | 0.430 | 0.1309 | 0.1738 | 0.00101 |
| ½ | 22 | 0.028 | 0.444 | 0.1309 | 0.1411 | 0.00108 |
| ½ | 24 | 0.022 | 0.456 | 0.1309 | 0.1123 | 0.00113 |
| ⁹⁄₁₆ | 21 | 0.032 | 0.498 | 0.1473 | 0.183 | 0.00135 |
| ⅝ | 10 | 0.134 | 0.357 | 0.1636 | 0.703 | 0.00070 |
| ⅝ | 11 | 0.120 | 0.385 | 0.1636 | 0.647 | 0.00081 |
| ⅝ | 12 | 0.109 | 0.407 | 0.1636 | 0.605 | 0.00090 |
| ⅝ | 13 | 0.095 | 0.435 | 0.1636 | 0.540 | 0.00103 |
| ⅝ | 14 | 0.083 | 0.458 | 0.1636 | 0.481 | 0.00114 |
| ⅝ | 15 | 0.072 | 0.481 | 0.1636 | 0.430 | 0.00126 |
| ⅝ | 16 | 0.065 | 0.495 | 0.1636 | 0.390 | 0.00134 |
| ⅝ | 18 | 0.049 | 0.527 | 0.1636 | 0.301 | 0.00151 |
| ⅝ | 20 | 0.035 | 0.555 | 0.1636 | 0.220 | 0.00168 |
| ¾ | 10 | 0.134 | 0.482 | 0.1963 | 0.882 | 0.00127 |
| ¾ | 11 | 0.120 | 0.510 | 0.1963 | 0.810 | 0.00142 |
| ¾ | 12 | 0.109 | 0.532 | 0.1963 | 0.750 | 0.00154 |
| ¾ | 13 | 0.095 | 0.560 | 0.1963 | 0.670 | 0.00171 |
| ¾ | 14 | 0.083 | 0.584 | 0.1963 | 0.591 | 0.00186 |
| ¾ | 15 | 0.072 | 0.606 | 0.1963 | 0.522 | 0.00200 |
| ¾ | 16 | 0.065 | 0.620 | 0.1963 | 0.480 | 0.00210 |
| ¾ | 17 | 0.058 | 0.634 | 0.1963 | 0.429 | 0.00219 |
| ¾ | 18 | 0.049 | 0.652 | 0.1963 | 0.367 | 0.00232 |
| ¾ | 20 | 0.035 | 0.680 | 0.1963 | 0.267 | 0.00252 |
| ⅞ | 8 | 0.165 | 0.545 | 0.2297 | 1.251 | 0.00162 |
| ⅞ | 10 | 0.134 | 0.607 | 0.2297 | 1.060 | 0.00201 |

**Table 17.1.**[1] *Cont'd.*

| Outside diameter (in.) | B.W.G. gage | Wall thickness (in.) | Inside diameter (in.) | External surface (ft² per ft length) | Weight lb/ft | Internal area[a] (ft²) |
|---|---|---|---|---|---|---|
| ⅞ | 11 | 0.120 | 0.635 | 0.2297 | 0.968 | 0.00220 |
| ⅞ | 14 | 0.083 | 0.709 | 0.2297 | 0.702 | 0.00274 |
| ⅞ | 16 | 0.065 | 0.745 | 0.2297 | 0.562 | 0.00303 |
| ⅞ | 18 | 0.049 | 0.777 | 0.2297 | 0.432 | 0.00329 |
| ⅞ | 20 | 0.035 | 0.805 | 0.2297 | 0.314 | 0.00353 |
| 1 | 8 | 0.165 | 0.670 | 0.2618 | 1.471 | 0.00245 |
| 1 | 10 | 0.134 | 0.732 | 0.2618 | 1.240 | 0.00292 |
| 1 | 11 | 0.120 | 0.760 | 0.2618 | 1.130 | 0.00315 |
| 1 | 12 | 0.109 | 0.782 | 0.2618 | 1.040 | 0.00334 |
| 1 | 13 | 0.095 | 0.810 | 0.2618 | 0.920 | 0.00358 |
| 1 | 14 | 0.083 | 0.834 | 0.2618 | 0.813 | 0.00379 |
| 1 | 15 | 0.072 | 0.856 | 0.2618 | 0.714 | 0.00400 |
| 1 | 16 | 0.065 | 0.870 | 0.2618 | 0.650 | 0.00413 |
| 1 | 18 | 0.049 | 0.902 | 0.2618 | 0.500 | 0.00444 |
| 1 | 20 | 0.035 | 0.930 | 0.2618 | 0.361 | 0.00472 |
| 1¼ | 7 | 0.180 | 0.890 | 0.3272 | 2.057 | 0.00432 |
| 1¼ | 8 | 0.165 | 0.920 | 0.3272 | 1.912 | 0.00462 |
| 1¼ | 10 | 0.134 | 0.982 | 0.3272 | 1.597 | 0.00526 |
| 1¼ | 11 | 0.120 | 1.010 | 0.3272 | 1.450 | 0.00556 |
| 1¼ | 12 | 0.109 | 1.032 | 0.3272 | 1.328 | 0.00581 |
| 1¼ | 13 | 0.095 | 1.060 | 0.3272 | 1.172 | 0.00613 |
| 1¼ | 14 | 0.083 | 1.084 | 0.3272 | 1.040 | 0.00641 |
| 1¼ | 16 | 0.065 | 1.120 | 0.3272 | 0.823 | 0.00684 |
| 1¼ | 17 | 0.058 | 1.134 | 0.3272 | 0.738 | 0.00701 |
| 1¼ | 18 | 0.049 | 1.152 | 0.3272 | 0.629 | 0.00724 |
| 1¼ | 20 | 0.035 | 1.180 | 0.3272 | 0.454 | 0.00759 |
| 1½ | 10 | 0.134 | 1.232 | 0.3927 | 1.980 | 0.00828 |
| 1½ | 12 | 0.109 | 1.282 | 0.3927 | 1.640 | 0.00896 |
| 1½ | 14 | 0.083 | 1.334 | 0.3927 | 1.280 | 0.00971 |
| 1½ | 16 | 0.065 | 1.370 | 0.3927 | 0.996 | 0.01024 |
| 2 | 11 | 0.120 | 1.760 | 0.5236 | 2.450 | 0.01689 |
| 2 | 13 | 0.095 | 1.810 | 0.5236 | 1.933 | 0.01787 |
| 2½ | 9 | 0.148 | 2.200 | 0.6540 | 3.820 | 0.02640 |

[a] The values shown in ft² for the internal area also represent the volume in cubic feet per foot of pipe length.

# REFERENCES

1. Platecoil Catalog No. 5-63, Platecoil Division, Tranter Manufacturing, Inc., Lansing, Michigan, 1900.

# 18    Vapor–Liquid Separators

## VERTICAL DRUMS[1]

1. From Table 18.1, choose a value for $R_{dv}$. $R_{dv}$ is based on operating results reported in published papers and by manufacturers. Vapor–liquid separators and knock-out drums serve different functions. A vapor–liquid separator is a drum where entrainment is generated. It should have a lower $R_{dv}$ than a knock-out drum, where residual entrainment from another source is further reduced.
2. Calculate $V_{load}$ as

$$V_{load} = (\text{cfs vapor}) \left( \frac{\rho_v}{\rho_1 - \rho_v} \right)^{1/2} \qquad (18.1)$$

3. Calculate the drum diameter $D$ as

$$D = \left( \frac{V_{load}}{0.178 \, R_{dv}} \right)^{1/2} \qquad (18.2)$$

4. Use Figure 18.1 to determine the disengaging height $H_d$.
5. Liquid volume in the drum is such that the holding time should be 2–5 min. Select a holding time $\theta$ and calculate the liquid depth $H_L$.

$$H_L = \frac{0.17 \, Q_L \theta}{D^2} \qquad (18.3)$$

**Table 18.1.**[3]   Values for $R_{dv}$

|  | $R_{dv}$ |
| --- | --- |
| With mesh | |
|   Usual process applications | 1.54 |
|   If pressure surges | 1.15 |
| Without mesh | |
|   Vapor–liquid separators | 0.2–0.44 |
|   Knock-out drums | 0.88 |

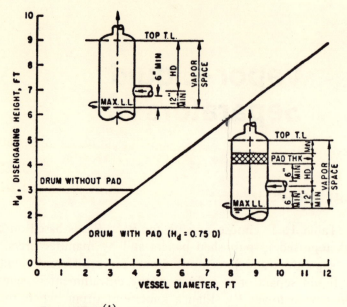

**Figure 18.1.**[1]  Recommended disengaging height.

## HORIZONTAL DRUMS*

1. Determine a trial diameter.
   a. Enter Figure 18.2 on the scale at the left with the holding time. Generally, a holding time of $7\frac{1}{2}$–10 min is recommended.
   b. Connect the time with the liquid space. Base your first trial on an 86% full drum (less if the minimum liquid level is other than the bottom of the drum).
   c. Mark the intersection with index line A.
   Reenter the nomograph at the right with the flowrate.
   d. Move horizontally to the flow index line, i.e., the gph line. If your units are gph, omit this step. You are starting at the flow index line.
   e. Draw a line from the flow index line to the mark on index line A. Mark where this line cuts index line B.
   f. Using an economic $L/D$ ratio, draw a line from the $L/D$ scale through the mark on index line B. Extend the line till it cuts the diameter scale.
   g. Adjust the diameter to suit a commercial head size (even 6-in. increments, except for 2 ft 8 in). A line drawn from this adjusted diam-

*After Ref. 2, with permission.

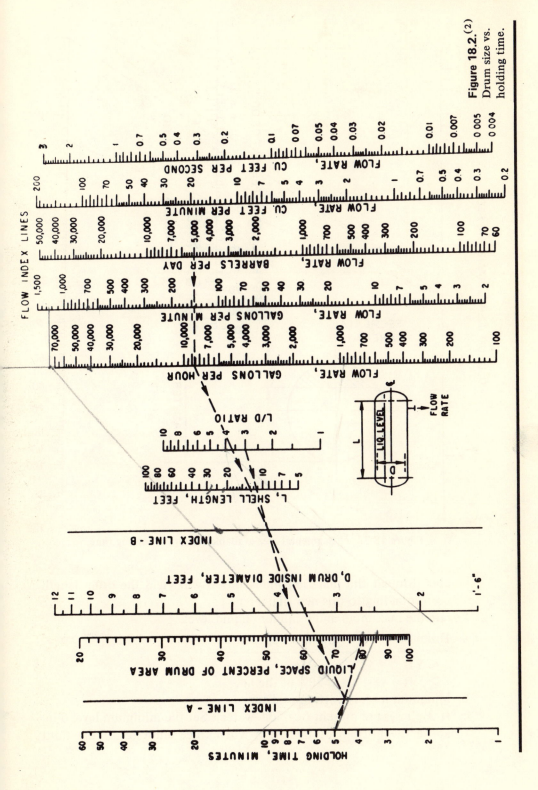

**Figure 18.2.**[2]
Drum size vs.
holding time.

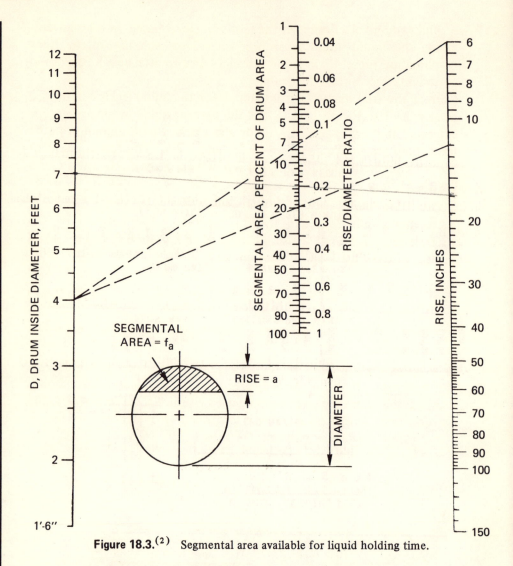

**Figure 18.3.**[2]   Segmental area available for liquid holding time.

eter through the mark on index line B will yield the drum length corresponding to this new diameter.

2. Locate the maximum and minimum liquid levels.

   a. 

| Drum i.d. | Minimum vapor space |
|---|---|
| ≤ 5 ft | 12 in. |
| > 5 ft | 20% of the i.d. |

   b. Minimum level is bottom of drum.

   c. If 3 phases are present (e.g., oil–water): Set the minimum level 6 in. above the bottom of the drum when trace amounts of the third phase are involved, and set the minimum level based on rules governing the liquid–liquid separation when large amounts of the third phase are present. In some instances, use an attached boot to keep the interface out of the drum.

3. Correct the trial diameter from step 1 for the liquid levels determined in step 2. Use Figure 18.3 to find the percent liquid in the drum. For example, if there is a 12-in. vapor space and a 6-in. minimum liquid level, their respective areas are 19.5% and 7.2%. This leaves 73.3% of the area for liquid holdup. Begin a new trial at step 1.b with 73.3% (vs. 86%).

4. Check for adequate reduction of entrainment.

   a. Assume a value for $R_{dh}$ of 0.167.
   b. Determine $L$ from Equation (18.4). Use Equation (18.1) to obtain $V_{load}$. Use Figure 18.3 to determine $f_a$.

$$L = \frac{V_{load}}{0.178 \, D^2} \left( \frac{a}{f_a R_{dh}} \right) \qquad (18.4)$$

where $f_a$ is the fraction of the drum occupied as a vapor, expressed as a decimal.

5. If $L$ is less than or equal to the trial length, stop here—the trial size is the final size. If the length is greater than the trial length, two choices exist. Either this new length can be used (if the $L/D$ ratio is still within the economic range), or a new trial length and diameter must be found.

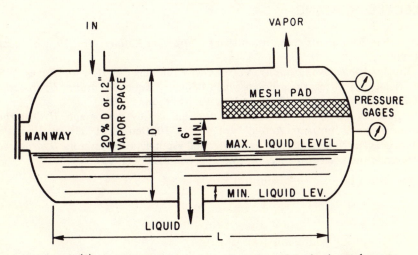

**Figure 18.4.**[2] General arrangement of mesh pad in horizontal vapor-liquid separator.

## NOMENCLATURE

$a$     Vapor space (ft)

$D$     Inside diameter of drum (ft)

$f_a$     % of drum area occupied as vapor space—determine from Figure 18.3

$H_d$     Disengaging height in vertical drum (ft)

$H_L$     Depth of liquid in vertical drum (ft)

$L$     Length of drum from tangent line to tangent line (ft)

$Q_L$     Liquid feedrate to the drum (gpm)

$R_{dh}, R_{dv}$     Vapor velocity in a drum divided by a base velocity (dimensionless)

$V_{load}$     Vapor flowrate through the drum (cfs)

$\rho_v$     Vapor density (lb/ft$^3$)

$\rho_l$     Liquid density (lb/ft$^3$)

$\theta$     Residence time in the drum (min)

## REFERENCES

1. A. D. Scheiman, Size vapor-liquid deparators quicker by nomograph, *Hydrocarbon Processing and Petroleum Refiner*, Vol. 42, No. 10, 165–168 (1963).
2. A. D. Scheiman, Nomographs to size horizontal vapor–liquid separators, *Hydrocarbon Processing and Petroleum Refiner*, Vol. 43, No. 5, 155–160 (1964).
3. J. F. Kuong, *Applied Nomography*, Vol. 3, Gulf Publishing Co., Houston, Texas, 1969, pp. 105–106.

## SELECTED READING

G. D. Kerns, New charts speed drum sizing, *Petroleum Refiner*, Vol. 39, No. 7, July (1960).

J. F. Kuong, *Applied Nomography*, Vol. 3, Gulf Publishing Co., Houston, Texas, 1969.

E. R. Neimeyer, Check these points when designing knockout drums, *Hydrocarbon Processing and Petroleum Refiner*, Vol. 40, No. 6, June (1961).

A. D. Scheiman, Size vapor–liquid separators quicker by nomograph, *Hydrocarbon Processing and Petroleum Refiner*, Vol. 42, No. 10, October (1963).

A. D. Scheiman, Nomographs to size horizontal vapor–liquid separators, *Hydrocarbon Processing and Petroleum Refiner*, Vol. 43, No. 5, May (1964).

# 19    Vessel Design

The design of pressure vessels in the United States is closely governed by the ASME Boiler and Pressure Vessel Code Section VIII, Division 1.

A quick, convenient guide to this code (including the summer 1975 addenda) is provided in Figure 19.1.[1] The purpose of this guide is to illustrate some of the types of pressure vessel construction which are provided for under Section VIII, Division 1, of the ASME Code and to furnish direct reference to the applicable rule in the Code. In the event of a discrepancy, the rules in the current edition of the Code shall govern. This should be used only as a quick reference. The current edition of the Code should always be referenced. Table 19.1 provides a further guide to the Code.

Figure 19.2 provides a convenient method for estimating the shell thickness of a pressure vessel. This figure should be used only for estimating, and never for final design.

## FLANGES: PRESSURE-TEMPERATURE RATING

Pressure ratings for 150 and 300 lb flanges as a function of temperature and various materials of construction are given in Table 19.2 and 19.3.

### Conditions and Limitations*

1. Products used within the jurisdiction of the ASME Boiler and Pressure Vessel Code and the USA Standard Code for pressure piping are subject to the maximum temperature and stress limitations upon the material and piping stated therein.
2. The ratings at $-20$–$100°F$ given for the materials covered shall also apply at lower temperatures. The ratings for low temperature service of the cast and forged materials listed in ASTM A352 and A350 shall be taken the same as the $-20$–$100°F$ ratings for carbon steel. Some of the materials listed in the rating tables undergo a decrease in impact resistance at temperatures lower than $-20°F$ to such an extent as to be

---

*After Ref. 3, with permission.

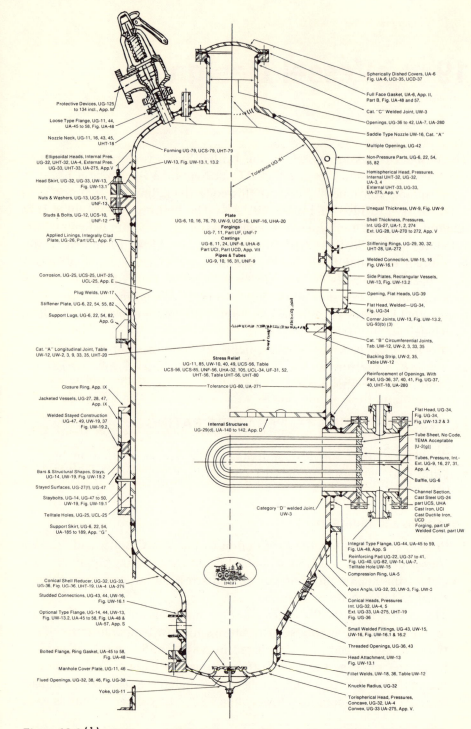

**Figure 19.1.**[1] Guide to ASME boiler and Pressure Vessel Code, Section VIII, Division I.

**Table 19.1.**[(1)]   General Notes,[ a] ASME Code

| | | | | | |
|---|---|---|---|---|---|
| Quality Control System | U-2, UA-900 | Design Temperature | UG-20 | Nameplates, Stamping & Reports | UG-115 to 120, App. W |
| Material—General | UG-5, 10, 11, 15 | Design Pressure | UG-21 | | |
| (a)  Plate | UG-6 | Loadings | UG-22 | Non-Destructive  Examination—As  required  per | |
| (b)  Forgings | UG-7 | Stress—Max. Allowable | UG-23 | Subsection and Part | |
| (c)  Castings | UG-8 | Manufacturer's Responsibility | UG-90 | (a)  Radiography | UW-51, 52 |
| (d)  Pipe & Tubes | UG-9 | Inspector's Responsibility | UG-90 | (b)  Ultrasonic | App. U |
| (e)  Bolts & Studs | UG-12 | Pressure Tests | UG-99, 100, 101, | (c)  Magnetic Particle | App. VI |
| (f)  Nuts & Washers | UG-13 | | UW-50, UCI-99, UCD-99 | (d)  Liquid Penetrant | App. VIII |
| (g)  Rods & Bars | UG-14 | Low Temperature Service | UG-84, UW-2 | Porosity Charts | App. IV |
| | | Quick Actuating Closures | UG-35 | Code Jurisdiction over piping | U-1(e) |
| | | Service Restrictions | UW-2, UCI-2, UCD-2 | Minimum Thickness of Plate | UG-16 |

[a]Reproduced with permission of the Hartford Steam Boiler Inspection and Insurance Company.

unable to safely resist shock loadings, sudden changes of stress or high stress concentrations. Therefore, products that are to operate at temperatures below −20°F shall conform to the rules of the applicable Codes under which they are to be used.

3. The pressure–temperature ratings in the tables apply to all products covered by this standard. Valves conforming to the requirements of this standard must, in other respects, merit these ratings. All ratings are the maximum allowable nonshock pressures (psig) at the tabulated temperatures (°F) and may be interpolated between the temperatures shown. The primary service pressure ratings (150, 300) are those at the head of the tables and shown in boldface type in the body of the tables. Temperatures (°F) shown in the tables, used in determining these rating tables, were temperatures on the inside of the pressure retaining structure. The use of these ratings require gaskets conforming to the requirements of Introductory Note 6.10 of USA Standard Specification B16.5-1961. The user is responsible for selecting gaskets of dimensions and materials to withstand the required bolt loading without injurious crushing, and suitable for the service conditions in all other respects.

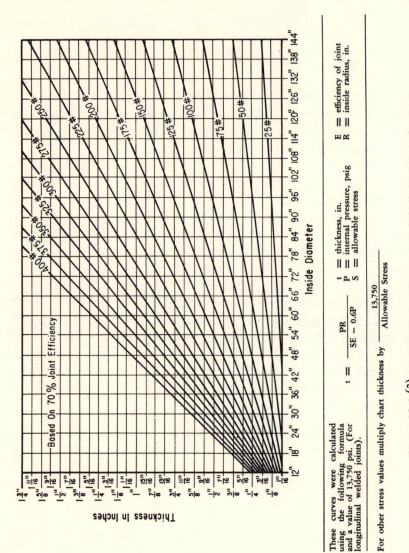

These curves were calculated using the following formula and a value of 13,750 psi. (For longitudinal welded joints).

$$t = \frac{PR}{SE - 0.6P}$$

For other stress values multiply chart thickness by $\frac{13,750}{\text{Allowable Stress}}$

$t$ = thickness, in.
$P$ = internal pressure, psig
$S$ = allowable stress

$E$ = efficiency of joint
$R$ = inside radius, in.

**Figure 19.2.**[2]  ASME Code pressure vessel shell thickness chart.

**Table 19.2.**[1]  150-lb Flange Pressure–Temperature Ratings: Subject to Stipulations in Text (All Pressures psig)

| Service Temperature Deg F | Carbon Steel | Carbon Moly | Cr-Mo ½-½ | Cr-Mo 1-½ | Cr-Mo 1¼-½ | Cr-Mo 2-½ | Cr-Mo 2¼-1 | Cr-Mo 3-1 | Cr-Mo 5-½ | Cr-Mo 5-½-Si | Cr-Mo 9-1 | 304 | 347 & 321 | 316 | 310 | 304L | 316L | Service Temperature Deg F |
|---|---|---|---|---|---|---|---|---|---|---|---|---|---|---|---|---|---|---|
| -20 to 100[2] | | | | | | | | 275 | | | | | | | | | | -20 to 100[2] |
| 150 | | | | | | | | 255 | | | | | | | | | | 150 |
| 200 | | | | | | | | 240 | | | | | | | | | | 200 |
| 250 | | | | | | | | 225 | | | | | | | | | | 250 |
| 300 | | | | | | | | 210 | | | | | | | | | | 300 |
| 350 | | | | | | | | 195 | | | | | | | | | | 350 |
| 400 | | | | | | | | 180 | | | | | | | | | | 400 |
| 450 | | | | | | | | 165 | | | | | | | | | | 450 |
| 500 | | | | | | | | 150 | | | | | | | | | | 500 |
| 550 | | | | | | | | 140 | | | | | | | | | | 550 |
| 600 | | | | | | | | 130 | | | | | | | | | | 600 |
| 650 | | | | | | | | 120 | | | | | | | | | | 650 |
| 700 | | | | | | | | 110 | | | | | | | | | | 700 |
| 750 | | | | | | | | 100 | | | | | | | | | | 750 |
| 800 | | | | | | | | 92 | | | | | | | | | | 800 |
| 850 | 82[1] | | | | | | | 82 | | | | | | | | — | 82 | 850 |
| 875 | 75[1] | 75[1] | | | | | | 75 | | | | | | | | — | — | 875 |
| 900 | 70[1] | 70[1] | | | | | | 70 | | | | | | | | — | — | 900 |
| 925 | 60[1] | 60[1] | | | | | | 60 | | | | | | | | — | — | 925 |
| 950 | 55[1] | 55[1] | | | | | | 55 | | | | | | | | — | — | 950 |
| 975 | 50[1] | 50[1] | | | | | | 50 | | | | | | | | — | — | 975 |
| 1000 | 40[1] | 40[1] | | | | | | 40 | | | | | | | | | | 1000 |

Hydrostatic Shell Test Pressure  425

[1] See Introductory Note 1 on Boiler Code and Pressure Piping Code limitations.

[2] See Introductory Note 2 for low temperature-pressure ratings including other materials.

**Table 19.3.**[1]  300-lb Flange Pressure–Temperature Ratings: Subject to Stipulations in Text (All Pressures psig)

| Service Temperature Deg F | Material | | | | | | | | | | | | | | | | | Service Temperature Deg F |
|---|---|---|---|---|---|---|---|---|---|---|---|---|---|---|---|---|---|---|
| | Carbon Steel | Carbon Moly | Cr-Mo 1/2-1/2 | Cr-Mo 1-1/2 | Cr-Mo 1 1/4-1/2 | Cr-Mo 2-1/2 | Cr-Mo 2 1/4-1 | Cr-Mo 3-1 | Cr-Mo 5-1/2 | Cr-Mo 5-1/2-Si | Cr-Mo 9-1 | 304 | 347 & 321 | 316 (TYPES) | 310 | 304L | 316L | |
| −20 to 100[2] | | | | | | 720 | | | | | | 615 | | 720 | | 515 | 515 | −20 to 100[2] |
| 150 | | | | | | 710 | | | | | | 585 | | 710 | | 510 | 515 | 150 |
| 200 | | | | | | 700 | | | | | | 550 | | 700 | | 505 | 515 | 200 |
| 250 | | | | | | 690 | | | | | | 520 | | 690 | | 465 | 495 | 250 |
| 300 | | | | | | 680 | | | | | | 495 | | 680 | | 430 | 475 | 300 |
| 350 | | | | | | 675 | | | | | | 470 | | 675 | | 395 | 435 | 350 |
| 400 | | | | | | 665 | | | | | | 450 | | 665 | | 360 | 395 | 400 |
| 450 | | | | | | 650 | | | | | | 430 | | 650 | | 340 | 380 | 450 |
| 500 | | | | | | 625 | | | | | | 410 | | 625 | | 320 | 360 | 500 |
| 550 | | | | | | 590 | | | | | | 395 | | 590 | | 310 | 350 | 550 |
| 600 | | | | | | 555 | | | | | | 380 | | 555 | | 300 | 335 | 600 |
| 650 | | | | | | 515 | | | | | | 370 | | 515 | | 290 | 325 | 650 |
| 700 | 470 | 480 | 480 | 485 | 485 | 480 | 485 | 480 | 485 | 480 | 485 | 355 | 495 | | 490 | 280 | 310 | 700 |
| 750 | 425 | 445 | 445 | 450 | 450 | 445 | 450 | 445 | 450 | 445 | 450 | 340 | 470 | | 465 | 275 | 300 | 750 |
| 800 | 365 | 410 | 410 | 415 | 415 | 410 | 415 | 410 | 415 | 410 | 415 | 330 | 450 | | 440 | 265 | 290 | 800 |
| 850 | 300[1] | 370 | 370 | 385 | 385 | 370 | 385 | 370 | 385 | 370 | 385 | 320 | 425 | | 415 | | 280 | 850 |
| 875 | 260[1] | 355[1] | 355 | 365 | 365 | 355 | 365 | 355 | 365 | 355 | 365 | 315 | 415 | | 400 | | | 875 |
| 900 | 225[1] | 335[1] | 335 | 350 | 350 | 335 | 350 | 335 | 350 | 335 | 350 | 310 | 400 | | 390 | | | 900 |

| | | | | | | | | | | | | | | | | | |
|---|---|---|---|---|---|---|---|---|---|---|---|---|---|---|---|---|---|
| 950 | 155[1] | **300**[1] | **300** | 315 | 315 | **300** | 315 | **300** | 315 | **300** | 315 | 305 | 380 | | 365 | | | 950 |
| 975 | 120[1] | 280[1] | 280 | **300** | **300** | 280 | **300** | 275 | **300** | 250 | **300** | **300** | 370 | | 350 | | | 975 |
| 1000 | 85[1] | 215[1] | 215 | 255 | 265 | 215 | 265 | 240 | 250 | 190 | 290 | **300** | 355 | | 340 | | | 1000 |
| 1025 | | | | 215 | 230[1] | 180 | 235 | 215 | 215 | 155 | 240 | 295 | 345 | | 325 | | | 1025 |
| 1050 | | | | 170 | 190[1] | 145 | 200 | 190 | 180 | 120 | 190 | 290 | 335 | | 315 | | | 1050 |
| 1075 | | | | 135 | 165[1] | 120 | 170[1] | 165 | 145 | 105 | 150 | 275 | 325 | | **300** | | | 1075 |
| 1100 | | | | 95 | 135[1] | 95 | 145[1] | 135 | 115 | 85 | 115 | 255 | 310 | | 290 | | | 1100 |
| 1125 | | | | 75[1] | 110[1] | 75 | 125[1] | 115 | 95 | 75 | 95 | 225 | **300** | | 270 | | | 1125 |
| 1150 | | | | 55[1] | 85[1] | 60 | 105[1] | 95 | 75 | 60 | 75 | 195 | 260 | 290 | 250 | | | 1150 |
| 1175 | | | | 45[1] | 65[1] | 50 | 85[1] | 70 | 65 | 50 | 65 | 175 | 215 | 260 | 225 | | | 1175 |
| 1200 | | | | 35[1] | 40[1] | 40 | 70[1] | 50 | 50 | 40 | 50 | 155 | 170 | 235 | 205 | | | 1200 |
| 1225 | | | | | | | | | | | | 135 | 140 | 205 | 185 | | | 1225 |
| 1250 | | | | | | | | | | | | 110 | 115 | 180 | 165 | | | 1250 |
| 1275 | | | | | | | | | | | | 100 | 95 | 160 | 140 | | | 1275 |
| 1300 | | | | | | | | | | | | 85 | 75 | 135 | 120 | | | 1300 |
| 1325 | | | | | | | | | | | | 75 | 65 | 115 | 100 | | | 1325 |
| 1350 | | | | | | | | | | | | 60 | 50 | 95 | 80 | | | 1350 |
| 1375 | | | | | | | | | | | | 55 | 45 | 80 | 70 | | | 1375 |
| 1400 | | | | | | | | | | | | 50 | 40 | 70 | 55 | | | 1400 |
| 1425 | | | | | | | | | | | | 40 | 35 | 60 | 45 | | | 1425 |
| 1450 | | | | | | | | | | | | 35 | 30 | 50 | 40 | | | 1450 |
| 1475 | | | | | | | | | | | | 30 | 30 | 45 | 30 | | | 1475 |
| 1500 | | | | | | | | | | | | 25 | 25 | 35 | 25 | | | 1500 |
| Hydrostatic Shell Test Pressure | 1100 | | | | | | | | | | | 925 | 1100 | | | 775 | | |

[1] See Introductory Note 1 on Boiler Code and Pressure Piping Code limitations.

[2] See Introductory Note 2 for low temperature-pressure ratings including other materials.

## REFERENCES

1. Quick Reference Guide to the ASME Boiler and Pressure Vessel Code Section VIII, Division 1, Hartford Steam Boiler Inspection and Insurance Company, Hartford, Connecticut, 1976.
2. Platecoil Product Data Manual No. 5-63, Platecoil Division, Tranter Inc., Lansing, Michigan, 1974, p. 74.
3. American Standard Steel Pipe Flanges and Flanged Fittings (USA B16.5−1961), The American Society of Mechanical Engineers, 345 E. 47th Street, New York, New York.

## SELECTED READING

ASME Boiler and Pressure Vessel Code Section VIII, Division 1, American Society of Mechanical Engineers, United Engineering Center, 345 E. 47th Street, New York, 1977.

# Index